Ulrich Krupp

Fatigue Crack Propagation in Metals and Alloys

1807–2007 Knowledge for Generations

Each generation has its unique needs and aspirations. When Charles Wiley first opened his small printing shop in lower Manhattan in 1807, it was a generation of boundless potential searching for an identity. And we were there, helping to define a new American literary tradition. Over half a century later, in the midst of the Second Industrial Revolution, it was a generation focused on building the future. Once again, we were there, supplying the critical scientific, technical, and engineering knowledge that helped frame the world. Throughout the 20th Century, and into the new millennium, nations began to reach out beyond their own borders and a new international community was born. Wiley was there, expanding its operations around the world to enable a global exchange of ideas, opinions, and know-how.

For 200 years, Wiley has been an integral part of each generation's journey, enabling the flow of information and understanding necessary to meet their needs and fulfill their aspirations. Today, bold new technologies are changing the way we live and learn. Wiley will be there, providing you the must-have knowledge you need to imagine new worlds, new possibilities, and new opportunities.

Generations come and go, but you can always count on Wiley to provide you the knowledge you need, when and where you need it!

William J. Pesce
President and Chief Executive Officer

Peter Booth Wiley
Chairman of the Board

Ulrich Krupp

Fatigue Crack Propagation in Metals and Alloys

Microstructural Aspects and Modelling Concepts

WILEY-VCH Verlag GmbH & Co. KGaA

The Author

Prof. Dr.-Ing. Ulrich Krupp
Faculty of Engineering and Computer Science
FH Osnabrück – University of Applied Sciences
Albrechtstrasse 30
49076 Osnabrück
Germany

Library of Congress Card No.: applied for

British Library Cataloguing-in-Publication Data
A catalogue record for this book is available from the British Library.

Bibliographic information published by the Deutsche Nationalbibliothek
The Deutsche Nationalbibliothek lists this publication in the Deutsche Nationalbibliografie; detailed bibliographic data are available in the Internet at http://dnb.d-nb.de.

Printed in the Federal Republic of Germany
Printed on acid-free paper

Typesetting Fotosatz Detzner, Speyer
Printing betz-druck GmbH, Darmstadt
Bookbinding Litges & Dopf Buchbinderei GmbH, Heppenheim
Wiley Bicentennial Logo Richard J. Pacifico

ISBN: 978-3-527-31537-6

To my children
Lukas and Lorenz

Foreword

Although it was understood very early that fatigue was a matter of creating and propagating a crack to the point of ultimate failure of a specimen or structural component, indeed long before "crystallization" theories of fatigue eventually died from natural causes, no significant measurements of crack propagation kinetics existed before the early 1960's. Theories of crack growth existed before then, of course, highly speculative and based on unrealistic models. Only after appropriate fracture mechanics parameters were developed to allow the driving forces for crack growth to be expressed for a wide variety of component geometries did predictive capability greatly improve for fatigue design purposes. The traditional approach of designing against fatigue, known as the "total-life" approach, worked adequately on the whole before the arrival of linear elastic fracture mechanics, mainly because it was backed by extensive (and expensive) experimentation. It also has undergone improvements, for examples in counting cycles during variable load cycling and in strain (rather than stress)-based predictions, and continues in wide use today, even in certain aerospace applications where usually fracture mechanics considerations now dominate. The total-life approach can be valid in situations where inspections for cracks are difficult or are economically prohibitive. The success of fracture mechanics in applications to aircraft probably arises from the necessity to use high stresses in aircraft and where relatively infrequent very high stress cycles promote early crack initiation, even in the absence of significant manufacturing defects, on which some fracture mechanicians rely to justify the approach.

Like most books on fatigue, the aim of Dr. Krupp's book is to help to make better fatigue life predictions. Although the title of his book implies an emphasis on fracture mechanics, and does focus on fracture mechanics (and almost exclusively on metals), it contains an elaborate chapter on crack initiation, and it also deals with the underlying physics of fatigue mechanisms. It is a well-balanced book in other respects as well, in the referencing between early and recent publications, between fundamental aspects such as dislocation considerations and modeling, and particularly in the balance between mechanics and materials considerations. It covers the range from elementary dislocation behavior to recent advances in modeling. It is thus suited to a wide range of students from different backgrounds and developmental stages; veterans of fatigue studies will find it interesting. In

Fatigue Crack Propagation in Metals and Alloys: Microstructural Aspects and Modelling Concepts. Ulrich Krupp

ISBN: 978-3-527-31537-6

that it contains detailed treatments of microstructural effects on short crack growth behavior, the book goes far to bridge gaps of understanding between material scientists and mechanical engineers. Thus, Dr Krupp's worthy book represents another important step in the evolutionary development of books dealing with fatigue.

It is a pleasure to acknowledge a highly productive collaboration with Dr. Krupp during his sabbatical visit to the University of Pennsylvania. His Penn colleagues gained much from his industry and insights, of which his book is an excellent example.

November 2006

Campbell Laird
Stockbridge, Michigan

Preface

As long as machines are moving we have to care about fatigue and fatigue-crack propagation. What we don't know about fatigue damage mechanisms is hidden in safety factors. However, in terms of light-weight design and economic engineering design safety factors need to be chosen as small as possible, i.e., it has to be accepted that technical materials exhibit microstructural changes and crack propagation during service life. This book outlines shortly the present state in fatigue-life prediction paying particular attention to fracture mechanics concepts. Today we know that interactions between a growing fatigue crack with local features of the material's microstructure are most relevant for understanding and modeling fatigue damage in the fatigue-limit range under high-cycle fatigue (HCF) and very-high-cycle fatigue (VHCF) conditions. It is hoped that this book gives students in advanced courses in materials engineering and researchers in industry and academia a picture of the sum of implications of short-crack propagation in fatigue life, helping them to interpret failure cases and to identify strategies for mechanism-based service-life prediction and the design of fatigue-damage-resistant material concepts.

Beside a literature review this book contains the results of research work the author carried out at the University of Siegen, Germany, and the University of Pennsylvania, Philadelphia, USA. The complete work was encouraged and strongly supported by Prof. Dr. Hans-Jürgen Christ, holder of the chair of materials science and testing at the University of Siegen. Working with such an outstanding scientist and teacher was most enjoyable and I am greatly indebted not only for his great effort in supporting my scientific career but also for his friendship. I was lucky also to have the chance to closely collaborate with the Institute of Mechanics, Prof. Dr. Claus-Peter Fritzen. This collaboration, which is greatly acknowledged, allowed me to combine the experimental work with complex modeling concepts.

A very strong impact on the research presented in this book was the chance to work together and to discuss the results with Prof. Dr. Charles J. McMahon, Jr. and Prof. Dr. Campbell Laird at the University of Pennsylvania. Thanks to their hospitality and friendship my family and me spent a great and unforgettable time in Philadelphia and for that I wish to express my gratitude.

Of course, this book was only possible by the huge amount of work carried out by my former colleagues and PhD students at the University of Siegen, Dr. Wil-

Fatigue Crack Propagation in Metals and Alloys: Microstructural Aspects and Modelling Concepts. Ulrich Krupp

ISBN: 978-3-527-31537-6

helm Floer, Dr. Alexander Schick, Dr. Olaf Düber, Dr. Boris Künkler, Helge Knobbe, Philipp Köster, and Arne Ohrndorf and the great support by my colleagues at the University of Pennsylvania, Dr. William M. Kane and Dr. Alex Radin. Beside their invaluable scientific contribution to this book it was and is their friendship, which made work really enjoyable and for which I am very thankful.

My special thanks goes to Prof. Dr. Hans-Jürgen Christ, Prof. Dr. Campbell Laird, Prof. Dr. Charles J. McMahon, Jr., and Dr. Boris Künkler for critically reading parts of the manuscript and to Wiley's editing and publishing team for their patience and support during the completion of this book.

However, without financial support successful research is hardly possible. I was lucky that I received several grants from the German research foundation, Deutsche Forschungsgemeinschaft, and a personal fellowship from the Alexander von Humboldt foundation allowing me to live and work in the United States. This support is gratefully acknowledged.

Finally I am deeply grateful to my family, especially my grandmother Johanna Schlaadt, my parents Hiltrud and Peter and my two children Lukas and Lorenz. Without their help and their love I never would have come so far and I wish to dedicate this book to them.

Contents

Foreword *VII*

Preface *IX*

Symbols and Abbreviations *XV*

1 Introduction *1*

2 Basic Concepts of Metal Fatigue and Fracture in the Engineering Design Process *3*
2.1 Historical Overview *3*
2.2 Metal Fatigue, Crack Propagation and Service-Life Prediction: A Brief Introduction *10*
2.2.1 Fundamental Terms in Fatigue of Materials *12*
2.2.2 Fatigue-Life Prediction: Total-Life and Safe-Life Approach *15*
2.2.3 Fatigue-Life Prediction: Damage-Tolerant Approach *19*
2.2.4 Methods of Fatigue-Life Prediction at a Glance *24*
2.3 Basic Concepts of Technical Fracture Mechanics *25*
2.3.1 The *K* Concept of LEFM *27*
2.3.2 Crack-Tip Plasticity: Concepts of Plastic-Zone Size *31*
2.3.3 Crack-Tip Plasticity: The *J* Integral *34*

3 Experimental Approaches to Crack Propagation *39*
3.1 Mechanical Testing *39*
3.1.1 Testing Systems *39*
3.1.2 Specimen Geometries *43*
3.1.3 Local Strain Measurement: The ISDG Technique *46*
3.2 Crack-Propagation Measurements *48*
3.2.1 Potential-Drop Concepts and Fracture Mechanics Experiments *49*
3.2.2 *In Situ* Observation of the Crack Length *54*
3.3 Methods of Microstructural Analysis and Quantitative Characterization of Grain and Phase Boundaries *56*
3.3.1 Analytical SEM: Topography Contrast to Study Fracture Surfaces *56*

Fatigue Crack Propagation in Metals and Alloys: Microstructural Aspects and Modelling Concepts. Ulrich Krupp

ISBN: 978-3-527-31537-6

3.3.2 SEM Imaging by Backscattered Electrons and EBSD *58*
3.3.3 Evaluation of Kikuchi Patterns: Automated EBSD *62*
3.3.4 Orientation Analysis Using TEM and X-Ray Diffraction *63*
3.3.5 Mathematical and Graphical Description of Crystallographic Orientation Relationships *65*
3.3.6 Microstructure Characterization by TEM *70*
3.3.7 Further Methods to Characterize Mechanical Damage Mechanisms in Materials *72*
3.4 Reproducibility of Experimentally Studying the Mechanical Behavior of Materials *74*

4 Physical Metallurgy of the Deformation Behavior of Metals and Alloys *75*
4.1 Elastic Deformation *76*
4.2 Plastic Deformation by Dislocation Motion *80*
4.3 Activation of Slip Planes in Single- and Polycrystalline Materials *90*
4.4 Special Features of the Cyclic Deformation of Metallic Materials *94*

5 Initiation of Microcracks *99*
5.1 Crack Initiation: Definition and Significance *99*
5.1.1 Influence of Notches, Surface Treatment and Residual Stresses *100*
5.2 Influence of Microstructual Factors on the Initiation of Fatigue Cracks *101*
5.2.1 Crack Initiation at the Surface: General Remarks *101*
5.2.2 Crack Initiation at Inclusions and Pores *102*
5.2.3 Crack Initiation at Persistent Slip Bands *104*
5.3 Crack Initiation by Elastic Anisotropy *107*
5.3.1 Definition and Significance of Elastic Anisotropy *107*
5.3.2 Determination of Elastic Constants and Estimation of the Elastic Anisotropy *109*
5.3.3 FE Calculations of Elastic Anisotropy Stresses to Predict Crack Initiation Sites *113*
5.3.4 Analytical Calculation of Elastic Anisotropy Stresses *116*
5.4 Intercrystalline and Transcrystalline Crack Initiation *119*
5.4.1 Influence Parameters for Intercrystalline Crack Initiation *119*
5.4.2 Crack Initiation at Elevated Temperature and Environmental Effects *123*
5.4.3 Transgranular Crack Initiation *126*
5.5 Microstructurally Short Cracks and the Fatigue Limit *127*
5.6 Crack Initiation in Inhomogeneous Materials: Cellular Metals *129*

6 Crack Propagation: Microstructural Aspects *135*
6.1 Special Features of the Propagation of Microstructurally Short Fatigue Cracks *135*
6.1.1 Definition of Short and Long Cracks *136*

6.2 Transgranular Crack Propagation *139*
6.2.1 Crystallographic Crack Propagation: Interactions with Grain Boundaries *139*
6.2.2 Mode I Crack Propagation Governed by Cyclic Crack-Tip Blunting *145*
6.2.3 Influence of Grain Size, Second Phases and Precipitates on the Propagation Behavior of Microstructurally Short Fatigue Cracks *149*
6.3 Significance of Crack-Closure Effects and Overloads *153*
6.3.1 General Idea of Crack Closure During Fatigue-Crack Propagation *153*
6.3.2 Plasticity-Induced Crack Closure *156*
6.3.3 Influence of Overloads in Plasticity-Induced Crack Closure *160*
6.3.4 Roughness-Induced Crack Closure *161*
6.3.5 Oxide- and Transformation-Induced Crack Closure *162*
6.3.6 $\Delta K^{*}/K^{*}_{max}$ Thresholds: An Alternative to the Crack-Closure Concept *163*
6.3.7 Development of Crack Closure in the Short Crack Regime *164*
6.4 Short and Long Fatigue Cracks: The Transition from Mode II to Mode I Crack Propagation *171*
6.4.1 Development of the Crack Aspect Ratio a/c *173*
6.4.2 Coalescence of Short Cracks *179*
6.5 Intercrystalline Crack Propagation at Elevated Temperatures: The Mechanism of Dynamic Embrittlement *181*
6.5.1 Environmentally Assisted Intercrystalline Crack Propagation in Nickel-Based Superalloys: Possible Mechanisms *181*
6.5.2 Mechanism of Dynamic Embrittlement as a Generic Phenomenon: Examples *187*
6.5.3 Oxygen-Induced Intercrystalline Crack Propagation: Dynamic Embrittlement of Alloy 718 *192*
6.5.4 Increasing the Resistance to Intercrystalline Crack Propagation by Dynamic Embrittlement: Grain-Boundary Engineering *197*

7 Modeling Crack Propagation Accounting for Microstructural Features *207*
7.1 General Strategies of Fatigue Life Assessment *207*
7.2 Modeling of Short-Crack Propagation *211*
7.2.1 Short-Crack Models: An Overview *211*
7.2.2 Model of Navarro and de los Rios *218*
7.3 Numerical Modeling of Short-Crack Propagation by Means of a Boundary Element Approach *226*
7.3.1 Basic Modeling Concept *226*
7.3.2 Slip Transmission in Polycrystalline Microstructures *230*
7.3.3 Simulation of Microcrack Propagation in Synthetic Polycrystalline Microstructures *232*
7.3.4 Transition from Mode II to Mode I Crack Propagation *236*

7.3.5 Future Aspects of Applying the Boundary Element Method to Short-Fatigue-Crack Propagation 239
7.4 Modeling Dwell-Time Cracking: A Grain-Boundary Diffusion Approach 242

8 Concluding Remarks 251

References 255

Subject Index 281

Symbols and Abbreviations

Roman Symbols

a	[m]	crack length
a/a_0	[m]	lattice constant/equilibrium atomic spacing
a_0	[m]	starter crack length
a_0, a_1	[m]	transition crack length (nonpropagation condition determined by $\Delta\sigma_D$)
a_2	[m]	transition crack length (nonpropagation condition determined by LEFM)
a_{coales}	[m]	crack depth of coalescent cracks
a_{det}	[m]	detectable crack length
a_{eff}	[m]	effective crack length
a_{ij}	[–]	elements of misorientation matrix
a_{th}	[m]	critical crack length
a_{crit}	[m]	critical final crack length
A	[–]	anisotropy factor
A'	[–]	cyclic strength coefficient
b	[–]	fatigue strength exponent
$\vec{b}$	[m]	Burgers vector, displacement vector
$\vec{b}_k$	[m]	vector of displacement due to a load σ_k
b_n^i	[m]	normal component of displacement of the ith element
b_t^i	[m]	tangential component of displacement of the ith element
B_{net}	[m]	effective specimen width
c	[–]	fatigue ductility exponent
c	[m]	half surface crack length
c	[m]	crack length including plastic zone
c_0	[–]	concentration of an embrittling species at crack tip
c_{ij}	[MPa]	elements of the stiffness tensor $\underline{\underline{C}}$
c_σ/c_δ	[–]	critical concentrations accounting for σ_{max} or δ_{crit}, respectively
C	[1/MPa]	compliance (dependent on ρ and Φ)

Fatigue Crack Propagation in Metals and Alloys: Microstructural Aspects and Modelling Concepts. Ulrich Krupp

ISBN: 978-3-527-31537-6

$\underline{\underline{C}}$	[MPa]	stiffness tensor
$C_{v,sat}$	[–]	vacancy saturation concentration
COD	[m]	crack-opening displacement
CTOD	[m]	crack-tip-opening displacement
d	[m]	lattice spacing
d	[m]	grain size
$\overline{d}$	[m]	mean grain size
d/d_i	[m]	barrier spacing (mean value/individual values)
d	[m]	notch radius
d_0	[m]	crystal size
$\overline{d}_\alpha$	[m]	mean α-ferrite cluster size
$\overline{d}_{\alpha\alpha}$	[m]	mean grain size in α-ferrite cluster
$\overline{d}_{\alpha\gamma}$	[m]	mean cluster size
$\overline{d}_\gamma$	[m]	mean γ-austenite cluster size
$\overline{d}_{\gamma\gamma}$	[m]	mean grain size in α-austenite cluster
D	[m²/s]	(effective) diffusion coefficient
D_{GB}	[m²/s]	grain-boundary diffusion coefficient
D^0_{GB}	[m²/s]	grain-boundary diffusion coefficient (undamaged GB)
D_{SF}	[m²/s]	surface diffusion coefficient
e	[m]	extrusion height (PSB)
$\vec{e}_{GB}$	[–]	unit vector of GB trace on surface
$\vec{e}_r$	[–]	unit vector, specimen-coordinate system
$\vec{e}_x$	[–]	unit vector, crystal-coordinate system
E	[MPa]	Young's modulus
E'	[MPa]	Young's modulus variable
E_b	[J]	bonding energy
E_c	[J/m²]	energy of GB cohesion
E_{Voigt}, E_{Reuss}	[MPa]	Young's modulus of the concepts of Voigt and Reuss, respectively
f	[–]	dislocation density distribution function
f	[Hz]	frequency
$f_{D\sigma}, f_{D\delta}$	[–]	Kachanov-type damage functions
f_{PSB}	[–]	volume fraction of persistent slip bands (PSBs)
f^α, f^γ	[–]	relative volumes of α-ferrite or γ-austenite, respectively
$f^{\alpha cont}, f^{\gamma cont}$	[–]	relative continuous volumes of α-ferrite or γ-austenite, respectively
$f^{\alpha sep}, f^{\gamma sep}$	[–]	relative separated volumes of α-ferrite or γ-austenite, respectively
F^{sep}	[–]	degree of separation
G	[MPa]	shear modulus
G	[MPa]	influence function
G_{Ic}	[J/m]	elastic–plastic fracture toughness
h, k, l	[–]	Miller indices for crystallographic plane (*hkl*)

J	[J/m^2]	J integral
J_c	[J/m^2]	critical value of the J integral
k	[MPa]	second-order constant of elasticity
k	[J/K]	Boltzmann constant ($k = 1.38 \times 10^{23}$ J/K)
k/k'	[MPa m$^{1/2}$]	Hall–Petch constant (related to σ_a or τ_a)
k_c	[MPa m$^{1/2}$]	cyclic Hall–Petch constant
k_c^α/k_c^γ	[MPa m$^{1/2}$]	cyclic Hall–Petch constant of α ferrite or γ austenite, respectively
$k_c^{\alpha\gamma}$	[MPa m$^{1/2}$]	cyclic Hall–Petch constant (only phase boundaries)
k_p''	[g^2 cm^{-4} s^{-1}]	parabolic rate constant (mass change due to metal oxidation)
k_σ, k_δ	[–]	exponents of grain boundary cohesion
K	[MPa m$^{1/2}$]	stress intensity factor
K_c	[MPa m$^{1/2}$]	critical stress intensity factor, fracture toughness
K_{cl}	[MPa m$^{1/2}$]	crack-closure stress intensity factor
K_{Ic}	[MPa m$^{1/2}$]	fracture toughness, loading mode I
K_{max}	[MPa m$^{1/2}$]	maximum value of stress intensity factor during fatigue cycles
K_{max}^*	[MPa m$^{1/2}$]	threshold value for K_{max} (two-parameter approach)
K_{min}	[MPa m$^{1/2}$]	minimum value of stress intensity factor during fatigue cycles
$K_{op,max}$	[MPa m$^{1/2}$]	maximum steady-state crack-opening stress intensity factor
K_{crit}^m	[MPa m$^{1/2}$]	threshold of microstructural stress intensity factor
K^m	[MPa m$^{1/2}$]	microstructural stress intensity factor
K_{op}	[MPa m$^{1/2}$]	crack-opening stress intensity factor
l_1, l_2, l_3	[–]	direction cosine between crystal axis and loading direction
$\vec{l}_e$ (l_x, l_y)	[–]	unit vector of observed slip trace
$\vec{l}_e^{th}$ (l_x^{th}, l_y^{th})	[–]	unit vector of theoretically possible slip trace
m^*, m_i^*	[–]	orientation factor (mean value, individual values)
$\underline{m}_1$, $\underline{m}_2$, $\underline{m}_3$	[–]	rotation matrices
$\underline{M}$	[–]	misorientation matrix, rotation matrix
M_S	[–]	Schmid factor
M_{Sa}	[–]	Sachs factor
M_T	[–]	Taylor factor
n	[–]	monotonic strain-hardening coefficient
n	[–]	Norton exponent (power-law creep)
n'	[–]	cyclic strain-hardening coefficient
n_i^c	[–]	critical a/c ratio in grain i
n_i^s	[–]	start value for a/c ratio in grain i
n_{BCS}	[–]	a/c ratio (BCS crack)
n_{NR}	[–]	a/c ratio (model of Navarro and de los Rios)
N	[–]	number of cycles
N_f	[–]	number of cycles to fracture

$N^{\alpha\alpha}/N^{\gamma\gamma}$	[–]	number of grain boundaries within α or γ phase, respectively
$N^{\alpha\gamma}$	[–]	number of αγ phase boundaries
p	[bar]	pressure
P	[N]	load
P	[m²/s]	grain-boundary diffusion ($P = s\delta D_{GB}$)
P_{SWT}	[MPa]	Smith–Watson–Topper damage parameter
Q	[J]	activation energy
r_0	[m]	distance between grain boundary and dislocation source
r_p	[m]	plastic zone size ahead of crack tip
$\vec{r}_{RD}$, $\vec{r}_{TD}$, $\vec{r}_{N}$,	[m]	direction vectors of the specimen coordinate system
r_α/r_γ	[–}	block parameter of the α ferrite or γ austenite phase, respectively
R	[–]	stress ratio, $R = \sigma_{min}/\sigma_{max}$
R	[J/(mol K)]	universal gas constant ($R = 8.31$ J/(mol K))
R	[Ω]	electric resistance
$\vec{R}$	[–]	rotation vector
R_a	[m]	mean surface roughness
s	[–]	grain boundary segregation factor
$\vec{s}$	[m]	dislocation line vector
s_{ij}, s_{ij}^i	[1/MPa]	elements of the compliance matrix
S_{crit}	[MPa]	critical stress for dislocation-source activation
$\underline{\underline{S}}$	[1/MPa]	single-crystal compliance tensor ([001]-coordinate system)
$\underline{\underline{S}}^i$	[1/MPa]	compliance tensor for grain i
t	[s]	time
T	[°C][K]	temperature
u, v, w	[–]	Miller indices for crystallographic direction [uvw]
U	[–]	effective stress intensity ratio
U	[J]	atomic distance potential
v	[m/s]	dislocation velocity
v	[m/s]	crack-propagation velocity da/dt
w	[J/m²]	strain energy density
W	[J]	work
$w_{el,eff}$	[J/m²]	effective elastic energy density
W_{el}^0	[J]	elastic potential energy (loaded structure without crack)
W_S	[J]	energy of surface formation
w_{pl}	[J/m²]	plastic energy density
W_{pl}	[J]	plastic work
Z	[J/m²]	cyclic J integral (= ΔJ)
Z_{eff}	[J/m²]	effective cyclic J integral

Greek Symbols

Symbol	Unit	Meaning
α	[grad]	opening angle of Kossel cone ($\alpha = 180° - \Theta$)
β	[–]	notch factor
γ	[–]	shear strain
γ	[–]	chemical activity coefficient
$\dot{\gamma}$	[1/s]	slip velocity
γ_{ij}	[–]	shear-strain components, $i, j = x, y, z, i \neq j$
γ_{el}	[J/m²]	specific ideal-elastic fracture work
γ_{GB}	[J/m²]	grain boundary energy
γ_{SF}	[J/m²]	specific surface energy
γ_{pl}	[J/m²]	specific plastic fracture work
Γ	[m]	integration line
Γ	[–]	accumulated plastic deformation / gross shear strain in a material
Γ_i	[–]	orientation function of grain i
δ	[m]	opening (atomic spacing) within the cohesive zone ahead of crack tip
δ	[m]	grain boundary width ($\delta \approx 0.5$ nm)
$\dot{\delta}_{cr}$	[m/s]	crack-opening velocity due to creep
Δa	[m]	crack-propagation increment
$\Delta \vec{b}_k$	[m]	displacement difference between load levels σ_k and σ_{k+1}
ΔCTOD	[m]	cyclic crack-opening displacement
ΔCTSD	[m]	cyclic crack-tip slide displacement
ΔG	[J]	activation energy
ΔJ	[J/m²]	cyclic J integral (= Z)
ΔK	[MPa m$^{1/2}$]	range of the stress intensity factor
ΔK^*	[MPa m$^{1/2}$]	threshold of ΔK (two-parameter approach)
ΔK_o	[MPa m$^{1/2}$]	threshold of ΔK
ΔK_{eff}	[MPa m$^{1/2}$]	effective range of the stress-intensity factor
$\Delta K_{th, eff}$	[MPa m$^{1/2}$]	effective threshold of ΔK
Δm	[kg]	mass change due to oxidation
ΔR	[Ω]	change in electric resistance
$\Delta \gamma_{pl, i}/2$	[–]	plastic shear-strain amplitude ith cycle
$(\Delta \gamma_{pl}/2)_{matrix}$	[–]	plastic shear-strain amplitude within matrix
$(\Delta \gamma_{pl}/2)_{PSB}$	[–]	plastic shear-strain amplitude within persistent slip band
$\Delta \varepsilon$	[–]	strain range
$\Delta \varepsilon/2$	[–]	strain amplitude
$\Delta \varepsilon_b$	[–]	Bauschinger strain
$\Delta \varepsilon_{pl}$	[–]	plastic-strain range
$\Delta \varepsilon_{pl}/2$	[–]	plastic-strain amplitude
$\Delta \Theta$	[grad]	tolerance range for special CSL boundaries (Σ values)

$\Delta\sigma$	[MPa]	stress range
$\Delta\sigma/2$	[MPa]	stress range amplitude
$\Delta\sigma_{FL}$	[MPa]	fatigue/endurance limit
$\Delta\sigma_{FL,i}$	[MPa]	local fatigue limit in grain i
$\Delta\sigma_{eff}$	[MPa]	effective stress range
$\Delta\sigma_{s}/2$	[MPa]	saturation stress amplitude
$\Delta\tau_{p}/2$	[MPa]	plateau stress amplitude
ε	[–]	total strain
$\underline{\varepsilon}$	[–]	strain tensor
ε_{B}	[–]	strain to fracture
ε_{el}	[–]	elastic strain
ε'_{f}	[–]	fatigue-ductility coefficient
ε_{ij}	[–]	strain tensor, $i, j = x, y, z$
$\dot{\varepsilon}_{GB}$	[1/s]	grain boundary sliding velocity
ε_{ii}	[–]	normal strain, $i = x, y, z$
ε_{m}	[–]	mean strain
ε_{pl}	[–]	plastic strain
$\dot{\varepsilon}_{st}$	[1/s]	steady-state creep rate
ζ	[–]	normalized coordinate (model of Navarro and de los Rios)
η_{KG}	[Ns/m^2]	grain boundary viscosity
Θ	[rad]	Bragg angle
Θ	[rad]	misorientation angle
Θ_{0}	[grad]	maximum misorientation of small-angle grain boundaries ($\Theta_0 = 15°$)
λ	[m]	wave length
λ	[m]	cell size of dislocation-cell structures
λ	[m]	diffusion length
μ	[J]	chemical potential
ν	[–]	Poisson ratio
ν_{Voigt}, ν_{Reuss}	[–]	Poisson ratio of the concepts of Voigt and Reuss, respectively
ξ	[rad]	twist misorientation between neighboring grains
Π	[J]	potential energy
ρ	[1/m^2]	dislocation density
ρ_{P}	[1/m^2]	dislocation density on parallel slip planes
ρ_{F}	[1/m^2]	forest dislocation density
σ	[MPa]	mechanical stress
$\underline{\sigma}$	[MPa]	stress tensor
σ_{0}	[MPa]	critical stress
$\sigma_{1}, \sigma_{2}, \sigma_{3}$	[MPa]	principal stresses (all shear stresses vanish)
σ_{a}	[MPa]	applied stress
σ_{cl}	[MPa]	crack-closure stress
σ'_{f}	[MPa]	fatigue strength coefficient
σ_{fr}	[MPa]	friction stress

σ_{ij}	[MPa]	stress tensor, $i, j = x, y, z$
σ_{ii}	[MPa]	normal stresses, $i = x, y, z$
σ_{m}	[MPa]	mean stress
$\sigma_{\max}$	[MPa]	maximum stress
$\sigma_{\min}$	[MPa]	minimum stress
σ^{i}_{nn}	[MPa]	stress normal to slip band (element i)
$\sigma^{i}_{\mathrm{nn,a}}$	[MPa]	applied normal stress (acting on element i)
σ_{op}	[MPa]	crack-opening stress
$\sigma_{\text{pile-up}}$	[MPa]	stress due to pileup at microstructural barrier
σ_{Y}	[MPa]	yield strength
σ_{Yc}	[MPa]	cyclic yield strength
$\sigma^{\alpha}_{\mathrm{Yc}}, \sigma^{\gamma}_{\mathrm{Yc}}$	[MPa]	cyclic yield strength of α ferrite or γ austenite, respectively
$\sigma^{\alpha\gamma}_{\mathrm{Yc}}$	[MPa]	cyclic yield strength (only $\alpha\gamma$ phase boundaries)
$\sigma^{\alpha\gamma}_{0\mathrm{c}}$	[MPa]	cyclic critical stress (only $\alpha\gamma$ phase boundaries)
$\sigma_{\mathrm{Y}i}$	[MPa]	yield strength in grain i
$\sigma_{\mathrm{Y(von\ Mises)}}$	[MPa]	von Mises equivalent stress
Σ	[–]	reciprocal fraction of coincident lattice sites of neighboring grains
τ	[MPa]	shear stress
τ_0	[MPa]	critical shear stress
τ_{a}	[MPa]	applied resolved shear stress (acting on slip band)
τ^{o}_{fr}	[MPa]	initial friction stress
$\tau_{\mathrm{fr}i}$	[MPa]	friction stress in grain i
τ_{ij}	[MPa]	shear stresses $i, j = x, y, z, i \neq j$
τ_{P}	[MPa]	Nabarro–Peierls stress
τ_{pass}	[MPa]	shear stress allowing dislocations to pass
τ_{s}	[MPa]	saturation stress
τ_{th}	[MPa]	theoretical critical shear stress
τ^{i}_{tn}	[MPa]	shear stress acting parallel to slip band (ith element)
$\tau^{i}_{\mathrm{tn,a}}$	[MPa]	resolved shear stress (acting on ith element)
τ_{Y}	[MPa]	(resolved) yield strength
$\varphi_1, \varphi_2, \Phi$	[rad]	Euler angles
ω	[m]	plastic zone size
Ω	[m^2]	atomic volume

Abbreviations

AES	Auger electron spectroscopy
ACPD	alternating current potential drop
AFM	atomic force microscopy
ASTM	American Society for Testing and Materials
CBED	convergent-beam electron diffraction
CCD	charge coupled device

CMOD	crack-mouth-opening displacement
CSL	coincidence site lattice
CSSC	cyclic stress–strain curve
CT	compact tension
DCPD	direct current potential drop
EBSD	electron backscattered diffraction
EPFM	elastic–plastic fracture mechanics
EFS	electrochemical fatigue sensor
FIB	focused ion beam
HCF	high-cycle fatigue
HRTEM	high-resolution transmission electron microscopy
ISDG	interferometric strain/displacement gauge
IST	incremental step test
LCB	low-cost beta
LCF	low-cycle fatigue
LEFM	linear elastic fracture mechanics
MFM	microstructural fracture mechanics
OIM™	orientation imaging microscopy
PSB	persistent slip band
SEM	scanning electron microscopy
RT	room temperature ($T = 298$ K)
SAD	selected area diffraction
SAGBO	stress-assisted (accelerated) grain boundary oxidation
SENB	single-edge-notched bend
SIT	silicon-intensified target
TEM	transmission electron microscopy
UHCF	ultrahigh-cycle fatigue
VHCF	very high-cycle fatigue

1
Introduction

One of my teachers in materials science once mentioned that the methodology of structural integrity belongs to the humanities rather than rational physics. Probably, there are only very few humanists interested in structural integrity. However, understanding materials fatigue and developing service-life prediction concepts became one of the major driving forces of the industrial and technological revolutions in the 19th and 20th centuries. Until today, choosing the right way in predicting a component's fatigue life is a matter of believing. Particularly in the high-cycle-fatigue (HCF) regime, there is mostly a factor of two and more between the predicted and the actual fatigue life. Even the question about the existence of a fatigue limit already raised 1860 by August Wöhler is not finally answered. Recent studies in the very-high-cycle-fatigue (VHCF) regime revealed that structures may fail well below the "fatigue limit" commonly defined as to occur at two millions of cycles (plain carbon steel).

The poor performance of fatigue-damage prediction methods is surprising when taking into account that elastic and plastic deformation of ductile materials is fairly well understood and that substantial progress has been achieved in the last twenty years in the fields of high-resolution materials characterization and powerful numerical simulation methods.

One of the central problems in structural integrity is the missing link between the microstructural dimensions of materials fatigue on the atomic length scale and the engineering design concepts for structures subjected to complex loading spectra. The moving together of design engineers and materials scientists is probably one of the main technical challenges of the 21st century. The steady need for smaller and lighter structures requires a reduction of existing safety factors and the revision of the total-life concept, which is based on the existence of a fatigue limit. It is the aim of this book to provide an overview of the general theoretical and experimental concepts of fatigue-life assessment and fracture mechanics (Chapters 2 and 3) as well as an insight into the fundamentals of elastic and plastic deformation (Chapter 4) and to review the current state in fatigue-crack-propagation research (Chapters 5 and 6). The focus is placed on a material's mesoscale, i.e., grain and phase distribution, and its interaction with crack initiation and propagation being responsible for the large scatter in fatigue life in the HCF and in the VHCF regimes. Depending on the material's strength, the phase of crack initiation and microcrack propagation determines up to 90% of fatigue life, while damage monitor-

Fatigue Crack Propagation in Metals and Alloys: Microstructural Aspects and Modelling Concepts. Ulrich Krupp

ISBN: 978-3-527-31537-6

ing using conventional nondestructive testing methods will not detect any technical cracks below 10^{-3} m dimension. It was shown that crack propagation on the micrometer scale is possible far below the threshold for long fatigue cracks. Hence, using conventional damage-tolerant life-prediction concepts based on fracture mechanics might be strongly nonconservative. Modern fatigue-damage simulation concepts accounting for the material microstructure and the transient development of crack closure effects (Chapter 7) can be considered as a baseline for the missing link between atomistic modeling and traditional concepts of engineering fatigue life assessment.

The situation becomes more complex when cycle-dependent fatigue damage is superimposed by time-dependent environmental effects, e.g., corrosion fatigue or stress-corrosion cracking. At elevated temperatures, interface diffusion of corrosive species becomes sufficiently fast to reduce the grain-boundary cohesion leading to intergranular crack propagation, when high tensile stresses act on the grain-boundary plane. This mechanism has been termed dynamic embrittlement and has been identified as a generic failure mechanism (Chapters 6 and 7).

Generally, fatigue damage is strongly dependent on local microstructural features like the grain size and geometry, the crystallographic orientation relationship, and the grain-boundary structure. This is proven and discussed by various examples and the implementation in numerical modeling concepts, which can be used not only for mechanism-based fatigue-life prediction but also for the derivation of trends for microstructure design, e.g., grain-boundary engineering.

2
Basic Concepts of Metal Fatigue and Fracture in the Engineering Design Process

2.1
Historical Overview

In the modern world, being characterized by a high degree of mobility, automation, telecommunication and computer networking, fatigue and fracture of engineering components have become an unpleasant but quite frequent experience of daily life. Even though the fatigue behavior of materials has been the subject of continuous research work for more than 150 years, there have been only a few spectacular failure cases in combination with technological development that really pushed the state of the art in fatigue research and service-life prediction: railroad accidents in the 19th century, aerospace technology and the invention of electron microscopy in the 20th century, and nanotechnology in the 21st century.

In ancient times, procedures to ensure the safety of a construction were draconic, e.g., while testing of bridges by overloads the civil engineer in charge had to stay below the construction. At least this could help to avoid carelessness during the design process. Later, people noticed that buildings having withstood the overload testing procedure might have a tendency to collapse even after years of being in use. Hence, time-dependent damage mechanisms like weathering effects or complex loading conditions had to be taken into consideration by improving the design, according to the trial-and-error method. For instance, palaces or aqueducts of the ancient Romans demonstrate how far engineering design techniques were developed 2000 years ago.

Systematic analysis of the mechanical failure behavior of metals probably started in the 15th century, when mastermind Leonardo da Vinci attributed his observation that mechanical strength of metal wires decreases when increasing their length to an increasing significance of defects in the material [1]. But it was not before the 18th century, when James Watt invented the steam machine and therefore put the cornerstone for industrialization and mobilization in place, that the focus of scientific interest was slowly directed towards the mechanical behavior of metals and alloys subjected to cyclic loading. Due to frequent accidents during continuous operation of stagecoaches and later steam engines (Fig. 2.1) engineers became aware of the fact that the service life of forged axles or shafts is limited.

Fatigue Crack Propagation in Metals and Alloys: Microstructural Aspects and Modelling Concepts. Ulrich Krupp

ISBN: 978-3-527-31537-6

Fig. 2.1 Steam engine of the New Hope and Ivyland Railroad, Pennsylvania, USA.

A railroad accident of particular tragedy occurred in May 1842, when a train fully occupied by 1500 people and pulled by two locomotives on its way from King Louis Philipe's birthday celebration in Versailles to Paris crashed as a consequence of a fractured axle. The affected locomotive immediately stopped so that the following engine and the carriages were piled up and squashed. Presumably, more than 50 people lost their lives during the Versailles disaster. The tragedy that is described in more detail by Smith [2] had become a milestone in fatigue research, even though the term "fatigue" itself was introduced later in 1854 by Braithwaite [3]. The first failure analyses of the failed axle were certainly rather speculative; the sudden fracture was attributed to time-dependent changes in the material's microstructure developing a certain kind of "crystalline texture" caused by vibrations, heat and magnetic induction. A remarkable contribution to the discussion of the reasons for the Versailles accident was provided by Wilam J. M. Rankine (1820–1872), who later became a leading scientist in the thermodynamics of the 19th century (Rankine process). Rankine excluded any time-dependent changes of the material. Based on the observation of the fracture surface, where damage seemed to have started by a regularly shaped plane pointing from the outer diameter (fatigue-crack initiation) to the center of the axle, he concluded that sudden fracture is the consequence of continuous decrease in the load-carrying cross section.

The first systematic and extensive experimental studies of the fatigue behavior of railroad axles were introduced by the Royal Prussian Obermaschinenmeister August Wöhler (1819–1914), laying the cornerstone for systematic approaches to fatigue-life prediction [4]. Wöhler himself carried out strain measurements during train journeys on a 5000-km reference route. He found particular high-strain peaks when passing over switches. To this day, the number of passed switches is a measure to determine an approximate value of the maximum allowable stresses in railroad design [5]. However, the famous Wöhler diagram (S–N diagram, see Fig. 2.2), which represents the applied stress amplitudes vs. number of cycles to fracture,

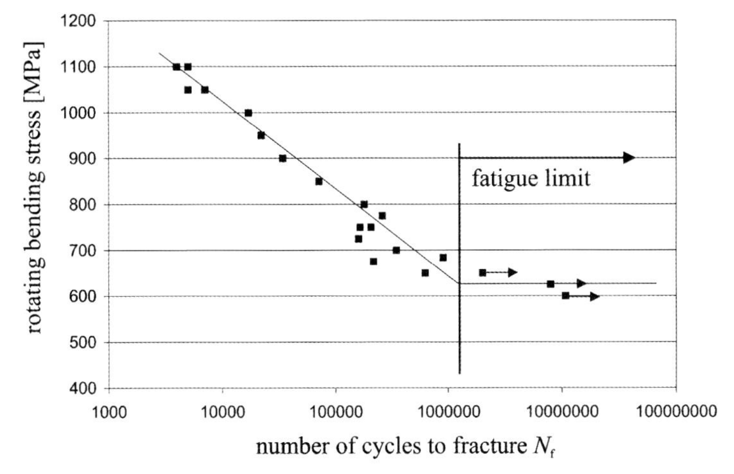

Fig. 2.2 Wöhler diagram of PH 15-5 precipitation-hardened steel (X3CrNiCuNb 15 5 4) showing rotating bending stress vs. number of cycles to fracture.

was not first introduced by Wöhler, who summarized his results in tables, but by Spangenberg, successor to Wöhler as head of the Mechanisch Technische Versuchsanstalt in Berlin.

In spite of the promising results of Wöhler, his co-workers and successors, the physical mechanisms of material fatigue were more or less completely unknown at the beginning of the 20th century. Due to the lack of high-resolution and high-depth-of-focus microscopy, observation of microstructural changes in materials was strongly limited. An exception was the outstanding work of Johann Bauschinger (1834–1893), professor for mathematics and mechanics at the Technical University Munich, who found and discussed a decrease in the yield strength after load reversal into plastic compression from initial plastic tension and vice versa (see [6] and Chapter 4).

The first direct microscopic observations of microstructural mechanisms during cyclic deformation of metals were made by Ewing and Humfrey at the University of Cambridge in 1903. They studied the polished surface of Swedish iron specimens subjected to rotation bending and found the formation and growth of slip bands in certain crystallites which eventually developed into fatigue cracks [7]. Even though fatigue research in the early 20th century was determined mainly by phenomenological and empirical studies, a few concepts remained of particular significance even for today's service-life assessment. Among them is the logarithmic Basquin relationship between the number of cycles to fracture N_f (corresponding to $2N_f$ load reversals) and the respective stress amplitude $\Delta\sigma/2$ of load reversals published in 1910 [8] (cf. Fig. 2.9):

$$\frac{\Delta\sigma}{2} = \sigma_f'\left(2N_f\right)^b \tag{2.1}$$

where σ_f' denotes the coefficient and b the exponent of fatigue strength. To account for a variation of the stress amplitudes during fatigue life of technical components, Palmgreen [9] in 1924, and later Miner [10] in 1945 with additional experimental support, introduced the hypothesis of linear damage accumulation, which is again an important tool of today's concepts of service-life assessment:

$$\sum_i \frac{N_i}{N_{fi}} = 1.0 \tag{2.2}$$

The summands of the damage sum given by Eq. (2.2) are the quotients of the actual numbers of cycles at a certain stress amplitude N_i and the respective number of cycles to fracture N_{fi}. However, one should be aware of the fact that the Miner rule is often a strong oversimplification. Depending on the load history, the damage sum can be higher or lower than one.

Besides the empirical Wöhler approach to predict fatigue life, a new concept had been introduced in the early 20th century renewing the basic idea of Leonardo da Vinci to attribute the susceptibility to fracture to material inhomogeneities. The new concept, termed fracture mechanics, focuses on the local analysis of such inhomogeneities, especially of cracks, by means of stress parameters [11] and energy parameters [12] correlating the problem of material failure by cracking with stress concentration at the crack tips. Later, the reduction of fatigue life by the influence of notches became a vital part of fatigue-life design. August Thum, the first holder of the chair of materials science at the Technical University Darmstadt, and Heinz Neuber, professor of engineering mechanics at the Technical University Munich, played a key role in establishing the field of structural integrity in materials engineering, e.g., by introducing the theory of notch stresses and fatigue notch factors [13, 14].

With the advent of World War II the number of fatigue-caused plane crashes increased tremendously giving rise to aircraft construction becoming the key driving force for metal-fatigue research. This was led to new microstructural dimensions by both the development of the dislocation theory by Polanyi [15] and Orowan [16], and by the invention of electron microscopy and its application to observe dislocations by Hirsch et al. [17]. The acceleration in the attainment of scientific knowledge in fatigue research is, last but not least, a consequence of improved methods to measure microstrains and crack propagation as well as of the introduction of servohydraulic mechanical testing systems in the late 1950s.

In the second half of the 20th century the main focus of fatigue research shifted towards the mechanisms of plastic deformation and the fracture mechanical treatment of fatigue-crack propagation. Independently of each other, Manson [18] and Coffin [19] proposed a simple relationship between the plastic strain amplitude $\Delta\varepsilon_{pl}/2$ and the number of cycles to fracture N_f, which is still used today to describe the low-cycle fatigue (LCF) regime of the strain vs. life curve (see Fig. 2.9):

$$\frac{\Delta\varepsilon_{pl}}{2} = \varepsilon_f'(2N_f)^c \tag{2.3}$$

where, by analogy to the Basquin relationship that is predominantly used to describe the high-cycle fatigue regime (HCF), $\varepsilon_{\mathrm{f}}'$ is the ductility coefficient and c the exponent of fatigue ductility.

Addressing the microstructural mechanisms of fatigue damage, Thompson et al. [20] found the significance of persistent slip bands (PSBs, see Section 5.1), while the phenomenon of the occurrence of striations in the fracture surface of fatigued metallic samples was firstly discussed by Zappfe and Worden in 1955 [21] and later in more detail by Forsyth and Ryder in 1960 [22]. The relationship between plastic deformation (blunting) ahead of a propagating fatigue crack and the formation of striations can be understood following the remarkable work of Laird [23] and Neumann [24] and is the subject of Section 6.2.2.

In spite of the significance of plastic deformation in the crack-tip zone, the linear-elastic-fracture-mechanics (LEFM) approach became an established theoretical tool to treat crack-propagation processes. In 1958, George R. Irwin introduced the stress-intensity factor K for static fracture analysis [25]

$$K = \sigma \sqrt{\pi a} Y \tag{2.4}$$

as an universal parameter to be used independent of component or sample dimensions since it depends at the same time on the crack length a and the mechanical stress σ (Y is a geometry function). Irwin's concept was later used by Paul C. Paris, during this time assistant professor at the University of Wahington in 1961, who took into account that a structure subject to fatigue has always flaws or defects that may grow by a rate da/dN, driven by the local stress-intensity range ΔK [26, 27]:

$$\frac{da}{dN} = C \Delta K^{n} \tag{2.5}$$

Even though this simple empiric equation, containing only two material-specific constants C and n, was initially not widely accepted and rejected by scientific journals, Paris' law has become the cornerstone of the use of fracture-mechanical methods in damage-tolerant service-life assessment in the aviation and automotive industries. Figure 2.3 shows examples of experimentally determined crack-propagation curves (after [28]) and their evaluation according to "Paris' law", Eq. (2.5), that is applicable only for stress-intensity range higher than the threshold value ΔK_{th} considered as a criterion for technical crack initiation.

However, in its original form, Paris' law is an oversimplification for many problems of fatigue-crack propagation, since many important parameters are neglected, e.g., the mean stress σ_{m} or crack-closure effects.

A significant modification for Paris' law arose form the work of Wolf Elber, carried out as a graduate student in the 1960s at the University of New South Wales. He noticed that the crack faces of a growing fatigue crack may have contact before completely unloading the sample from a tensile load [29, 30]. This kind of premature crack closure, which can be attributed to plasticity effects, roughness or surface oxidation, gives rise to a decrease in the crack-driving stress-intensity range to a value of ΔK_{eff} to be used instead of ΔK in Eq. (2.5). Elber's argument, which

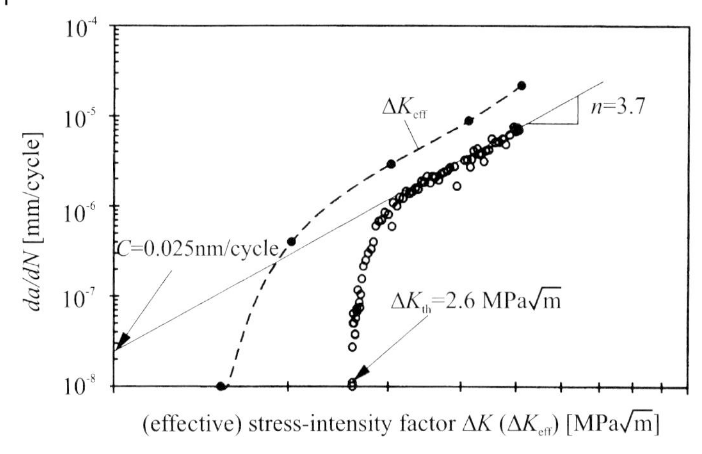

Fig. 2.3 Crack-propagation rate da/dN vs. the range of the stress-intensity factor ΔK for a X8019+12.5%SiC aluminum alloy at a stress ratio of $R = 0.1$ with (ΔK_{eff}) and without (ΔK) consideration of crack-closure effects (after [28]).

seems at a first glance to be very plausible, generated much, still ongoing, controversial discussion during the last three decades.

Even though LEFM became a well-established approach in particular for structural integrity concepts of low-weight constructions, at the beginning of the 1970s design questions were raised that could not be treated without taking substantial crack-tip plasticity into account. On the one hand, an absolutely safe service-life assessment had to be applied to tubes for nuclear power plants made of ductile austenitic steels; on the other hand, there was an increasing need to develop and use modern high-performance materials and to reasonably define their fatigue-strength limits (e.g., for the application of gas turbine blades at temperatures up to 1300 °C). Hence, elastic–plastic fracture mechanics (EPFM) approaches like the cyclic J integral and cyclic crack-tip opening (ΔCTOD) became important parameters to describe locally the driving forces to fatigue-crack propagation.

Furthermore, it turned out that the microstructural mechanisms at the crack tip had to be reconsidered when studying the early stages or the HCF regime of fatigue-crack growth. This is discussed in detail in Chapters 5 to 7. Pearson [31] was in 1975 among the first who found that short cracks of a length within the microstructural dimension of the alloy, and hence not detectable by conventional methods of nondestructive crack-detection, may grow at an unexpected high rate and at stress-intensity ranges below the threshold value ΔK_{th} for technical (long) crack initiation. This implies that the definition of service intervals according to long-crack behavior (Paris' law) can be nonconservative. The "abnormal" behavior of short-crack propagation has attracted much experimental and modeling research activity, but nevertheless it has not yet been accounted for in an explicit way within structural integrity concepts. Certainly, this is due to the fact that technical failures can hardly be attributed in a doubtless manner to short-crack effects.

Finally, the benefit of fatigue research is not exclusively the improvement of service-life prediction methods in engineering design. Beyond this, the mechanism-based quantification of fatigue-damage processes allows the tailor-made optimization and adaptation of new materials and processing routes. It is shown in Chapters 5 to 7 that the early phase of crack propagation can be influenced by adjusting a material's microstructure and, hence, to increase the fatigue life until the occurrence of technical crack initiation, which can comprise more than 90% of the total fatigue life until fracture [32, 33].

Two more recent severe accidents have shown that in spite of a tremendous number of scientific studies in the field of fatigue mechanism – already in 1969 more than 21 000 publications were counted [34], a number that probably has been increased to more than 100 000 in the meantime – today's dimensioning methods and technical regulations are not sufficient to eliminate any risk of fatigue damage. In early summer 1998 a broken wheel rim failed by fatigue caused the derail of a German ICE high-speed train. The fold-up of the following carriages and the eventual collision with a bridge that fall on top of the train made the accident with 101 casualties the worst one of German railroad history. The responsible fatigue damage process for this disaster is still the subject of investigations [35]. In fall 2001, a fully occupied airbus A300-600 crashed into New York's district of Queens, and more than 250 people were killed. First investigations of the accident revealed that fracture of the vertical stabilizer and the rudder made of carbon-fiber-reinforced composite was due to high turbulences in the wake of a preceding aircraft. Even though this accident could be attributed to an inappropriate manipulation of the rudder by the pilot leading to stresses beyond the design limit, these kind of catastrophic material failure support the necessity of ongoing research in the field of fatigue and fracture of engineering materials with new challenges:

- Development of methods to recognize dangerous microstructural changes, e.g., small cracks, during operation (damage monitoring).
- Analysis of time- and cycle-dependent failure mechanisms of new materials like composite materials, nanomaterials, substrate-coating systems, cellular metals, ceramics and composites (cf. Suresh's book on fatigue of materials [36]).
- Development of new, and improvement of existing, mechanism-based damage-prediction and service-life-prediction methods and their consequent application in industrial design processes and implementation into teaching of university engineering classes.

In spite of the many problems and failure cases still occurring due to fatigue of materials, one should take into consideration that every day planes, cars, trains and ships move millions of miles without any technical trouble. However, nowadays a successful flight over the Atlantic Ocean is not attributed to the excellent work in fatigue and fracture research and concepts of structural integrity.

2.2
Metal Fatigue, Crack Propagation and Service-Life Prediction: A Brief Introduction

Comparing failure of a ductile material during a tensile test with that during a fatigue test, one may notice that in the latter case fracture occurs suddenly without any preceding macroscopic indication, while the tensile specimen shows necking prior to fracture (Fig. 2.4). The course of the monotonic stress vs. strain curve during uniaxial deformation shows clearly the change from reversible elastic deformation, described by Hooke's law

$$\sigma = E\,\varepsilon \tag{2.6}$$

with the applied stress σ, the strain ε and the isotropic Young's modulus E, to irreversible plastic deformation when exceeding the yield strength σ_Y (often defined by a 0.2% plastic-strain offset $\sigma_{p0.2}$; Fig. 2.4).

In contrast to this, specimens subjected to fatigue (Wöhler experiments) may fail by fracture at stress levels far below the yield strength; i.e., within the macroscopically reversible regime. Accounting for this behavior is certainly of essential significance for dimensioning cyclically loaded technical components: data from monotonic tensile tests are of very limited utility. The question is: What is the reason for some kind of material damage at load levels for which one would expect just a reversible deflection of atoms from their equilibrium sites?

As has been known since Leonardo da Vinci's tensile tests, technical materials contain imperfections like surface roughness, notches, pores, inclusions and seg-

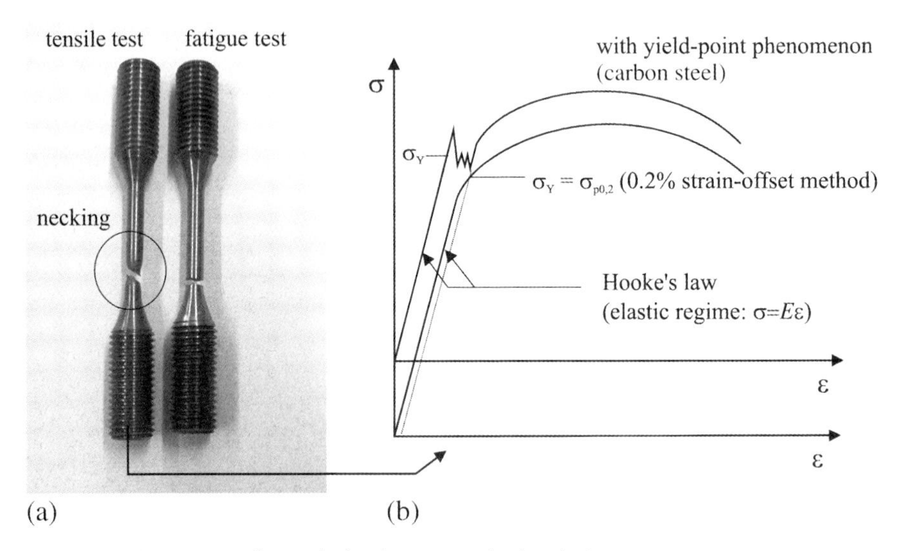

Fig. 2.4 (a) Comparison of two cylindrical specimens broken during a monotonic tensile test and broken during fatigue. (b) Stress vs. strain diagram for monotonic tensile tests.

regation of alloying elements, as well as a microstructure of grains of various crystallographic orientation, morphology and boundary structures. These imperfections may act as stress raisers, giving rise to local exceeding of the microscopic yield strength, and hence activation of plastic slip by generation and movement of dislocations (this is described in more detail in Chapter 4). The main origin of technical stress concentration is design notches, e.g., at crankshafts. Since at such locations a certain degree of plasticity has often to be tolerated due to economic factors, the stress state in the vicinity of such notches has to be carefully analyzed, e.g., by finite-element calculations (structural-stress approach) or measurements. The corresponding cyclic stress-strain behavior reflecting the situation in the root of a notch can be determined by LCF experiments applying the respective stress levels. Usually such experiments are terminated when the stiffness of the specimen shows a steep decrease due to the onset of instable crack propagation. To correlate damage evolution with the loading conditions of structures that already contain cracks, cyclic deformation tests are required on characteristically notched specimens to determine the development of the crack-propagation rate da/dN. The relationship between fatigue-life design of a notched component and the application of LCF and crack-propagation experiments is shown schematically in Fig. 2.5 and can be used for the design of cyclically loaded components.

From the materials science and design point of view, microstructural stress concentrators are of particular importance. That is why the main focus of Chapters 5 to 7 is placed on the influence of crystallographic orientation relationships of polycrystalline microstructures on the fatigue and crack-propagation properties of metals and alloys.

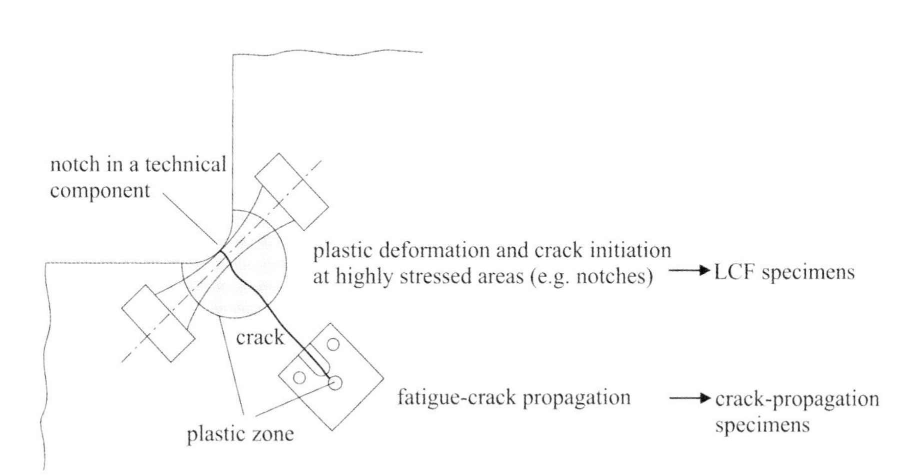

Fig. 2.5 Use of LCF and crack propagation specimens to represent the fatigue damage behavior in terms of cyclic deformation and fatigue crack propagation in the vicinity of a notch (after [36]).

2.2.1
Fundamental Terms in Fatigue of Materials

In any case, fatigue of materials is a consequence of a time-dependent stress–strain loading spectrum. In the most simple case, such a spectrum is described by a sine curve as shown in Fig. 2.6a. Each cycle can – in analogy to the monotonic stress–strain curve of the tensile test – be represented by a stress–strain hysteresis loop (Fig. 2.6b), where σ_{max} and σ_{min} denote the absolute value maximum and minimum stress of the cycle, $\sigma_m = (\sigma_{max}+\sigma_{min})/2$ and $\varepsilon_m = (\varepsilon_{max}+\varepsilon_{min})/2$ the mean stress and mean strain, respectively, $\Delta\sigma/2 = (\sigma_{max} - \sigma_{min})/2$ the stress amplitude, ε_{max} and ε_{min} the respective maximum and minimum strain, and $\Delta\varepsilon/2$ the strain amplitude. The index "pl" refers to the plastic part of the strain values, which are obtained by subtracting the elastic strain from the total strain; i.e., for the plastic-strain amplitude this yields $\Delta\varepsilon_{pl}/2 = \Delta\varepsilon/2 - (\Delta\sigma/2)/E$. A parameter of particular significance is the stress ratio $R = \sigma_{min} / \sigma_{max}$, e.g., for analyzing crack-closure effects. For fully reversed cycling ($\sigma_m = 0$) the stress ratio is $R = -1$, for cycling in pure tension compression ($\sigma_{min} = 0$) it is $R = 0$.

Since plotting the whole set of stress–strain hysteresis loops is not very useful – in the case of HCF 10^4 to $>10^6$ cycles need to be considered – the maximum, minimum and mean values are represented in cyclic-deformation curves vs. the number of cycles or the cumulative plastic strain $\varepsilon_{pl,cum}$. The schematic in Fig. 2.7 reveals how transient processes are represented by cyclic deformation curves, e.g., cyclic softening or hardening, cycle-dependent changes in the mean stress by cyclic creep or cyclic relaxation. Cyclic hardening is due to the increasing degree of irreversibility and blocking of dislocation motion (Chapter 4) and can be observed in a pronounced way during fatigue of initially annealed metals. Cyclic softening may occur during fatigue of work-hardened materials or due to Lüders-band formation in carbon steel; i.e., dislocations initially blocked by build-up of interstitial carbon will break free like avalanches once a certain plastic strain level is exceeded, leading to a subsequent decrease in the stress level.

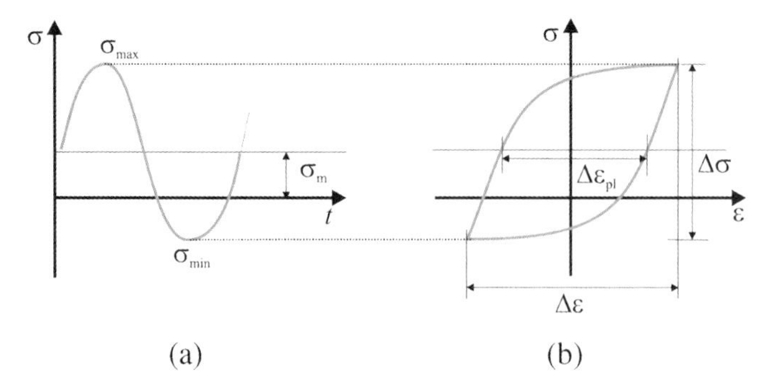

Fig. 2.6 Schematic representation of (a) sinusoidal fatigue loading with a mean stress σ_m as stress vs. time plot and (b) corresponding stress–strain hysteresis loop with definition of the main parameters (see text for details).

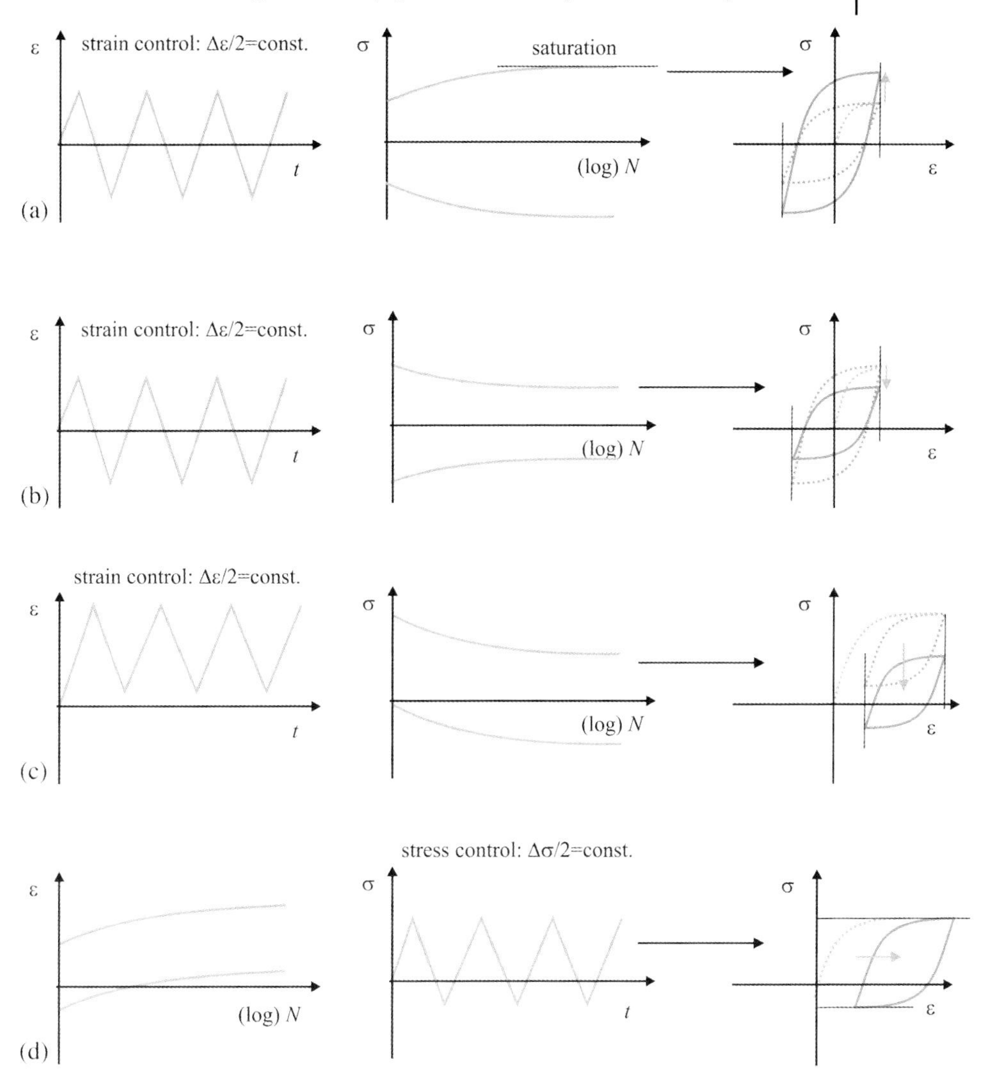

Fig. 2.7 Transient processes during cyclic deformation: (a) cyclic strengthening, (b) cyclic softening, (c) cyclic relaxation and (d) cyclic creep.

In many cases, the transient regime is followed by a steady-state saturation regime (cf. Fig. 2.7), during which the stress–strain hysteresis loops exhibit almost no changes. By plotting the respective stress–strain data pairs of several fatigue tests of different load levels into a stress vs. strain diagram one obtains the cyclic stress–strain curve (CSSC), the pendant to the monotonic stress–strain curve of the tensile test. In the case of an absence of a steady-state saturation regime the data of cycle $0.5N_f$ should be used. Since the cyclic stress–strain curve represents the influence of transient processes, it is commonly used in service-strength analysis. The course of the cyclic stress–strain curve can be mathematically described in analogy to the three-parameter approach for the monotonic stress–strain curve of Ramberg and Osgood [37], simply by using the stress and strain-range values:

$$\frac{\Delta\varepsilon}{2} = \frac{\Delta\sigma}{2E} + \left(\frac{\Delta\sigma}{2A'}\right)^{1/n'} \tag{2.7}$$

Here, A' is the cyclic strength coefficient and n' the cyclic exponent of strain hardening (n' varies for most metals and alloys between 0.1 and 0.2 [36]). Assuming Masing behavior [38]; i.e. the increasing branches of hysteresis loops for various load levels must be coincident, the stress–strain hysteresis loops can be approximated. This relationship is shown in Fig. 2.8.

Normally, analysis of cyclic-deformation behavior of materials by means of Wöhler diagrams and stress–strain hysteresis loops does not imply the mechanisms of crack initiation and crack propagation. There are specific experimental methods available to include fatigue-crack initiation and propagation into service-strength analysis (cf. Chapter 3), e.g., replica techniques, *in situ* observation of cracks by optical microscopy or potential-drop measurements on standardized notched specimens. The respective fracture-mechanical approaches are introduced in Sections 2.2.3 and 2.3 and further discussed in Chapter 7 regarding the applicability to propagation mechanisms of microstructurally short fatigue cracks.

At this point it should be mentioned that the present overview on methods of service-strength analysis and fatigue-life prediction is a coarse and incomplete one. A comprehensive introduction and more details can be found in the books by Radaj [39], Haibach [40] or Naubereit [41] the latter in compressed form but with several examples given to be used with a computer.

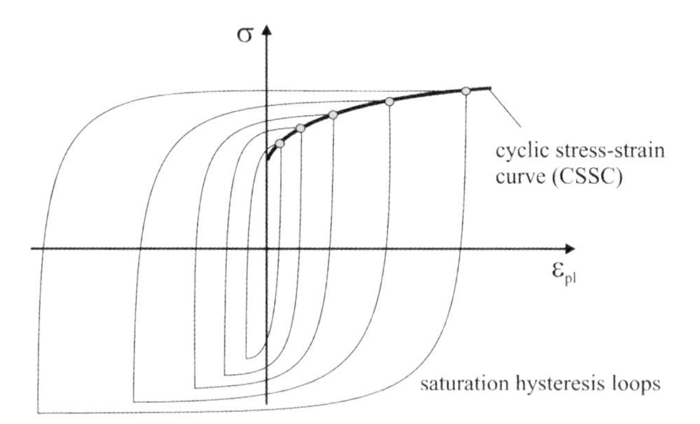

Fig. 2.8 Schematic representation of the cyclic stress–strain curve (CSSC) obtained from hysteresis loops of the saturation regime of fatigue tests with different stress amplitudes.

2.2.2
Fatigue-Life Prediction: Total-Life and Safe-Life Approach

By means of stress- or strain-based Wöhler diagrams the approximate number of cycles can be predicted up to which a component can be used until a failure criterion is reached, either defined as complete fracture, or by exceeding a critical crack length or maximum strain, according to dimensioning of the whole system.

The stress-based Wöhler diagram is used for fatigue-life prediction according to the nominal-stress approach. The nominal stress is obtained by multiplying mechanical stresses, determined experimentally or by finite-element calculations, with an influence factor accounting for variations in, e.g., the type of loading, frequency, temperature, etc. The data points to create a Wöhler diagram can be derived from cyclic-loading tests either on components or standardized specimens. To accommodate the strong stochastic scattering of the results the application of statistical approaches is advisable, e.g., the $\arcsin\left(\sqrt{P}\right)$ method to obtain a scatter band of failure probabilities [42] or the staircase method to obtain reliable values for the fatigue limit [43].

The strain-based Wöhler diagram can be described mathematically by superimposing Basquin's equation (Eq. 2.1) for the HCF regime ($N_f > 10^4$ cycles) and Manson's and Coffin's equation (Eq. 2.3) for the LCF regime ($N_f \leq 10^4$ cycles) as shown in Fig. 2.9. Taking the detrimental effect of a mean stress after Morrow [44] into account, the superimposition yields the following relationship between the total strain amplitude and the number of cycles to technical crack initiation N_f:

$$\frac{\Delta\varepsilon}{2} = \frac{\sigma_f' - \sigma_m}{E}\left(2N_f\right)^b + \varepsilon_f'\left(2N_f\right)^c \tag{2.8}$$

The strain-based Wöhler diagram can be approximated by data from monotonic tensile tests (see [39]) using the ultimate tensile strength σ_{UTS} and the true strain to fracture ε_f = ln(original specimen cross section/cross section after fracture):

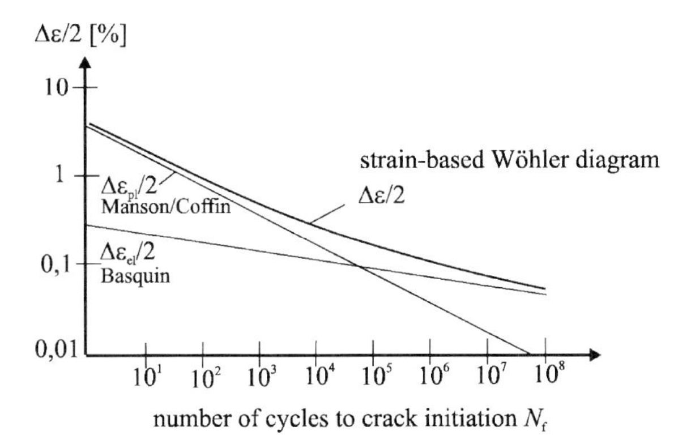

Fig. 2.9 Example of a strain-based Wöhler diagram (for 0.2%C plain steel) (after [39]).

$$\Delta\varepsilon = 3.5\frac{\sigma_Z}{E}N_f^{-0.12} + \varepsilon_f^{0.6}N_f^{-0.6} \tag{2.9}$$

An alternative way to estimate the service life of technical components accounting for the effect of notches as stress raisers is the local-strain approach (notch stress concept). Using Neuber's rule (cf., e.g., [14, 36]), the local deviation $\Delta\sigma_k$ from the nominal fully reversed stress range $\Delta\sigma$ and the corresponding strain-range deviation within the notch $\Delta\varepsilon_k$ are represented by means of a fatigue notch factor k:

$$k = \frac{\sqrt{\Delta\sigma_n \Delta\varepsilon_n E}}{\Delta\sigma} \quad \Rightarrow \quad \Delta\sigma_n \Delta\varepsilon_n = (k\Delta\sigma)^2/E = \text{const} \tag{2.10}$$

Equation (2.10) can be graphically represented as hyperbola (Fig. 2.10). The intersection X between this hyperbola and the cyclic stress–strain curve of the material defines the local stress–strain state at the notch tip (σ_n and ε_n, respectively) for the nominal maximum stress σ_{max}. The intersection point X serves as the origin for load reversal to the nominal compressive stress σ_{min} by $\Delta\sigma = \sigma_{min} - \sigma_{max}$. The new stress–strain state at the notch tip corresponds to the intersection between an experimental stress–strain hysteresis loop and the Neuber hyperbola applied to $\Delta\sigma_n$ and $\Delta\varepsilon_n$ with respect to the origin X. This approach to determine the cycle-dependent stress–strain behavior at the notch tip is shown schematically in Fig. 2.10.

Combining the constitutive equation for the cyclic stress–strain curve (Eq. 2.7) with the right-hand part of Eq. (2.10) yields

$$\frac{(k\Delta\sigma)^2}{E} = \Delta\varepsilon_n \Delta\sigma_n = \frac{(\Delta\sigma_n)^2}{E} + \Delta\sigma_n\left(\frac{\Delta\sigma_n}{A'}\right)^{1/n'} \tag{2.11}$$

Hence, the number of cycles to crack initiation can be estimated, provided the cyclic stress–strain curve, the Wöhler diagram and the fatigue notch factor are known.

Besides the actual local strain range within a notch, there are several additional parameters of the loading spectrum and the material microstructure that signifi-

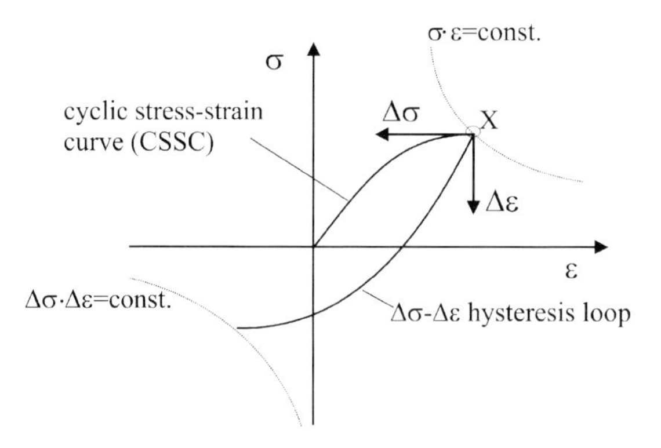

Fig. 2.10 Schematic representation of the application of Neuber's rule according to Eq. (2.10) (after [36]).

cantly influence the fatigue life. Most important are effects due to an applied mean stress, load history (overloads) or crack closure. Staying with the formalism of the local-strain approach and the strain-based Wöhler diagram for fully reversed cyclic deformation ($R = -1$), these effects can be taken into account by means of damage parameters. Probably the most common damage parameter is the one derived in 1970 by Smith et al. [45]:

$$P_{SWT} = \sqrt{\sigma_{max} \, {}^{\Delta\varepsilon}\!/_{2} \, E} \tag{2.12}$$

The damage parameter P_{SWT} contains the total-strain amplitude $\Delta\varepsilon/2$ and the maximum stress σ_{max}, assuming that fatigue damage is limited to nominal tensile stresses $\sigma_{max} > 0$. Further damage parameters that have been published include the effective stress amplitude σ_{eff} accounting for crack-closure effects that reduce the stress range during which local fatigue damage mechanisms are active [46] or the influence of a pre-deformation that alters a material's microstructure [47, 48]. The combination of the damage parameter with the mathematical expression of the strain-based Wöhler diagram (Eq. 2.8) yields the damage-parameter Wöhler diagram, e.g., for the Smith–Watson–Topper parameter in the form

$$P_{SWT} = \sqrt{\sigma_f'^2 (2N_f)^{2b} + \varepsilon_f' \sigma_f' (2N_f)^{b+c}} \tag{2.13}$$

Instead of uniform cyclic loading conditions, most technical components experience complex loading spectra during service life. In spite of the problem to consider history effects in a mechanism-based way, the damage accumulation approach using the Miner rule (Eq. (2.2)) or modifications of it is the most common way of service-life prediction for non-uniform cyclic loading of technical components, e.g., wheels of automobiles or trucks. To analyze fatigue-loading spectra to implement them in fatigue-life assessment concepts, counting methods are used. In the most simple way this can be done by assigning the stochastic load distribution vs. time or number of cycles to discrete load ranges that can be easily treated by the Miner rule. For instance, exceeding of predefined maximum and/or minimum values by the stress–strain course are counted (single parametric approaches, cf. [39]). Dual parametric approaches consider both the range as well as the respective mean value of the stress–strain course. A well-established method that accounts for "memory rules" (Masing behavior) is the rainflow-counting approach according to Matsuishi and Endo [49] (see also Dowling and Socie [50] or, in a more current modification, Anthes [51]). To replace a random loading spectrum by a set of complete stress–strain hysteresis loops, the actual strain course vs. time is represented in a vertical way according to Fig. 2.11a. Starting "rain" flows off all strain amplitudes as if they were roofs of a pagoda according to the following rules. The rain flow is stopped (*i*) when it reaches a load value where the opposite load minimum falls below the starting minimum; i.e., $|\varepsilon_H|$ is below $|\varepsilon_0|$ (rain flow coming from ε_0 is stopped at ε_C), (*ii*) when it reaches a load value where the opposite load maximum is higher than the starting maximum; i.e., $|\varepsilon_T|$ is above $|\varepsilon_N|$ (rain flow coming from ε_M is stopped at ε_Q), or (*iii*) when it is interrupted by rain coming

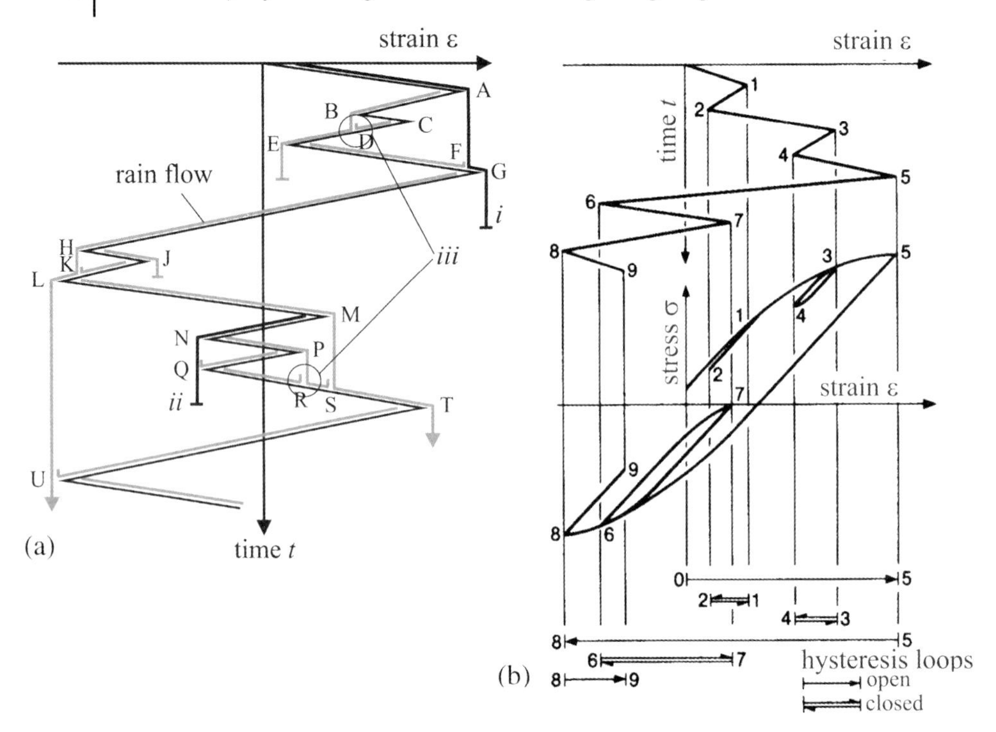

Fig. 2.11 Application of the rain-flow counting method: (a) rain flow along a random loading spectrum; (b) construction of stress–strain hysteresis loops (after [39]).

from the pagoda roof above, e.g., at ε_D or ε_R. By implementing these rules as computer algorithms in fatigue-life assessment programs one can replace random loading spectra together with the experimental cyclic stress–strain curve stepwise as hysteresis loops as shown in Fig. 2.11b. Only when a hysteresis loop is complete will it be counted as one loading cycle of a certain class, corresponding to the partitioning of the loading spectrum. Again, these loading classes can be treated by the damage-accumulation approach after Palmgreen and Miner (Eq. 2.22).

In many cases, three-dimensional structural integrity problems should not be simplified by applying a uniaxial-loading approach, since real loading and damage conditions are determined by complex superimposition of cyclic torsion, bending and compression–tension. To treat such kinds of multiaxial fatigue, the concept of the critical plane has become established (see [36]). The critical plane represents the area where the service-life-relevant damage mechanisms are active, depending on the degree of exceeding of certain criteria that are defined by the mechanical load combination (normal and shear stresses within the critical plane).

Of course, design engineers do not solely rely on such comparatively simple methods of service-life prediction, which are based on several fatigue experiments on standardized specimens. In general, technical components having a vital function undergo tailor-made cyclic-loading tests (*safe-life* design). For instance, in automobile chassis design multiaxial servohydraulic testing devices apply loading sequences that correspond to test runs under extreme weather, road and driving-style conditions. The wings of Airbus aircrafts are tested as complete structures by means of 25 servohydraulic jigs and more than 1000 strain sensors to simulate takeoff under twice their limit load. Applying such overloads on real components is required to make sure that the design process based on simplified assumptions yields a product that guaranties a nearly absolute safe operation during the predicted lifetime or inspection interval.

If there are technical restrictions for *safe-life* design, e.g., limited maximum weight of the component or unpredictable loading spectrum, *fail-safe* design is an alternate option. Here, it must be ensured that the complete structure will not loose its safe operability in the case of component failure.

2.2.3
Fatigue-Life Prediction: Damage-Tolerant Approach

Involvement of the mechanisms of crack initiation and propagation with approaches of service-life prediction seems to be plausible and obvious at a first glance, in order to account for the physical damage evolution. However, the interactions between the complex microstructure of technical materials and the mechanisms of crack initiation and early propagation are not fully understood yet and hence, the development of physically based models that can be used for service-life prediction of technical structures is still extremely challenging. More recent developments based on the experimentally gained knowledge of the interactions mentioned above (see also Chapters 5 and 6) are reviewed in Chapter 7.

The following section is focused on the rather empirical Paris law of fatigue-crack propagation, which has been introduced by Eq. (2.5), and some of its modifications. Even though, these approaches consider the microstructure only by material parameters they have become increasingly established in lightweight design in the aerospace and automotive industry.

The damage-tolerant design approaches postulate that generally each component contains imperfections in the form of cracks. However, these cracks are able to propagate only when a certain critical length is exceeded. This behavior is represented for most materials by the characteristic sigmoidal relationship between the logarithm of the crack-propagation rate da/dN and the range of the stress-intensity factor ΔK (see Fig. 2.12) where the linear part can be described by Paris' law (Eq. 2.5). The latter includes the concept of transferability of LEFM; i.e., by combining the stress range $\Delta\sigma$, the crack length a and the geometry function Y for a crack in a component within the parameter ΔK, the individual dimensions of a component are eliminated. Hence, the linear part of the $\log(da/dN)$ vs. $\log(\Delta K)$ curve is described by Paris' law as follows:

$$\frac{da}{dN} = C\Delta K^n = C\left(\Delta\sigma\sqrt{\pi a}Y\right)^n \qquad (2.14)$$

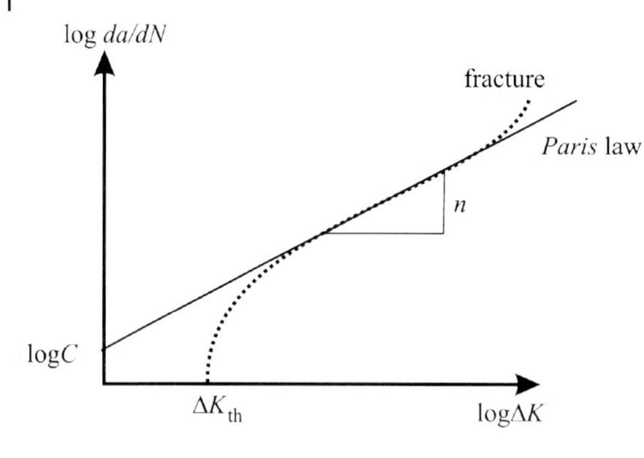

Fig. 2.12 Schematic representation of the crack propagation rate da/dN vs. the range of the stress-intensity factor ΔK in a double-logarithmic plot with the Paris law represented as a straight line (cf. Fig. 2.3).

Only when ΔK exceeds regime I in Fig. 2.12, i.e., the threshold value of the range of the stress-intensity factor ΔK_{th} (corresponding to a critical combination between nominal stress range and crack length), is an existing crack considered as technical crack that can propagate. It will be shown later that this assumption is not valid for microstructurally short fatigue cracks. When K_{max} reaches the fracture toughness K_c (see Section 2.3) stable fatigue-crack propagation switches to unstable crack propagation (regime III in Fig. 2.12), causing immediate failure by ductile rupture. Figure 2.12 shows a schematic representation of the technical fatigue-crack-propagation characteristics while Table 2.1 provides examples of crack-propagation data for some materials.

Assuming a starter crack of length a_0 and a critical final crack length a_{crit}, which corresponds to the failure condition of a technical component, then, under the simplifying condition that $Y = \text{const.}$, the number of cycles to fracture N_f can be calculated by separating the variables da and dN in Eq. (2.14) and subsequent integration:

$$\int_0^{N_f} C\left(\Delta\sigma\sqrt{\pi}Y\right)^n dN = \int_{a_0}^{a_{crit}} \frac{1}{\left(\sqrt{a}\right)^n}\, da \tag{2.15}$$

Table 2.1 Crack propagation data for various technical materials.

$da/dN = C\Delta;K^n$	C	n	ΔK_{th} (MPa m$^{1/2}$)
Steels [39]	5.79×10^{-11}	2.25	$\approx 1.7 + \sqrt{d}$ (d = grain size) [105]
Aluminum alloys [39]	9.82×10^{-12}	3	7 (Al bronze [105])
Titanium alloys [39]	3.56×10^{-15}	4	2.9 ($R = 0.7$), 6.1 ($R = 0.1$), β-Ti [90]
Approximation da/dN [39]	$da/dN \approx 8000\ (\Delta K/E)^{3.4}$ E = Young's modulus		–

$$N_{\mathrm{f}} = \frac{2\left(a_0^{1-n/2} - a_{\mathrm{crit}}^{1-n/2}\right)}{(n-2)C\left(\Delta\sigma\sqrt{\pi}Y\right)^n}, n \neq 2 \quad (2.16)$$

For the special case that the exponent n is equal to 2, Eq. (2.16) becomes a rather simple fatigue-life equation that shows the significance of the most relevant parameters, particularly that of the starter crack length a_0:

$$N_{\mathrm{f}} = \frac{1}{\pi C(\Delta\sigma Y)^2} \ln \frac{a_{\mathrm{crit}}}{a_0} \quad (2.17)$$

Hence, early detection of propagating starter cracks is essential for application of the damage-tolerant design approach. The most common technique of nondestructive crack analysis is ultrasonic detection having a resolution limit of approximately 0.5 mm for regular inspection intervals. By means of an analytical or numerical solution of Paris' law (cf. Eqs. 2.14 to 2.17) in combination with some kind of safety factor, the time or number of cycles to failure is calculated for which a crack propagates from the detectable length a_{det} to a maximum tolerable length a_{tol}. Based on the result of this calculation the inspection interval is determined. In the case that a crack is detected during an inspection, the duration of the following inspection interval is determined on the basis of the time or number of cycles that is required to grow the crack from the actual measured length to the maximum tolerable length. Only when cracks that exceed the maximum tolerable length are detected, the component has to be shut down and replaced. This procedure of damage-tolerant design is shown schematically in Fig. 2.13.

After the publishing of Paris' law by Paris and Erdogan [27], it was heavily criticized since it neither accounts for load-history effects, e.g., the influence of overloads, nor for mean-stress or crack-closure effects. Furthermore, it is questionable

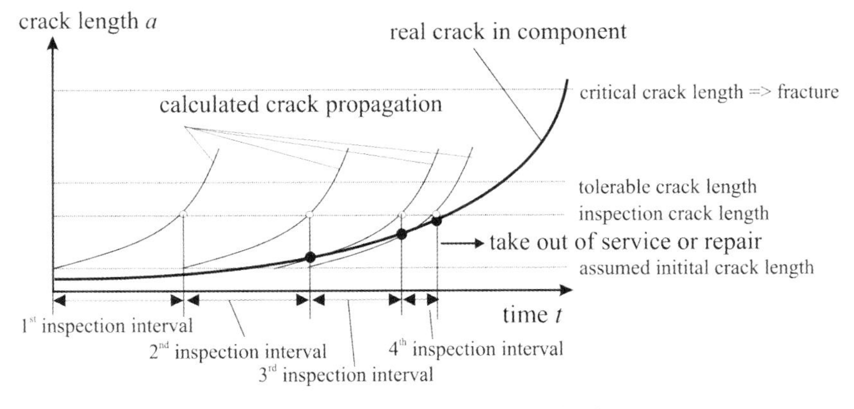

Fig. 2.13 Procedure of damage-tolerant design based on calculated inspection intervals. The inspection intervals correspond to the time to grow the crack to a critical length, which is calculated for the assumed initial crack length a_0 (inspection intervals 1 and 2) or the actual real crack length a (inspection intervals 3 and 4).

if the linear-elastic K concept is suitable to describe fatigue-crack propagation in ductile materials that is driven by plastic deformation ahead of the crack tip.

Nevertheless, the impressive simplicity and the amazingly good correlation with experimental da/dN data meant Paris' law became the most frequently applied equation in damage-tolerant design research. Certainly, many modifications of Paris' law have been developed during the last few decades. Without being a complete coverage, a few of them are introduced in the following section and later in Chapter 7.

To extend the Paris–Erdogan approach from its restriction to stable fatigue-crack growth (regime II, linear part in Fig. 2.12) Foreman et al. [52] suggested an equation that (1) brings the crack-propagation rate da/dN to infinite when approaching the fracture toughness K_c and (2) accounts for the mean stress by including the stress ratio R:

$$\frac{da}{dN} = \frac{C_F \Delta K^{n_F}}{(1-R)K_c - \Delta K} \tag{2.18}$$

The index "F" in Eq. (2.18) is used to clarify that the constants C_F and n_F are not of the same values as the respective constants C and n in Eq. (2.5). There are several further approaches to accommodate the sigmoidal shape of the log(da/dN) vs. log(ΔK) curve in Fig. 2.12, e.g., from Weertman [53] or Klesnil and Lukas [54]. Emphasis should be put on the one suggested by McEvily [55], taking into account not only the mean stress by the maximum stress-intensity factor K_{max} and the fracture toughness K_c, but also the threshold range of the stress intensity factor ΔK_{th}:

$$\frac{da}{dN} = C\left(\Delta K - \Delta K_{th}\right)^2 \frac{\Delta K}{K_c - K_{max}} \tag{2.19}$$

A significant impetus for the mechanism-based understanding of fatigue-crack propagation was due to Elber's observation of crack-closure effects [29, 30]. Several plasticity-, roughness-, transformation- or oxide-induced effects may lead to a volume increase in the wake of a propagating crack, giving rise to premature contact of the crack faces when unloading a sample, and hence reducing the crack driving force to an effective range of the stress intensity factor $\Delta K_{eff} = K_{max} - K_{op}$. Below the crack-opening stress intensity factor K_{op} the crack is closed and the respective residual range of the stress intensity factor $\Delta K_{res} = \Delta K - \Delta K_{eff}$ does not contribute to crack advance. Instead of the crack-opening stress σ_{op} or crack-opening stress-intensity factor K_{op}, respectively, it is also common to use the crack-closure stress σ_{cl} or K_{cl}, respectively, to account for differences in the loading and in the unloading regime of a complete fatigue cycle (see Section 6.3). To take crack-closure effects in consideration it is merely required to replace ΔK by ΔK_{eff} in Paris' law:

$$\frac{da}{dN} = C\Delta K_{eff}^{n} \tag{2.20}$$

For a very simplified application of the ΔK_{eff} concept it may be assumed that no intrinsic material threshold for crack initiation exists; i.e., crack propagation sets in as soon as the effective range of the stress intensity becomes $\Delta K_{eff} > 0$. By defining

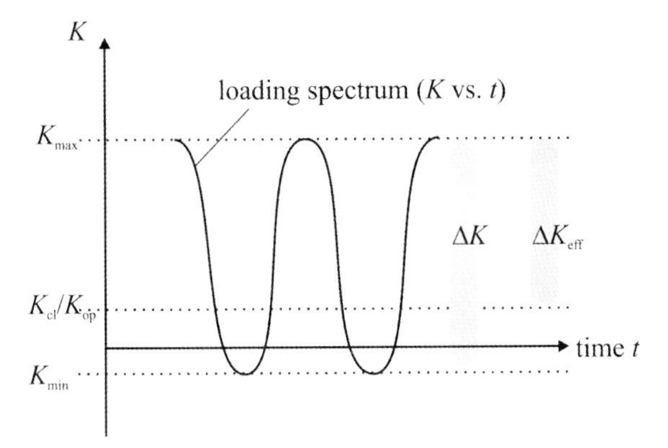

Fig. 2.14 Reduction of the range of the stress intensity factor to an effective range due to crack-closure effects (for details, see Section 6.3).

an effective stress-intensity ratio $U=\Delta K_{eff}/\Delta K$ the influence of mean stress is included according to

$$\Delta K_{eff} = U\Delta K = \frac{K_{max} - K_{op}}{K_{max} - K_{min}} \cdot \Delta K = \overbrace{\left(\frac{1}{1-R} - \frac{K_{op}}{\Delta K} \right)}^{U} \cdot \Delta K \tag{2.21}$$

The threshold value of the range of the stress intensity factor follows from Eq. (2.21) for $U = 0$:

$$U = \frac{1}{1-R} - \frac{K_{op}}{\Delta K_{th}} \Rightarrow \Delta K_{th} = K_{op}(1-R) \tag{2.22}$$

Figure 2.15 (after [56]) shows by a visualization of Eqs. (2.20) to (2.22) the influence of the stress ratio. In agreement with experimental observations an increase in the stress ratio leads to a decrease in the contribution of crack closure, and hence to a decrease in the threshold range of the stress intensity factor ΔK_{th}. Even though this treatment of crack closure is an oversimplification, it shows clearly the main effects of changing the mean stress.

There are basically two reasons why the applicability of the methods of LEFM to predict the propagation fatigue cracks is limited to "long" cracks:

1. As shown in Section 2.3, the stress-intensity factor K of LEFM describes the elastic stress field in the vicinity of a crack tip. However, initiation and early propagation of fatigue cracks in ductile materials are governed by dislocation movement in the plastic zone around the crack tip, the extension of which can only be neglected for long cracks.
2. In the early stages of the fatigue-damage process the length of short cracks is of the dimension of the material's characteristic microstructural features, like the grain size or the size of precipitates. Since in this dimension the material properties have a pronounced anisotropy, the material cannot be treated as a sort of continuum, which is one of the basic assumptions of LEFM.

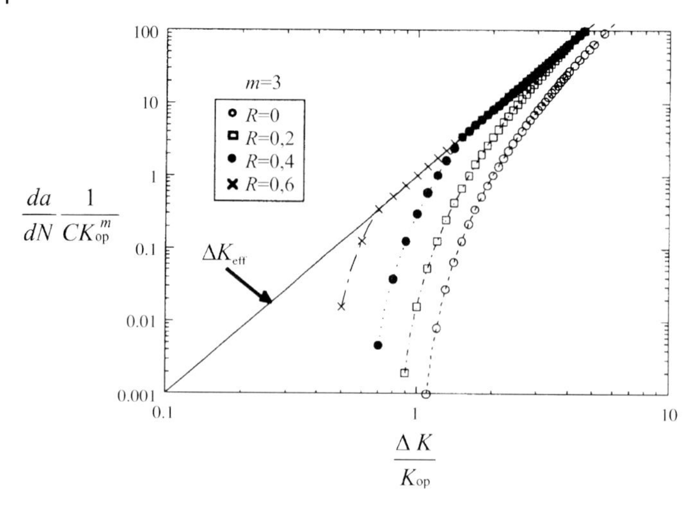

Fig. 2.15 Dimensionless crack propagation vs. range of stress intensity factor curves for various stress ratios R according to Eq. (2.21) (after [56]).

These limitations can be partially overcome by applying the methods of elastic–plastic fracture mechanics (EPFM), which correlate the crack-propagation rate with the elastic–plastic deformation around the crack tip. The most commonly used concepts that are outlined in Section 2.3 include the cyclic J integral, the range of the plastic crack-tip opening displacement (ΔCTOD), and the yield-strip approaches. In addition to this, the anisotropy of the mechanical properties of crystalline, ductile materials is taken into account by microstructural fracture mechanics (MFM). Some methods to treat microcracks are addressed in Chapter 7.

2.2.4
Methods of Fatigue-Life Prediction at a Glance

Figure 2.16 shows stepwise the way from material selection to the application of a component. Besides the pure mechanical loading effects of homogeneous materials, particular attention has to be paid to microstructure alterations due to joining (brazing, welding, etc.) or coating techniques, as well as to environmentally assisted degradation (stress-corrosion cracking, corrosion–creep interactions, etc.). It is a challenge for future research in the field of material fatigue to include such kinds of complex conditions into mechanism-based fatigue-life prediction methods.

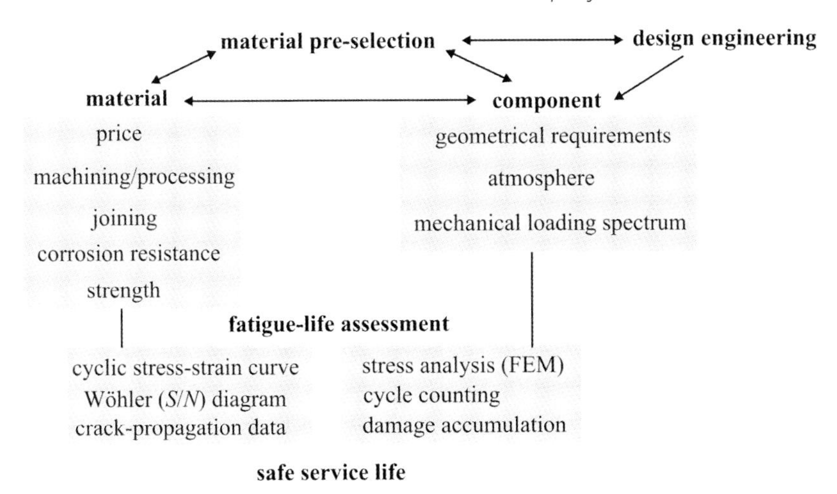

Fig. 2.16 Fatigue life prediction in the engineering design process.

2.3
Basic Concepts of Technical Fracture Mechanics

It is beyond all question that it is impossible to pack the whole field of fracture mechanics into an introductory sub-section of this book. However, the fracture-mechanics approach is a key issue for understanding the concepts of the experimental and theoretical work discussed in the following sections, and hence the most striking issues and important terms relevant to fatigue-crack propagation need to be briefly outlined.

In addition to the cited original work, textbooks on fracture mechanics are recommended, e.g., those of the following authors: Anderson [56], Broek [57], Gross [58], and Schwalbe [59].

In general, fracture-mechanics research deals with the question as to which way the driving force for crack advance can be mathematically described. Figure 2.17 shows that this question can be answered in different ways depending on the length scale for which the fracture problem is treated.

Including the atomic length scale in the simulation of the mechanical behavior of technical material systems by means of molecular- or dislocation-dynamic approaches is still extremely difficult due to a lack of potential data to be obtained, e.g., as embedded-atom-method potentials (EAM), and closed solutions for dislocations in polycrystals. Nevertheless, there is no doubt that these kinds of *ab initio* methods are going to attract considerable attention when dealing with nanostructured materials or thin films. The state of the art in challenges and limits of *ab initio* models is reviewed, e.g., in [60, 61].

Basically, separation of atoms is the fundamental mechanism of crack propagation. It requires both overcoming the cohesive strength between the atomic layers and the generation of new surfaces. Griffith [12] applied a simple energy balance

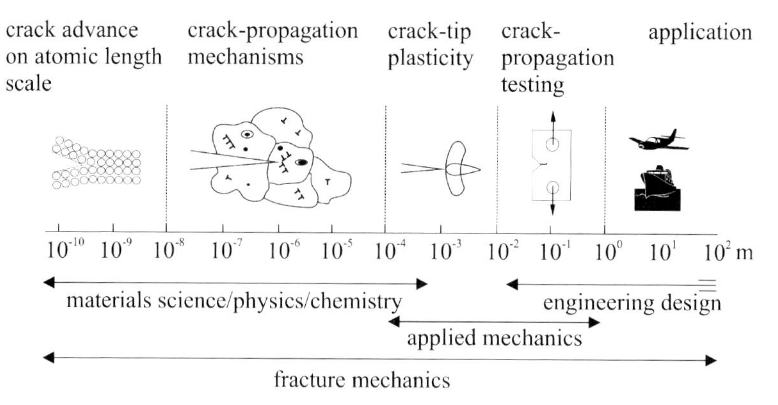

Fig. 2.17 Various length scales, scientific disciplines and modeling approaches relevant for the analysis of cracks.

for total energy W_{tot} including the formation energy of new surface W_s with the specific surface energy γ_s on the one hand and the release of elastic potential energy W_{el} due to crack advance by da in a plate of unit thickness $B = 1$ (see Fig. 2.18) on the other hand.

$$\frac{dW_{tot}}{da} = \overbrace{\frac{dW_{el}}{da}}^{-G} + \frac{dW_s}{da} = -\frac{2\pi\sigma^2 a}{E} + 4\gamma_s = 0 \tag{2.23}$$

The fracture criterion for ideally brittle materials can be defined as the situation when the total energy decreases; i.e., $dW_{tot}/da = 0$. This is the case when, for an applied stress value σ_f and an increase in crack length by da, the release in elastic po-

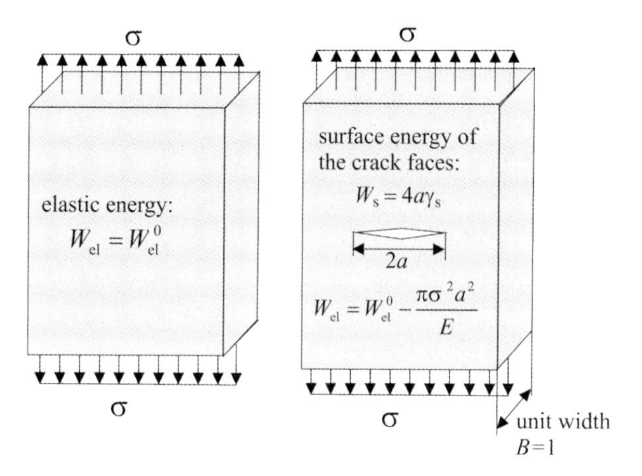

Fig. 2.18 Schematic representation of the energy balance for a Griffith crack in a plate of unit thickness.

tential energy (energy release rate G) overcomes the energy required to form two new surfaces:

$$\sigma_{\mathrm{f}} = \sqrt{\frac{2E\gamma_{\mathrm{s}}}{\pi a}} \tag{2.24}$$

Griffith's energy balance provides a simple but clear insight in the fracture mechanism in ideally brittle materials. To also apply it to crack propagation in ductile metals and alloys, the energy balance has to be modified according to Orowan [62] and Irwin [63] by means of replacing the ideal elastic fracture work $2\gamma_{\mathrm{s}}$ in Eq. (2.24) by a value for the elastic ideal plastic fracture toughness $G_{\mathrm{c}} = 2(\gamma_{\mathrm{el}} + \gamma_{\mathrm{pl}})$.

However, of most technical relevance as a quantitative criterion for crack propagation accounting for the local situation at the crack tip is the K concept of LEFM which is introduced briefly in the following section.

2.3.1 The *K* Concept of LEFM

In general, a crack tip can be loaded under three different modes: assuming a planar crack as shown in Fig. 2.19, a normal load with respect to the crack plane is considered as mode I (opening mode), a shear load in the direction of crack advance as mode II (in-plane shear mode) and a shear load in the direction perpendicular to crack advance as mode III (anti-plane shear mode). It is worth mentioning that most technical loading cases are a superimposition of different modes, called mixed-mode loading.

According to Irvin's analysis [25], the stress tensor σ_{ij} ($i, j = x, y, z$) near the crack tip ($r \ll a$) depending on the angle Θ with respect to the crack plane and the distance r from the crack tip (Fig. 2.20) is given by

$$\sigma_{ij} = \frac{1}{\sqrt{2\pi r}}\left[K_{\mathrm{I}} f_{ij}^{\mathrm{I}}(\Theta) + K_{\mathrm{II}} f_{ij}^{\mathrm{II}}(\Theta) + K_{\mathrm{III}} f_{ij}^{\mathrm{III}}(\Theta)\right] \tag{2.25}$$

taking the stress intensity factors K, and dimensionless functions $f(\Theta)$ for the three different modes into account.

Assuming plane-stress conditions; i.e. $\sigma_{zz} = \tau_{xz} = \tau_{yz} = 0$, for the sake of simplicity the following equilibrium conditions are valid for the volume element in Fig. 2.20:

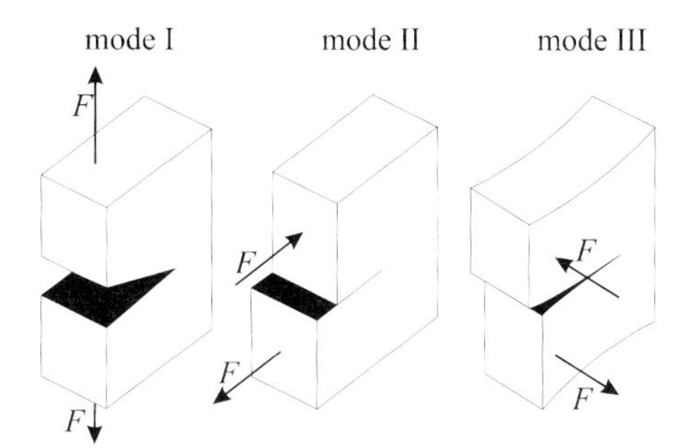

Fig. 2.19 Schematic representation of the three fundamental crack-opening modes: (a) mode I (opening), (b) mode II (in-plane shear), (c) mode III (anti-plane shear).

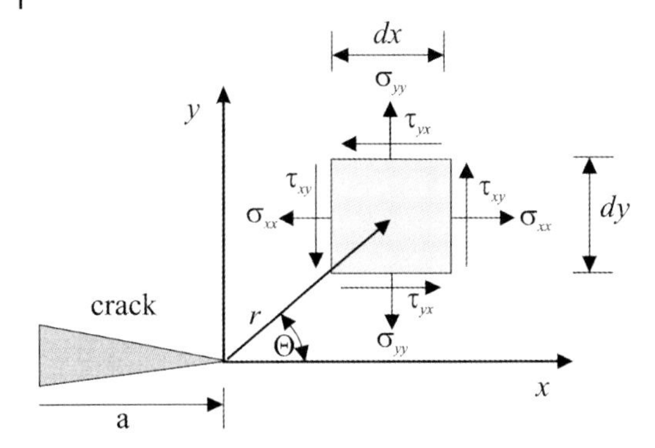

Fig. 2.20 Definition of the coordinate system and the stress directions for the crack tip stress field.

$$\frac{\partial \sigma_{xx}}{\partial x} + \frac{\partial \tau_{xy}}{\partial y} = 0, \quad \frac{\partial \sigma_{yy}}{\partial y} + \frac{\partial \tau_{xy}}{\partial x} = 0 \tag{2.26}$$

The strain values resulting from the displacements u in x direction and v in y direction are given by

$$\varepsilon_{xx} = \frac{\partial u}{\partial x}, \quad \varepsilon_{yy} = \frac{\partial v}{\partial y}, \quad \gamma_{xy} = \frac{\partial u}{\partial y} + \frac{\partial v}{\partial x} \tag{2.27}$$

and have to fulfill the compatibility condition; i.e., neither holes nor interpenetration can occur in the material:

$$\frac{\partial^2 \varepsilon_{xx}}{\partial y^2} + \frac{\partial^2 \varepsilon_{yy}}{\partial x^2} = \frac{\partial^2 \gamma_{xy}}{\partial x \partial y} \tag{2.28}$$

The strains can be correlated with the stresses by means of Hooke's law, that is in the case of isotropic linear elastic behavior described by the Young's modulus E, the Poisson ratio ν and the shear modulus G and given as

$$E\varepsilon_{xx} = \sigma_{xx} - \nu\sigma_{yy}, \quad E\varepsilon_{yy} = \sigma_{yy} - \nu\sigma_{xx}, \quad G\gamma_{xy} = \tau_{xy} \tag{2.29}$$

Hence, an elastic-deformation problem is fully described by (1) the equilibrium equations (Eq. 2.26), (2) the strain equations (Eq. 2.27), (3) the compatibility condition (Eq. 2.28) and (4) the constitutive material equation (Hooke's law, Eq. 2.29), and can be solved for respective boundary conditions either numerically by means of the finite-element method (FEM; mesh generation for complete structures) or the boundary element method (BEM; mesh generation restricted only to surfaces, cf. Chapter 7), or analytically by means of the real Airy stress function in combination with complex stress functions, after Westergard [64] (cf. [56] for a more detailed and comprehensive description).

For the stress field *near* the crack tip one obtains the so-called Sneddon equations for the Griffith crack:

$$\begin{bmatrix} \sigma_{xx} \\ \sigma_{yy} \\ \tau_{xy} \end{bmatrix} = \sigma_a \sqrt{\frac{a}{2r}} \cos\left(\frac{\Theta}{2}\right) \begin{bmatrix} 1-\sin\left(\frac{\Theta}{2}\right)\sin\left(3\frac{\Theta}{2}\right) \\ 1+\sin\left(\frac{\Theta}{2}\right)\sin\left(3\frac{\Theta}{2}\right) \\ \sin\left(\frac{\Theta}{2}\right)\cos\left(3\frac{\Theta}{2}\right) \end{bmatrix} - \begin{pmatrix} \sigma_a \\ 0 \\ 0 \end{pmatrix} \tag{2.30}$$

Far away from the crack tip, the shear stress τ_{xy} and the normal stress σ_{xx} in x direction vanish, while the normal stress σ_{yy} in y direction approaches the applied nominal stress σ_a. At the crack tip the stresses in Eq. (2.30) increase to infinity due to the $1/\sqrt{r}$ singularity.

The stress-intensity factor K is a measure for the intensity of the stress increase in the stress field at the crack tip. Since the stress state is assumed to be elastic, K must be proportional to the applied nominal stress σ_a. Comparing Eq. (2.25), now given for mode I by

$$\begin{bmatrix} \sigma_{xx} \\ \sigma_{yy} \\ \tau_{xy} \end{bmatrix} = \frac{K_I}{\sqrt{2\pi r}} \cos\left(\frac{\Theta}{2}\right) \begin{bmatrix} 1-\sin\left(\frac{\Theta}{2}\right)\sin\left(3\frac{\Theta}{2}\right) \\ 1+\sin\left(\frac{\Theta}{2}\right)\sin\left(3\frac{\Theta}{2}\right) \\ \sin\left(\frac{\Theta}{2}\right)\cos\left(3\frac{\Theta}{2}\right) \end{bmatrix} \tag{2.31}$$

and Eq. (2.30) and taking into account that the nonsingularity term can be neglected ($\Theta = 0$), the stress-intensity factor K_I for mode I loading conditions (and in the same way for mode II and III) can be expressed as

$$K_I = \sigma_a \sqrt{\pi a} \tag{2.32}$$

Unfortunately, a self-contained analytical solution is only possible for the special case of an infinite plate with a center crack. For technical cases it turns out that using geometry functions Y, which depend on characteristic dimensional parameters, e.g., the ratio of the crack length to specimen width ratio a/W, is most practical to accommodate for deviations from the expression given by Eq. (2.31). Detailed overviews of such functions are given by Tada et al. [65] or Murakami [66]. Figure 2.22 shows some selected examples.

According to [67] for a semi-elliptical surface crack, where the a/W ratio is negligible, the stress-intensity-factor distribution along the crack front (according to Fig. 2.22) can be approximated as

$$K_I = \sigma_a \sqrt{\frac{\pi a}{\Phi}} \left(\left(\frac{a^2}{c^2}\right) \sin^2\varphi + \cos^2\varphi \right)^{1/4} \quad \text{with} \quad \Phi \approx 1 + 1.464 + \left(\frac{a}{c}\right)^{1.65} \tag{2.33}$$

The K concept (K field) is valid only for a certain range of the distance r from the crack tip. For large r values; i.e., far away from the crack tip, the nonsingular terms of Eq. (2.30) cannot be neglected any more; for low r values the stress increase at

load case	stress-intensity factor
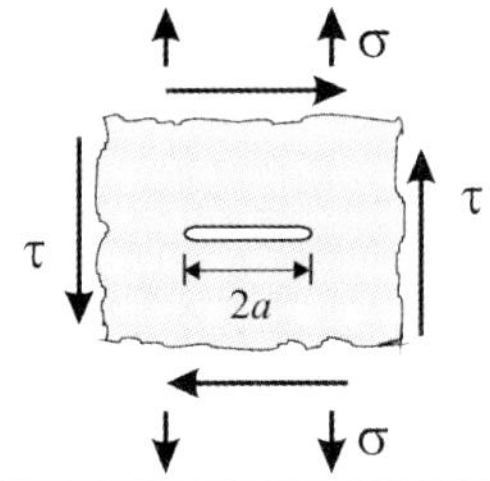	$\begin{Bmatrix} K_I \\ K_{II} \end{Bmatrix} = \begin{Bmatrix} \sigma \\ \tau \end{Bmatrix} \sqrt{\pi a}$
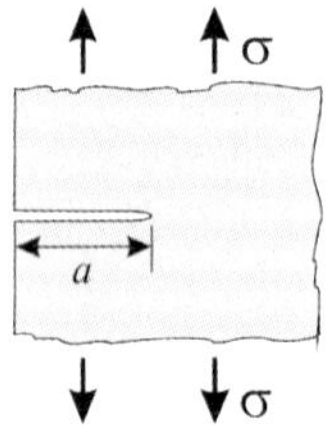	$K_I = 1.12\sigma\sqrt{\pi a}$
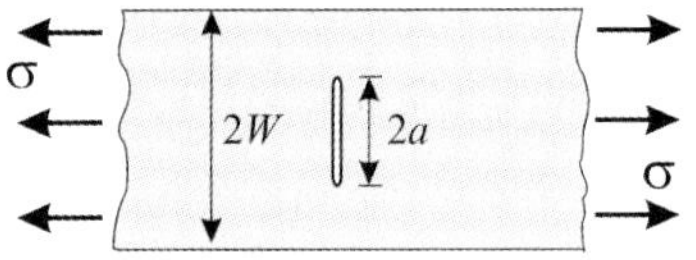	$K_I = \sigma\sqrt{\pi a}\, Y_I(a/W)$ $Y_I = \dfrac{1 - 0.025(a/W)^2 + 0.06(a/W)^4}{\sqrt{\cos(\pi a/2W)}}$
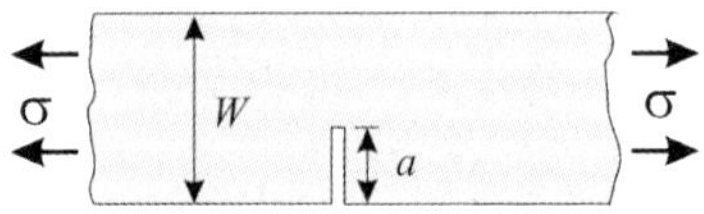	$K_I = \sigma\sqrt{\pi a}\, Y_I(a/W)$ $Y_I = \sqrt{\dfrac{2W}{\pi a}\tan\left\{\dfrac{\pi a}{2W}\right\}}\left[\dfrac{0.752 + 2.02\dfrac{a}{W} + 0.37\left\{1 - \sin\dfrac{\pi a}{2W}\right\}^3}{\cos\left\{\dfrac{\pi a}{2W}\right\}}\right]$
SENB specimen (4pt bending) 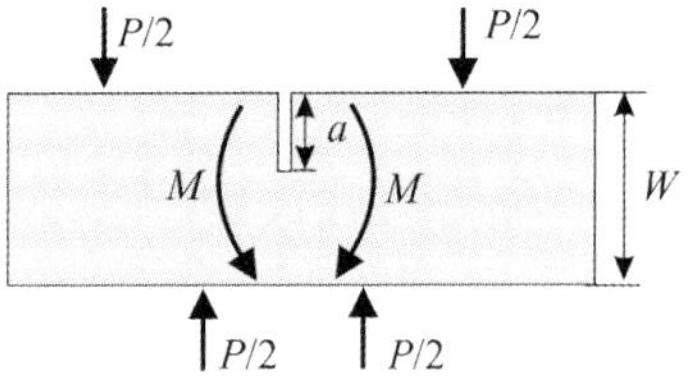	$K_I = \sigma\sqrt{\pi a}\, Y_I(a/W)$ $\quad \sigma = \dfrac{6M}{W^2 B}$ $Y_I = \sqrt{\dfrac{2W}{\pi a}\tan\left\{\dfrac{\pi a}{2W}\right\}}\left\{\dfrac{0.923 + 0.199\left[1 - \sin\left\{\dfrac{\pi a}{2W}\right\}\right]^4}{\cos\left\{\dfrac{\pi a}{2W}\right\}}\right\}$

Fig. 2.21 Examples of geometry functions of several crack geometries for calculation of stress-intensity factors.

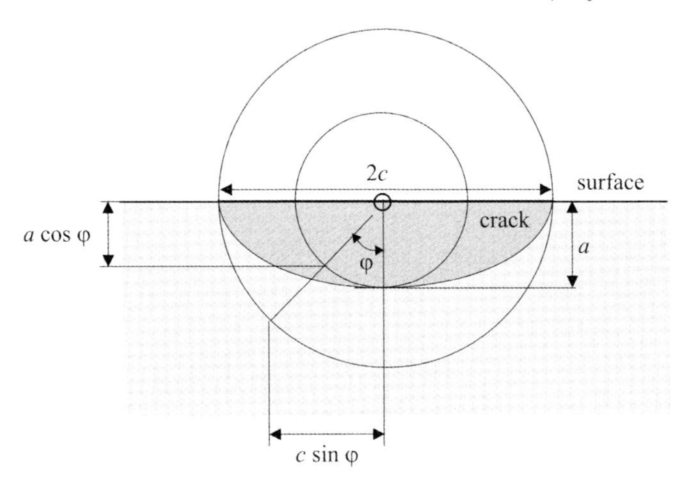

Fig. 2.22 Crack-geometry parameters for calculation of the stress-intensity factor distribution by means of Eq. (2.33).

the crack tip is limited by plastic yielding. Only if the plastic zone is sufficiently small compared to the *K* field is the stress-intensity factor a useful parameter for the treatment of fracture processes.

2.3.2
Crack-Tip Plasticity: Concepts of Plastic-Zone Size

Directly at the crack tip, Eq. (2.30) yields theoretical stress values of infinity. However, in a ductile metallic material the stress level cannot exceed the local yield stress, and therefore the actual gradient of the stress σ_{yy} ahead of the crack tip corresponds to the solid line in Fig. 2.23. To implement this into the *K* concept Irwin [68] introduced a correction of the physical crack length by a small distance δ in such a way that the amount of stress intensity cut off by the yield strength σ_Y (area I in Fig. 2.23) can be accommodated by the plastic deformation within the distance δ (area II in Fig. 2.23). A first estimate of the plastic zone size r^* can be obtained by Eq. (2.31) for $\Theta = 0$ and σ_{yy} to reach the yield strength σ_Y:

$$r^* = \frac{K_I^2}{2\pi\sigma_Y^2} \tag{2.34}$$

The condition that the areas I and II are of the same size ($A_1 = A_2$) can be expressed as

$$A_1 = \sigma_Y\left(r_p - r_p^*\right) = \int_0^{r^*} \frac{K_I}{\sqrt{2\pi r}}\,dr - \sigma_Y r_p^* = A_2 \tag{2.35}$$

which results in a larger plastic zone size of

$$r_p = \frac{K_I^2}{\pi\sigma_Y^2} \tag{2.36}$$

for plane-stress conditions.

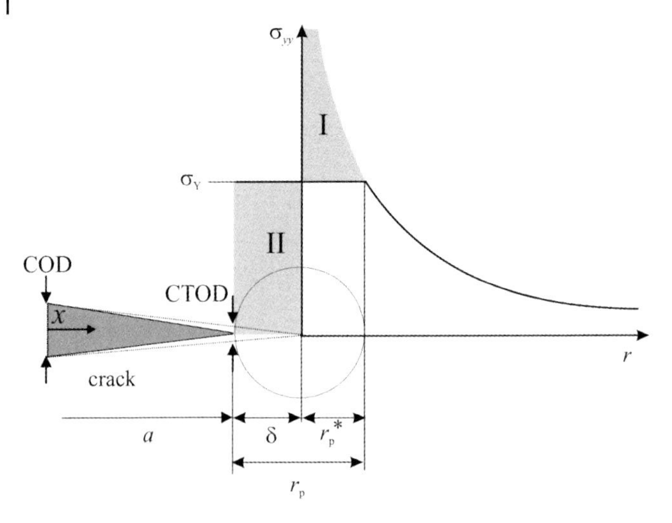

Fig. 2.23 Plastic zone correction according to Irwin [68].

Under plane-strain conditions, yielding is suppressed by a triaxial stress state at the crack tip. Yielding under triaxial stress can be accounted for by using the equivalent stress of Tresca σ_{Tr} being equal to the maximum shear stress $2\tau_{max} = \sigma_1 - \sigma_3$:

$$\sigma_{Tr} = \sigma_1 - \sigma_3 \tag{2.37}$$

or von Mises stress σ_{vM}, which accounts for the superimposition of the whole set of principal stresses σ_1, σ_2, and σ_3 (cf. Section 4.1):

$$2\sigma_{vM}^2 = (\sigma_1 - \sigma_2)^2 + (\sigma_2 - \sigma_3)^2 + (\sigma_1 - \sigma_3)^2 \tag{2.38}$$

Since for $\Theta = 0$ all shear stresses vanish, the main stresses correspond to the normal stresses σ_{xx}, σ_{yy} , σ_{zz} with $\sigma_1 = \sigma_{yy} = \sigma_2 = \sigma_{xx}$, and $\sigma_3 = \sigma_{zz} = \nu(\sigma_{xx} + \sigma_{yy})$. Hence, the plastic zone size for plane-strain conditions is given by

$$r_p = \frac{K_I^2}{\pi \sigma_Y^2}(1 - 2\nu)^2 \tag{2.39}$$

which is for a Poisson ratio of $\nu \approx 0.3$ (for most metallic materials) smaller by a factor of approximately 6 than for plane-stress conditions. Eventually, the plastic zone extension around the crack tip as a function of the angle Θ (see Fig. 2.20) can be expressed as follows:

$$r_p = \frac{K_I^2}{4\pi \sigma_Y^2}\left(\frac{3}{2}\sin^2\Theta + A(1 + \cos\Theta)\right) \tag{2.40}$$

Figure 2.24 shows qualitatively the three-dimensional plastic zone ahead of a planar crack.

As an alternative approach to treat plastic deformation at the crack tip, Dugdale [69] and in a similar form Barenblatt [70] proposed a yield-strip model, an idea that was further developed later by Bilby et al. [71] representing the yield strip ahead of

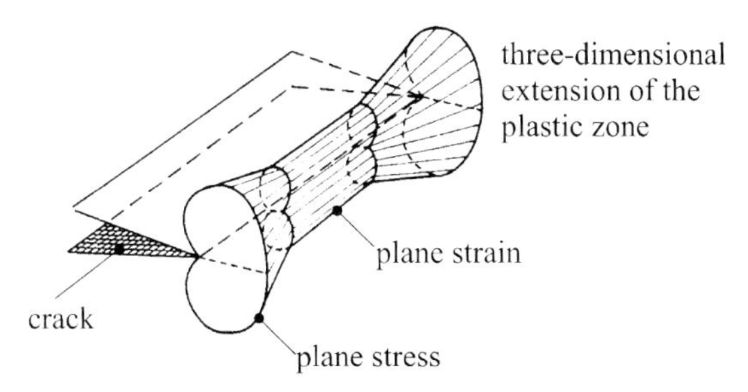

Fig. 2.24 Three-dimensional plastic zone ahead of a planar crack front (after [57]).

the crack tip by means of continuously distributed dislocations (BCS crack model). As the basis of several important models for microcrack propagation, the BCS approach is introduced in Section 7.2.

Analogous to Irwin's plastic zone correction, the Dugdale–Barenblatt model considers a crack, a length $2a$ of which is extended by a plastic zone of length ρ at both tips of the crack, according to Fig. 2.25.

It is assumed that the stress-intensity factor K_I due to the remote applied stress σ_a at the tip of the plastic zones

$$K_\mathrm{I} = \sigma_\mathrm{a}\sqrt{\pi(a+\rho)} \tag{2.41}$$

is superimposed by a "crack-closing" stress-intensity factor K_ρ, which is attributed to yielding in the plastic zone (yield strip, stress σ_Y) acting opposite to σ.

In general, a single load P acting at x assuming an infinite plate of unit thickness yields to the stress-intensity at the crack tips $+a$ and $-a$ [56, 58]:

$$K_{\rho,\pm a} = \frac{P}{\sqrt{\pi a}}\sqrt{\frac{a \pm x}{a \mp x}}\, a \tag{2.42}$$

When replacing both the load P by the yield criterion within the plastic zone, i.e., $P = -\sigma_\mathrm{Y}\,dx$, and the crack tip a by the plastic zone tip $a+\rho$, the total "crack-closing" stress-intensity factor for both yield strips is given by the integration of Eq. (2.42):

$$K_\rho = -\frac{\sigma_\mathrm{Y}}{\sqrt{\pi(a+\rho)}}\int_a^{a+r}\left(\sqrt{\frac{a+\rho+x}{a+\rho-x}} + \sqrt{\frac{a+\rho-x}{a+\rho+x}}\right)dx \tag{2.43}$$

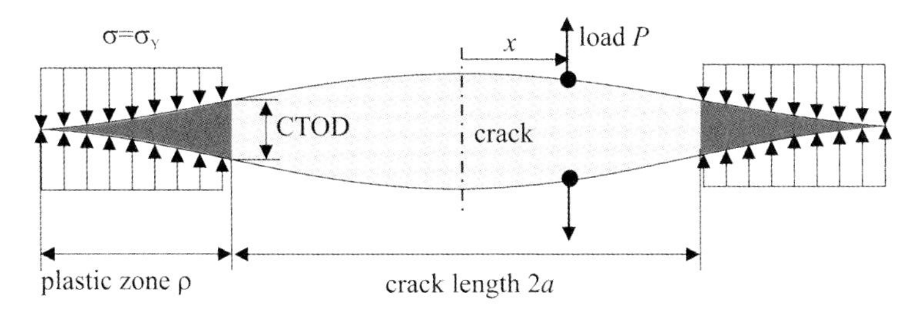

Fig. 2.25 Geometrical parameters of the yield-strip model after Dugdale and Barenblatt.

Solving this integral results in

$$K_\rho = -2\sigma_\mathrm{Y}\sqrt{\frac{a+\rho}{\pi}}\,\arccos\left(\frac{a}{a+\rho}\right) \tag{2.44}$$

Since the remote stress-intensity factor (Eq. 2.41) and the "crack-closing" stress-intensity factor must balance each other ($K_\rho + K_\mathrm{I} = 0$) the plastic-zone size is given by

$$\frac{a}{a+\rho} = \cos\left(\frac{\pi\sigma_\mathrm{a}}{2\sigma_\mathrm{Y}}\right) \tag{2.45}$$

Expressing the right-hand side of Eq. (2.45) by a Taylor series, where higher order terms are neglected, the plastic-zone size ahead of the crack tip can be estimated as

$$\rho = r_\mathrm{p} = \frac{\pi}{8}\left(\frac{K_\mathrm{I}}{\sigma_\mathrm{Y}}\right)^2 \tag{2.46}$$

By means of both the Irwin and the Dugdale–Barenblatt analyses, the so-called crack-tip-opening displacement (CTOD) as a measure of crack-tip blunting (see Figs. 2.23 and 2.24) can be determined [72]. From the elastic crack-tip stress analysis one obtains the crack-tip displacement field [56]. Substituting the plastic-zone size r_p given by Eq. (2.36) for the distance r from the crack tip one obtains for CTOD according to Irwin (plane-stress conditions)

$$\mathrm{CTOD} = \frac{4}{\pi}\frac{K_\mathrm{I}^2}{E\sigma_\mathrm{Y}} \tag{2.47}$$

Analogously, CTOD can be given in a more general form following the yield-strip approach as

$$\mathrm{CTOD} = \frac{1}{m}\frac{K_\mathrm{I}^2}{E'\sigma_\mathrm{Y}} \tag{2.48}$$

where $m \approx 1$ and $E' = E$ corresponds to plane stress and $m \approx 2$ and $E' = E/(1-\nu)^2$ to plane strain conditions.

Applying the plastic-zone-size approaches to cyclic-loading conditions one has to take into account that loading to K_max is followed by unloading by $-\Delta K$. Hence, the plastic-zone size for fully reversed unloading is r_p2 corresponding to $2\sigma_\mathrm{Y}$. This is shown schematically in Fig. 2.26. If σ_Y in Eqs. (2.36) and (2.46) is replaced by $2\sigma_\mathrm{Y}$ the compressive or cyclic plastic-zone size ρ^* can be estimated to be a factor of four smaller than the monotonic plastic zone. Consequently, ΔCTOD is a factor of two smaller than the monotonic CTOD.

2.3.3 Crack-Tip Plasticity: The *J* Integral

To analyze pronounced crack-tip plasticity, Rice [73] suggested the nonlinear energy-release rate, the so-called *J* integral. It is based on the change in the potential energy $d\Pi$ due to crack advance by da. The potential energy consists of the stored

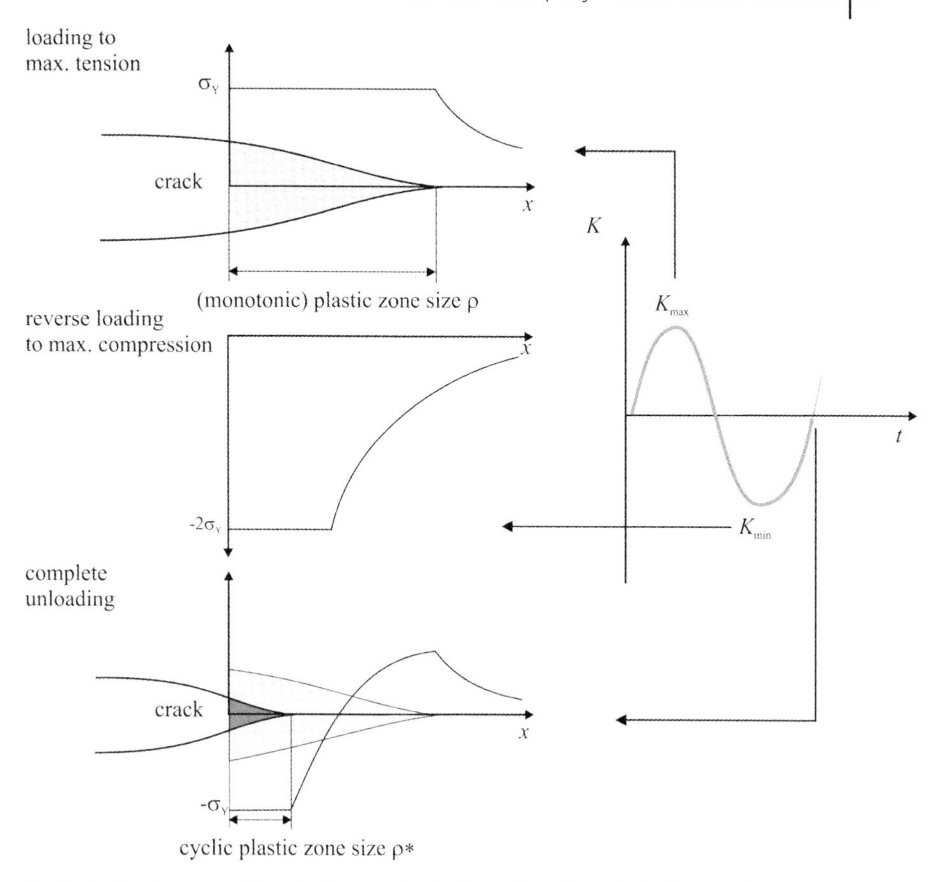

Fig. 2.26 Development of the plastic zone ahead of a crack tip for cyclic-loading conditions.

strain energy and the work W applied by a load P, leading to a displacement v of a cracked structure (see Fig. 2.27a):

$$J = -\frac{d\Pi}{da} = -\frac{d(U-W)}{da} = \left(\frac{\partial}{\partial a} \int_0^P v dP \right)_{\text{load control}} \tag{2.49}$$

The J integral can be given by a path-independent contour integral

$$J = \int_{\Gamma} \left(w dy - \underline{T} \frac{\partial \underline{u}}{\partial x} ds \right) \tag{2.50}$$

with strain-energy density

$$w = \int_0^{\varepsilon_{ij}} \sigma_{ij} d\varepsilon_{ij} \tag{2.51}$$

and $\underline{T}$ being the traction vector with the components of the stress tensor σ_{ij} normal to the integration path Γ, and $\underline{u}$ the respective displacement vector, as shown in Fig. 2.27b.

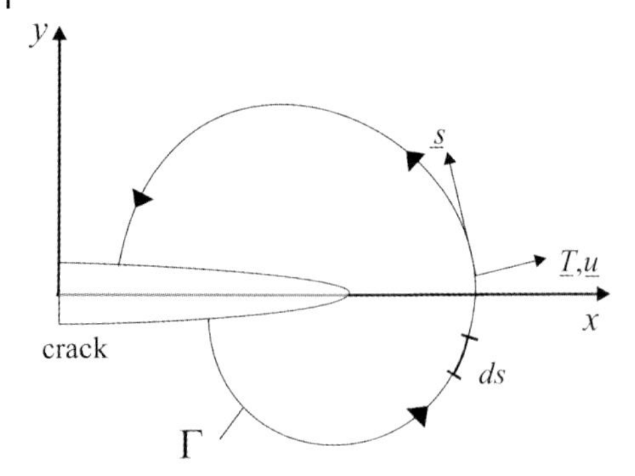

Fig. 2.27 Application and integration path of the J integral.

Basically, the J integral can be applied to linear elastic materials in a similar way to the stress-intensity factor K of the LEFM. Hutchinson [74] and Rice and Rosengreen [75] used the Ramberg–Osgood equation (cf. Eq. (2.7) for monotonic loading conditions) that describes elastic–plastic material behavior. Close to the crack tip, the elastic part of the strain can be neglected, and hence the stresses and strains ahead of the crack tip are given as

$$\sigma_{ij} = k_1 \left(\frac{J}{r} \right)^{\frac{1}{n+1}} \tag{2.52}$$

$$\varepsilon_{ij} = k_2 \left(\frac{J}{r} \right)^{\frac{n}{n+1}} \tag{2.53}$$

with k_1 and k_2 being proportionality constants referring to the so-called HRR singularity at the crack tip. In the case of linear-elastic mode I loading the following conversion for the stress intensity factor K_I, the J integral and the energy release rate G (cf. Eq. 2.23) is valid:

$$J = G = \frac{K_I^2}{E'} \tag{2.54}$$

with E' being equal to the Young's modulus E for plane-stress conditions and equal to $E/(1-\nu^2)$ for plane-strain conditions. More details of the practical use of the J integral are given in Section 3.2.

Dowling and Begley [76] were the first to apply the J integral concept as ΔJ, or Z integral as proposed by Wüthrich [77], to cyclic-loading conditions. It can be applied to fatigue-crack propagation in a similar way to Paris and Erdogan's law (Eq. 2.5) for the propagation rate of long cracks vs. the range of the linear-elastic stress intensity factor ΔK:

$$\frac{da}{dN} = C'(\Delta J)^{m'} \tag{2.55}$$

Various approaches to determine values for ΔJ are proposed in the literature; e.g., the area below the load–displacement hysteresis loop can be used according to [56]

$$\Delta J = Z = \frac{2}{Bb} \int_{v_{\min}}^{v_{\max}} \left(P - P_{\min}\right) dv \tag{2.56}$$

where B denotes the specimen width, b the unbroken ligament and v the actual load line displacement. An estimative solution for Z can be obtained by superimposing the linear elastic solution (as a function of the effective elastic-energy density $w_{el,eff}$, see Fig. 2.28) and the fully plastic solution (as a function of the plastic energy density w_{pl}, see Fig. 2.28) according to Heitmann et al. [78] as follows:

$$Z_{eff} = a\left(f_1 \cdot w_{el,eff} + f_2 \cdot w_{pl}\right) = 2{,}9 \cdot \frac{\Delta\sigma_{eff}^2}{2E} a + 2{,}5 \cdot w_{pl} a \tag{2.57}$$

with f_1 und f_2 being geometry functions (for details see [79]) and $\Delta\sigma_{eff}$ being the effective part of the stress range where the crack is open. An extension for high-temperature fatigue-crack propagation is suggested by Riedel [80], containing the Norton exponent n and the strain range of power-law creep $\Delta\varepsilon_{cr}$:

$$Z_{eff}^{*} \approx 2{,}9 \cdot \frac{\Delta\sigma_{eff}^2}{2E} a + 2{,}4\left(1 + \frac{3}{n}\right)^{-0{,}5} \Delta\sigma\Delta\varepsilon_{pl}\left[1 + \left(\frac{\Delta\varepsilon_{cr}}{\Delta\varepsilon_{pl}}\right)^{1+n'}\right] a \tag{2.58}$$

It is worth mentioning that high-temperature fatigue is strongly affected by the environment, particularly if hold times in tension are superimposed. This is discussed in more detail in Sections 6.5 and 7.4. As a general approach it is useful to subdivide environmentally assisted fatigue-crack propagation in a pure fatigue contribution $(da/dN)_{fat}$, a creep contribution $(da/dN)_{cr}$ and an environment contribution $(da/dN)_{env}$ (cf. [81, 82]):

$$\frac{da}{dN} = \left(\frac{da}{dN}\right)_{fat} + \left(\frac{da}{dN}\right)_{cr} + \left(\frac{da}{dN}\right)_{env} \tag{2.59}$$

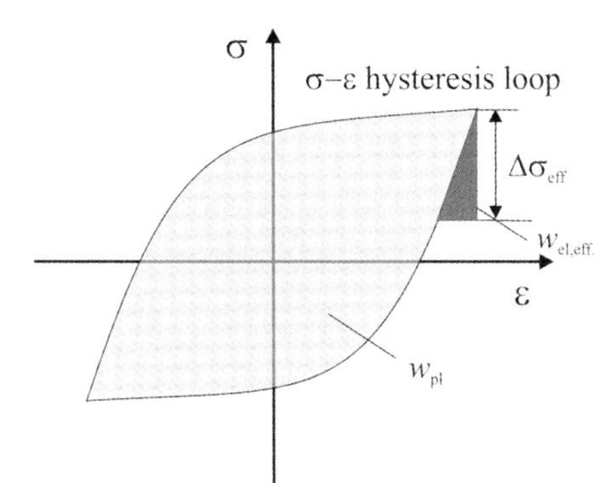

Fig. 2.28 Elastic and plastic energy densities obtained by stress–strain hysteresis loops to be used for the application of the J integral to cracks growing under cyclic-loading conditions.

3
Experimental Approaches to Crack Propagation

3.1
Mechanical Testing

Mechanical testing of materials covers a wide variety of experimental approaches, ranging from a simple standard tensile test to complex multiaxial testing of technical components under near-service conditions. There is no doubt that the latter yield very precise results if the relevant in-service loading conditions are sufficiently known. However, most structural components used in mechanical engineering have to work in an appropriate and safe way for many years, and therefore even small changes in design would require long-term test series. Hence, calculation methods are applied within industrial design processes that allow the assessment of the fatigue life of complex components based on data from standard material testing on simple specimen geometries. By doing this, design or material modifications are possible without the need for protracted component testing programs. In addition to this, fatigue tests on standardized specimens of simple cylindrical geometry provide more relevant information on fundamental mechanisms of damage evolution, since the local loading conditions can be determined easily.

3.1.1
Testing Systems

The most prominent approach to characterize the fatigue behavior of engineering materials are Wöhler tests, as discussed in Chapter 2. Here, standard cylindrical or flat specimens are loaded to a fixed stress or strain under torsion, push–pull or bend loading conditions. The data pairs, loading amplitude vs. respective number of cycles to fracture, yield the strain- or stress-based Wöhler diagram (see Figs. 2.2 and 2.9). The simplest testing method to obtain Wöhler (S/N) diagrams is the rotating-bending test (see Fig. 3.1a) that was already used for Wöhler's original analysis of railway axles. Rotating cylindrical specimens are loaded by an adjustable bending moment resulting in a sinusoidal fully reversed tension compression course, with the maximum load at the surface of the gauge length. As soon as the specimen is fractured the rotating–bending machine stops and the number of cycles to fracture can be read from the cycle counter.

Fatigue Crack Propagation in Metals and Alloys: Microstructural Aspects and Modelling Concepts. Ulrich Krupp

ISBN: 978-3-527-31537-6

A more detailed analysis of the fatigue behavior of materials is possible by using servohydraulic testing machines, which were introduced in the 1950s. Here, user-defined loading spectra can be applied by means of closed-loop controlling units. Modern state-of-the-art systems allow loading frequencies of up to approximately 100 Hz; in the case of a relatively new voice-coil-operated system even up to 1000 Hz (MTS 1000 Hz, see Fig. 3.1b). The working principle of servohydraulic testing systems is shown schematically in Fig. 3.2.

The core of a servohydraulic testing machine is the servo valve that controls electronically the flow of high-pressure hydraulic oil into the two separate chambers of the hydraulic actuator and hence its upward and downward movement (in the case of an uniaxial system). The test specimens are mounted between the machine's crosshead and the hydraulic actuator by means of various kinds of grips, e.g., hydraulically clamping grips or thread grips. This sounds simple, but in reality uniaxial clamping of the specimens is tricky and, particularly in the case of compression fatigue or compression–tension fatigue, essential to obtain reproducible results. To ensure uniaxial loading conditions, a careful alignment of the system by means of calibration specimens provided with several strain gauges or the use of universal joints (for self-alignment) is required. To avoid buckling during compression of, for example, thin-sheet samples, suitable supports must be used.

The servo valve is controlled by an analog or digital controller (control signal). The actual loading conditions at the specimen are measured by means of a load cell and by means of a strain gauge, which is either attached to the specimen or operates in an optical contactless way. In combination with an amplifier that provides

(a)

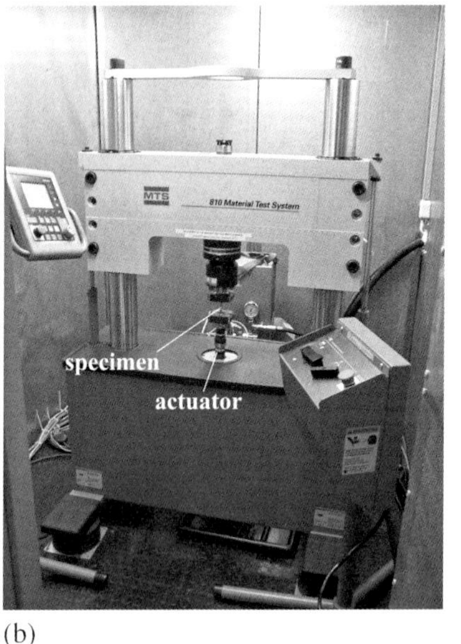

(b)

Fig. 3.1 Fatigue testing by means of (a) rotating–bending testing machine and (b) a modern servohydraulic push–pull testing system (MTS 1000 Hz).

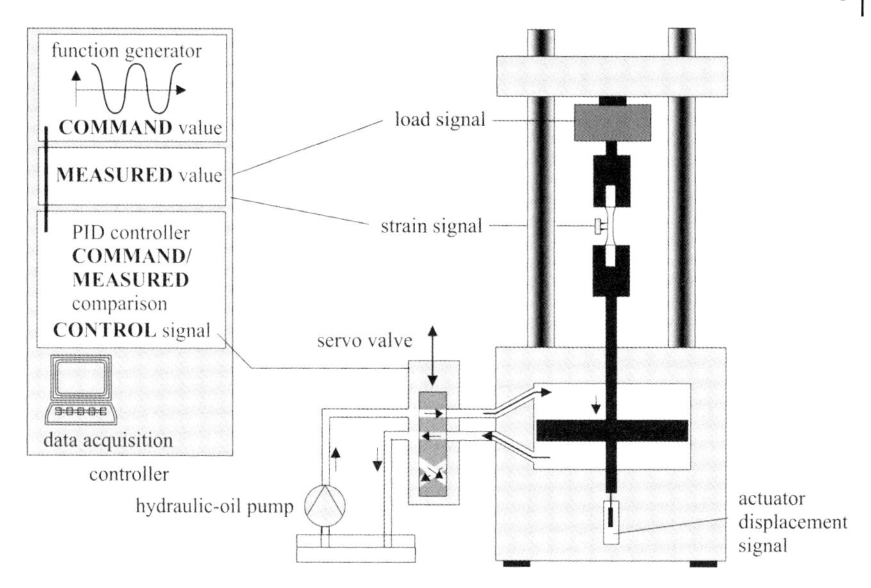

Fig. 3.2 Schematic representation of the working principle of servohydraulic materials testing systems.

the supply voltage and that amplifies the millivolt measurement signal, the actual measured signal is compared with the command signal, which is provided by a function generator. According to the respective deviations, the control signal is increased or decreased by means of a functional relationship, consisting of a proportional (P), an integrative (I) and a damping (D) contribution. The PID controlling performance is adjusted by stepwise optimizing the PID parameters for the given combination of testing system, material and specimen geometry. Details of the mode of operation of servohydraulic testing systems and the controlling systems can be found elsewhere (e.g., [83]).

When planning a series of fatigue tests, the question of which control parameter has to be used is the first one to be answered. If a test should be carried out under stress, strain or plastic-strain control depends on the technical application or the scientific objective. The easiest way is to apply a sinusoidal load signal and to measure the corresponding strain (stress control, $\Delta\sigma/2 = \text{const.}$, cf. Fig. 2.7). Plotting the reversal points of the stress–strain hysteresis loops yields the cyclic deformation curve, which reveals transient processes like cyclic softening or hardening. For most structural applications, certain load spectra are applied to complex components: This results in locally varying strain levels throughout the component. To simulate such a condition strain-controlled fatigue experiments are carried out ($\Delta\varepsilon/2 = \text{const.}$, cf. Fig. 2.7).

By means of the stress–strain data pairs, taken either from stable hysteresis loops (after the transient stage) or from the hysteresis loop at $N = 0.5N_f$, when a stable regime does not exist, the cyclic stress–strain curve (CSSC) can be set up. It is clearly less time-consuming to apply incremental step tests (ISTs) to obtain the

CSSC [84–86]. An IST involves loading blocks of about 30 cycles with gradually increasing and decreasing strain amplitudes, as is shown schematically in Fig. 3.3a. After reaching stable cyclic loading conditions in a block of cycles (cyclic saturation) the reversal points of the hysteresis loops for the various load steps yield directly the complete CSSC (see Fig. 3.3b). It should be pointed out here that, particularly for materials exhibiting pronounced transient behavior (cyclic hardening/softening), the IST may yield only an approximation for the CSSC obtained by single-step fatigue tests. For low strain amplitudes the stress amplitude measured by an IST is higher than for the corresponding single-step test, since it is influenced by the dislocation arrangement resulting from the preceding loading blocks. For high strain amplitudes it is the other way around, since the cumulative plastic strain during an IST is lower than that during the corresponding single-step tests.

To focus on the most damaging effect during low-cycle fatigue, which is the plastic strain occurring in the vicinity of notches of technical components, plastic-strain-controlled fatigue tests should be chosen (for details see [87]). Plastic-strain control reflects directly the fatigue-life-determining plastic-strain amplitude in the Manson–Coffin relationship, which was originally derived for high-temperature fatigue loading conditions (Eq. 2.3). To establish plastic-strain control, the elastic part $\varepsilon_{el} = \sigma/E$ has to be subtracted from the measured total strain signal ε_{tot}. In many cases, high-temperature fatigue systems are equipped with inductive or quartz lamp heaters instead of resistance furnaces, since they allow for a better accessibility of the specimens and the capability to cycle the temperature for thermal fatigue or thermomechanical fatigue testing. Such experiments are required to simulate the loading situation during start-up and shutdown processes of thermal power machines, e.g., aero-engines or land-based gas turbines, internal-combustion engines, etc., and to provide data for the corresponding fatigue-life assessment. It has been recognized that the phase shift between the load and the temperature cycles, the frequency and the hold times during high-temperature or thermo-

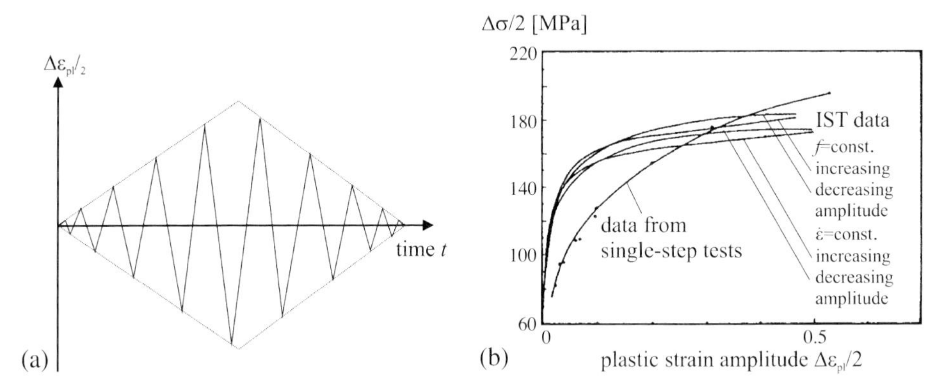

Fig. 3.3 (a) Course of the control signal (here ε_{pl}) for the incremental step test (IST). (b) Corresponding results: stress amplitude $\Delta\sigma/2$ vs. the respective steps in the plastic-strain amplitude $\Delta\varepsilon_{pl}/2$ compared with data from single-step experiments (after [85]).

(a)

(b)

Fig. 3.4 Experimental setup for high-temperature fatigue testing under vacuum or defined-atmosphere conditions: (a) for push–pull testing – specimen heated by induction coil; (b) for four-point bending – specimen heated by quartz lamps.

mechanical fatigue do strongly influence the fatigue-damage mechanisms by superimposing creep and high-temperature corrosion effects. To quantify such effects, experiments in varying gas atmospheres and in vacuum are required. Figure 3.4 shows examples of experimental setups for high-temperature fatigue testing.

3.1.2
Specimen Geometries

It probably does not make much sense to present the great variety of specimen geometries that are used for fatigue and fracture mechanics testing. In particular, scientific programs and/or the application of new multiaxial testing methods often require special specimen geometries, which makes comparability difficult in many cases. Data acquisition for industrial applications is typically based on international or national standards, e.g. ISO or ASTM, that determine specimen geometries, surface roughness and testing procedures.

Many of the results discussed in Chapters 5 and 6 were obtained by using nonstandardized specimens, where geometries and surface conditions were chosen according to the particular scientific objective. For instance, fatigue experiments for characterization of short-crack-propagation mechanisms were carried out on smooth specimens that were carefully electropolished (Fig. 3.5), allowing analytical scanning electron microscopy (SEM) of the specimen surface during periodic interruptions of the test. By means of a shallow notch in the gauge length

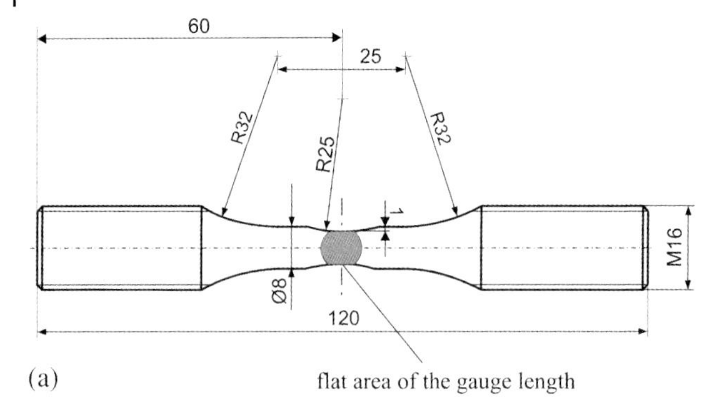

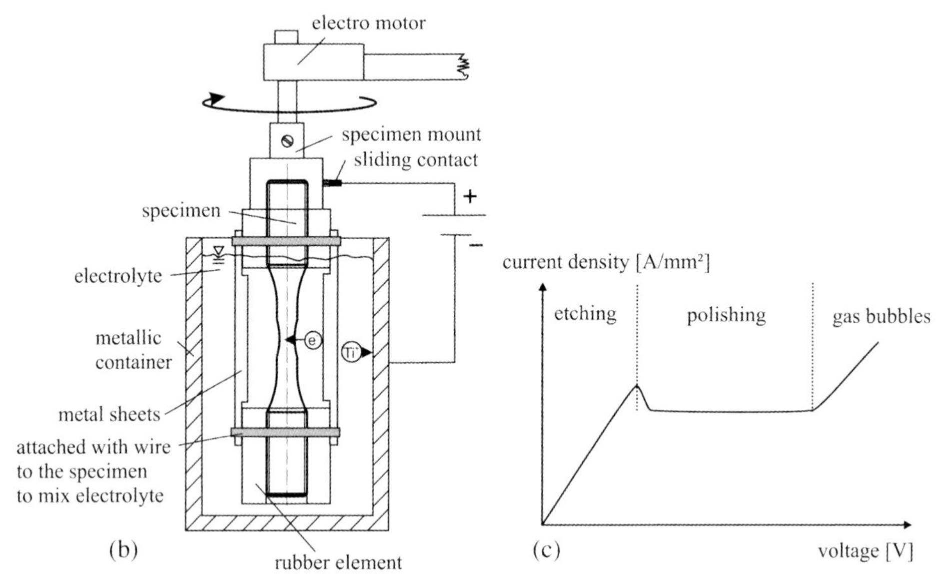

Fig. 3.5 (a) Shallow-notched specimen geometry that has been used for several of the short-crack-characterization experiments described in Chapters 5 and 6. (b) Schematic representation of an apparatus to electropolish cylindrical fatigue specimens according to the respective (c) current density–voltage curve (after [90]).

(see Fig. 3.5) the lateral extension of the fatigue-damage zone was limited without introducing a strong stress concentrator (stress concentration factor of 1.1). Figure 3.5b shows schematically an apparatus to electropolish cylindrical fatigue specimens. To ensure a sufficient supply of electrolyte, e.g., 58.8 vol.% methanol, 35.3 vol.% buthanol, 5.9 vol.% perchloric acid (60%) at $T=-30\,°C$, $U=10$–12 V and $I=0.8$–1.0A for the β-titanium alloy LCB (cf. Chapters 5 and 6), the specimens are rotated.

(a)

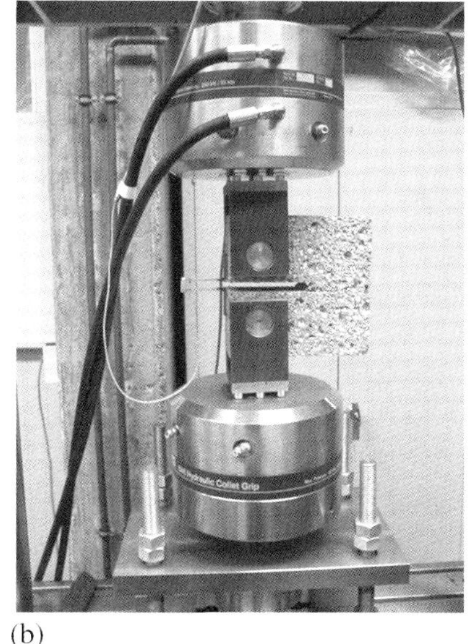

(b)

Fig. 3.6 Experimental setup for (a) fully reversed fatigue testing of open-cell Al sponge and (b) determination of fracture toughness K_{Ic} of a closed-cell Al foam.

A further example of using tailored specimen geometries is mechanical testing of cellular metals (cf. Section 5.6). Due to their inhomogeneity, relatively large samples have to be used to ensure a reasonable reproducibility; i.e., the diameter of the specimens should cover at least 10 pores. Figure 3.6 shows the setup for fatigue testing of an open-cell Al-9Si-3Cu sponge specimen that was glued into suitable grips (Fig. 3.6a), and fracture-toughness testing (K_{Ic}) of a closed-cell Al-7Si-Mg+15%SiC foam compact-tension specimen designed in accordance with ASTM 399-90 (see Section 3.2.1).

For measurement of the propagation of long fatigue cracks, notched specimens are typically used, e.g., single-edge notched bend (SENB) specimens under three- or four-point bending loading conditions. Figure 3.7 shows, as an example, the specimen geometry used to study crack propagation in a nickel-based superalloy under sustained-loading conditions at $T = 650\,°C$ in air and vacuum (see Section 6.5). A sharp notch was established by means of spark erosion (EDM, electro-discharge machining) followed either by fatigue pre-cracking applying a decreasing range of stress intensity ΔK or by the razorblade technique. In the latter case, a razorblade or alternatively a thin wire together with 1–3 μm diamond paste is drawn back and forth through the EDM starter notch. Such a sharpened crack is represented in Fig. 3.7b. To ensure a linear crack front, one should use side-grooved specimens (see Fig. 3.7a).

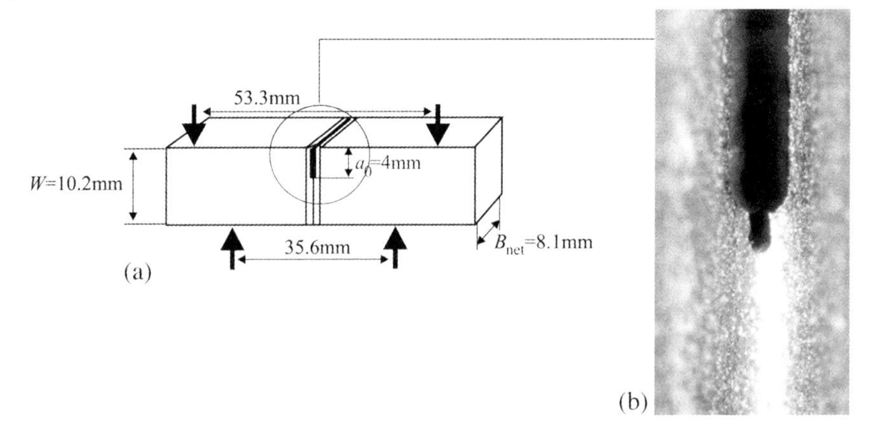

Fig. 3.7 (a) Geometry with loading points of a side-grooved SENB four-point bending specimen; (b) starter notch sharpened by a 50-μm Mo wire.

3.1.3
Local Strain Measurement: The ISDG Technique

A simple way to measure the uniaxial displacement of fatigue or tensile specimens is the application of an inductive plunger-type strain gauge, which is used in most mechanical testing machines to follow the position of the actuator. However, the actuator position is not a very useful measure for the true strain at the specimen, since it is influenced by the stiffness of all components of the testing machine that lie in the load path through the specimen. Hence, strain measurement has to be done either by external resistive or capacitive strain gauge devices, which have to be attached to the specimens (clip gauges), or by bonded electrical resistance strain gauges, which are glued onto the surface of samples or technical components. While the strain gauge devices are used to measure the uniaxial displacement between the two gauge edges, the bonded gauges are commercially available in various gauge sizes and patterns and can be used to characterize complete local strain states. The bonded electrical-resistance strain gauges, as shown in Fig. 3.8, consist of a photo-etched metal foil pattern (typically constantan or a NiCr alloy) mounted on a polymer backing (polyimide or glass-fiber-reinforced epoxy). Since they are small (order of magnitude approximately 1–10 mm^2) and flexible, they can be at-

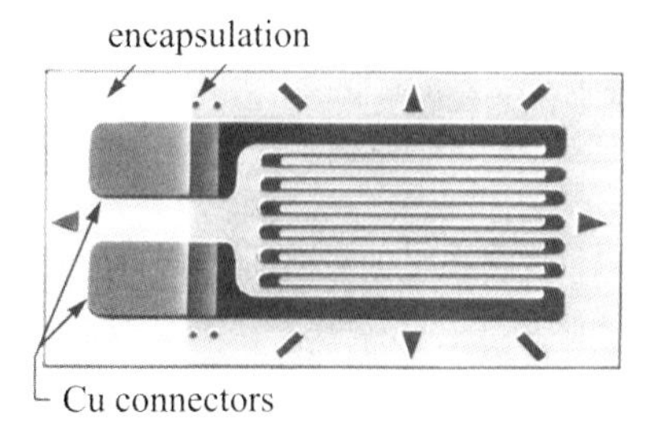

Fig. 3.8 Example of a metal foil pattern strain gauge.

tached even to small and curved surfaces. The main issue of the metal foil pattern is a sensitive relationship between the length change Δl and the corresponding change in the electric resistance $\Delta R/R = k\Delta l/l_0$, with k being a proportionality factor. Bonded electrical resistance strain gauges are commonly used for experimental stress–strain analysis on technical components as well as for local strain measurements, e.g., in the vicinity of notches or crack tips, as reported by Yu et al. [88].

Recent developments in image analysis and digital pattern recognition have led to a greatly increased resolution of optical strain analysis systems. Nowadays, displacement fields with a resolution smaller than 10 μm can be obtained. By means of two CCD cameras focused at different angles onto the specimen, such displacement fields can even be measured three-dimensionally.

Local strain measurements with a resolution of the order of nanometers can be carried out by means of an interferometric strain/displacement gauge (ISDG), which is perfectly suitable for precise crack-closure measurements (see Section 6.3). The technique, first proposed by Sharpe and Grandt [89] and later applied by several other authors [90–99], is based on the interference of laser light that is reflected by two markers on the specimen surface according to the following equation:

$$d \sin \Theta = m\lambda \tag{3.1}$$

In most cases, the laser light (for the experiments described in Section 6.3.7 a 5-mW Ne–He laser, $\lambda = 632$ nm, was used) is reflected by the flanks of two Vickers microhardness indentations that are applied with a spacing of $d \approx 50$–100 μm within the region of interest on the specimen surface. Such an experimental setup is shown in Fig. 3.9a. According to Fig. 3.9b, the flanks of the microhardness indentations are inclined by an angle of $\Theta = 42°$, which determines the positions of the two moveable photodiodes or high-resolution CCD cameras that record the interference pattern. A displacement of the two indentations by Δd results in two effects in the plane of the cameras at a distance of D: a change in the spacing of the interference maxima ΔY and a contraction or expansion of the complete pattern, leading to a shift of the maxima by dY (see Fig. 3.9c). Since only the latter can be measured with high accuracy, the use of two individual cameras is required to accommodate the pattern shift due to the shared displacement of the two indentations.

The resolution of the system is obtained by differentiating sin Θ in Eq. (3.1) by d:

$$\frac{\partial \sin \Theta}{\partial d} = \frac{\partial}{\partial d}\left(\frac{m\lambda}{d}\right) \Rightarrow \frac{\partial \Theta}{\partial d} = -\frac{m\lambda}{d^2 \cos \Theta} \tag{3.2}$$

Eliminating m by again applying Eq. (3.1) and using the relationship $dY = Dd\Theta$, one can obtain the resolution of an ISDG system from

$$\frac{\partial Y}{\partial d} = \frac{D \tan \Theta}{\lambda} \tag{3.3}$$

For a distance of $D = 300$ mm between the specimen surface and the cameras, an initial distance of the microhardness indentations of $d = 100$ μm and a minimum detectable shift of the interference pattern in the plane of the CCD cameras of

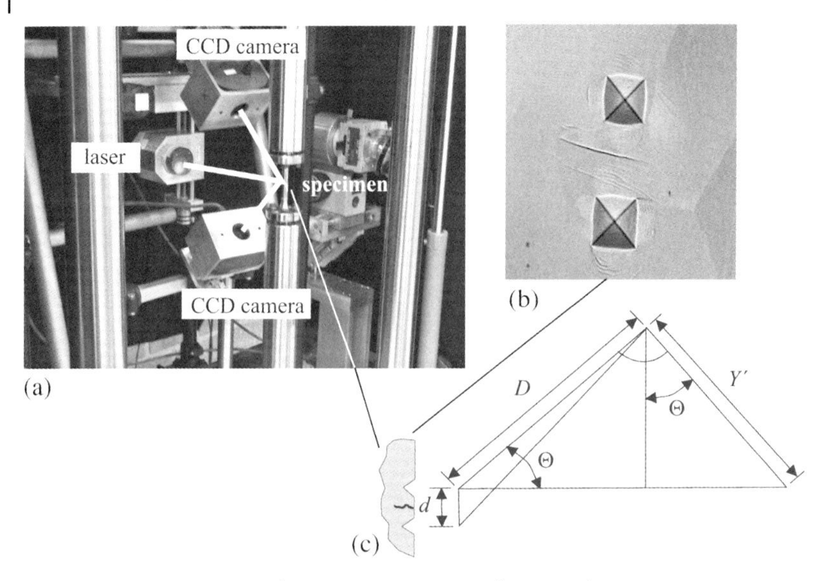

Fig. 3.9 Strain measurement by ISDG: (a) experimental setup within a servohydraulic testing system; (b) specimen surface with two microhardness indentations; (c) schematic representation of the geometrical parameters (for details see text).

$\delta Y_{\mathrm{min}} = 3.225$ μm, the theoretical resolution was calculated to be $\delta d_{\mathrm{min}} = 1.19$ nm [100]. In the literature, estimated values for the resolution are given by Akiniwa et al. [96] as of the order of 1 nm and by Sharpe and Grandt [89] as of the order of 0.5 nm. A more detailed error analysis of the CCD camera system in combination with FFT(fast Fourier transformation) pattern smoothening yielded a conservative estimate of the practical resolution of 23 nm [100].

A disadvantage of using Vickers microhardness indentations for ISDG analysis of crack-closure effects is the introduction of plastic deformation, which might influence the crack propagation under certain circumstances (cf. Fig. 3.9b). To exclude such effects the microhardness indentations have to be applied not before the crack front has already proceeded at least 50 μm from the gauge distance between the indentations. Alternatively, reflection points can be introduced by focused ion beam (FIB) milling or by gold markers as proposed by Fax et al. [101].

3.2
Crack-Propagation Measurements

Analogous to the various length scales of defects and cracks in engineering materials – from early flaw initiation on the nanometer scale to large cracks on the centimeter scale – there are a large variety of techniques for detecting fatigue damage and cracks. While cracks in components are typically detected by means of ultrasonic or eddy-current testing having a resolution of the order of 10^{-3} m, high-resolution transmission electron microscopy (HRTEM) allows the characterization of defects of the order of the atomic spacing (10^{-10} m). Of course, such high-resolution techniques are not suitable for failure analysis of technical components, since the overall volume of matter studied by TEM worldwide since its invention is well below 1 mm^3 [102]. Hence, when talking about measuring crack propagation one has to distinguish between practical crack-detection procedures to be applied during service intervals on technical components, considering cracks typically as large as 1 mm, and crack monitoring in research and development, considering even very small cracks at the micrometer and sub-micrometer length scale.

3.2.1 Potential-Drop Concepts and Fracture Mechanics Experiments

The most obvious and precise approaches to follow crack propagation during mechanical deformation are the observation of the crack tip by means of optical microscopy, SEM or the replica technique. However, these techniques become difficult when the data acquisition must be done automatically and/or when the experiments are carried out at high temperatures and/or in special environments.

In general, crack propagation leads to a decrease in stiffness of a component or a specimen. In particular, in the case of notched fracture mechanics specimens, like compact-tension specimens or SENB specimens, this effect is used for measuring the crack length by applying a clip strain gauge over the two flanks of the notched/cracked specimen (cf. Fig. 3.6 and ASTM E399-90). By very precisely determining the slope of the stress–strain hysteresis loops after the load-reversal points (single or multiple load reversals), the decrease in stiffness can be used as a measure of the damage evolution in cylindrical fatigue/tensile specimens [103].

A very effective and, if carefully calibrated, very sensitive method to measure crack propagation during monotonic or cyclic deformation (resolution ~ 1 μm) is provided by means of the potential-drop technique. As a propagating crack reduces the cross section of a specimen, the electric resistance across the crack path increases correspondingly. By using a stable power supply to feed a constant current of I = 20–50 A into the specimen, the increasing resistance can be measured by means of two wires attached close to the crack faces in combination with a nanovoltmeter, as shown schematically for a direct current potential-drop (DCPD) setup in Fig. 3.10a. Since the correlation between the crack length and the potential drop is not only nonlinear but also depends on several factors, such as the exact position of the wires, the specimen geometry, the crack aspect ratio and the temperature by superimposing a thermocouple effect [104], a thorough calibration of potential-drop systems is required. This is usually done by analyzing the fracture surface postmortem, either by (1) evaluating beach marks that can be generated, e.g., by high-frequency fatigue loading blocks or heat tinting, or (2) by brittle cracking of specimens at certain potential-drop signal levels and simply measuring the extent of the fatigue-crack fraction. Having a calibration curve for the correlation between potential drop and crack length (see Fig. 3.10b), the potential drop tech-

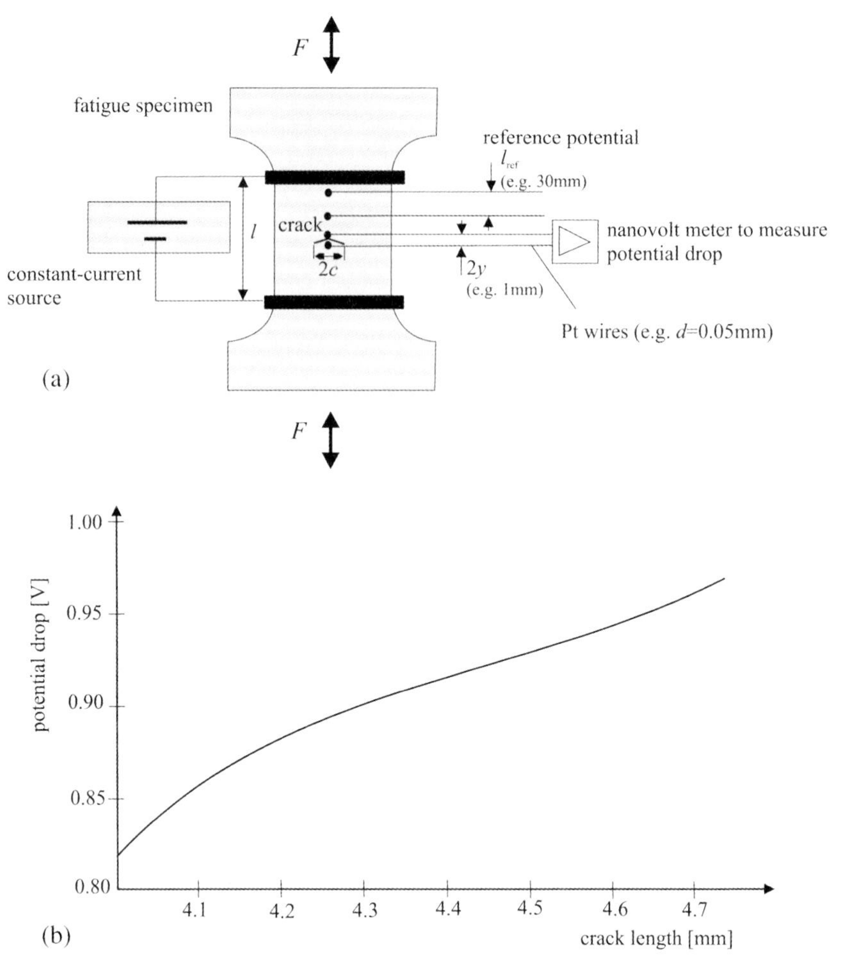

Fig. 3.10 (a) Schematic representation of the potential-drop technique; (b) calibration curve to correlate potential-drop data with the crack length (SENB specimen, ACPD).

nique can be used to monitor crack propagation for various loading conditions applied to the same specimen geometry. An even higher accuracy than that of the DCPD method can be obtained using a high-frequency alternating current power supply (ACPD method; typically 100–250 MHz, at $I = 5$ A). The main advantage is due to the "skin effect" resulting in a linear relationship between the actual crack length and the potential-drop signal [105, 106]. However, the high-frequency AC current makes the signal a function of the position of the leads (induced pick up).

An easy-to-handle alternative to potential-drop measurements are commercially available crack-measuring films. These films are glued on the surface of notched specimens (cf. Fig. 3.11) in such a way that they will crack along with the actual crack propagation in the specimen. The resistance of the foil increases correspondingly, yielding a measure for the crack length.

Crack advance is also a source of noise, and hence can be detected using noise sensors; i.e., applying the acoustic emission technique [107]. Since the actual crack signals are superimposed by a variety of unwanted signals, coming either from the testing environment or the specimen itself, both filtering the signal and calibration are tricky, making acoustic emission as a damage-monitoring system more suitable for scientific purposes than practical failure analyses.

As already mentioned in the preceding section, propagation of long cracks sets in as soon as the threshold value for the stress-intensity range ΔK_{th} is exceeded. The crack-propagation rate da/dN is well described by the empirical Paris–Erdogan relationship (Eq. 2.5) as a function of the applied stress-intensity range ΔK. Once the fracture toughness K_c is exceeded, unstable crack propagation results in failure of the specimen or technical component. Determination of ΔK_{th} and K_c, as well as of the constants C and m in Eq. (2.5), requires fracture-mechanics-type experiments. A common approach to obtain fatigue-crack threshold data is the load-shedding method [105]. For this purpose a notched and fatigue-pre-cracked specimen is loaded by a gradually decreasing range of stress-intensity factor ΔK until the crack-propagation rate falls below a pre-defined minimum value, which lies typically between $da/dN = 10^{-7}$ and 10^{-8} mm/cycle (cf. Fig. 3.11b and ASTM E399-90). The corresponding value for ΔK is then taken as the fatigue-crack threshold ΔK_{th}. To be sure that the obtained values are reproducible material data, the decrease rate for ΔK must be kept small. Otherwise residual stresses due to plastic deformation at the crack tip could retard or even stop further crack advance (cf. overload effects, Section 6.3). A well-established load-shedding procedure is the one proposed by Saxena et al. [108]:

$$C = \frac{1}{\Delta K}\frac{d(\Delta K)}{da} \tag{3.4}$$

with the constant C to be held constant throughout the test (C should be chosen between –0.02 and 0.1 mm^{-1}). This is easy to establish for computer-controlled tests, where the crack length is continuously monitored, e.g., by means of a crack-measuring film (see Fig. 3.11). To account for crack-closure effects, the influence of which depends strongly on the chosen stress ratio R, the strain at the back face of the specimen is measured by means of a bonded electrical-resistance strain gauge. The measured data of this strain gauge reveal different stiffness levels for the specimen corresponding to the situation where the crack is completely open or closed. To evaluate the constants C and m of the Paris–Erdogan equation, ΔK has to be gradually increased after determination of ΔK_{th}.

Of particular importance for damage-tolerant design of technical components is the fracture toughness K_c (cf. ASTM E399-90) or, more generally, the crack-resistance curve with the critical energy release rate G_c, characterizing the criterion for unstable crack propagation (cf. [56]). To account for elastic–plastic crack propagation, a critical value of the J integral, J_c, should be the appropriate parameter (ASTM E813-89).

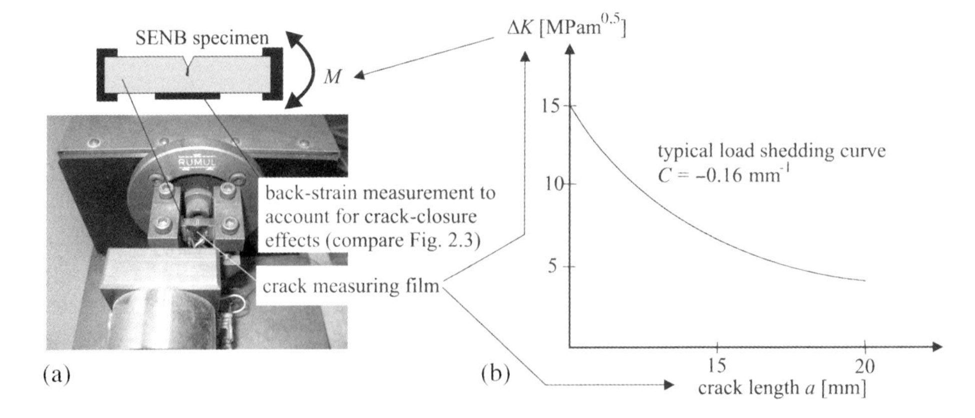

Fig. 3.11 Experimental procedure to determine ΔK_{th}, the constant C and the exponent m of Paris' law (Eq. 2.5): (a) resonance four-point bending system (Rumul cracktronic) for SENB specimens; (b) ΔK vs. a to be chosen for the load-shedding approach.

As an example, the procedure to determine the fracture toughness K_{Ic} (mode I loading conditions, see Fig. 3.6b) according to ASTM E399-90 is briefly introduced in the following. Again, a notched specimen is used, with the notch sharpened by fatigue pre-cracking at stress intensity values K well below $0.6\ K_{Ic}$. In addition to this, a size criterion has to be fulfilled that includes the actual K_{Ic} and hence has to be verified after the test. Referring to the schematic representation of a compact-tension (CT) specimen in Fig. 3.12, the specimen thickness B and its width W and the starter crack length a have to be chosen according to

$$B, a \geq 2{,}5\left(\frac{K_{Ic}}{\sigma_Y}\right)^2 \quad 0{,}45 \leq a/W \leq 0{,}55 \tag{3.5}$$

Hence, one should have an idea of the estimated value of the fracture toughness K_{Ic} prior to the actual test. After generation of the sharp fatigue pre-crack, the specimen is loaded monotonically to fracture, while recording the required load vs. the displacement of the crack flanks (see Fig. 3.12). Afterwards, a critical load P_Q is determined by following one of three alternative approaches, which are represented schematically in Fig. 3.12b.

- Approach 1: P_Q is the intersection point of a secant (P_5) having 95% of the original linear slope of the load–displacement curve.
- Approach 2: an intermediate regime of unstable crack propagation (pop-in) leads to a decrease in the load before the 5% deviation from the linear load–displacement curve is reached. Then, P_Q is the maximum load before the pop-in event.
- Approach 3: the specimen fails before reaching the 5% deviation. Then, P_Q is the maximum load reached during the test.

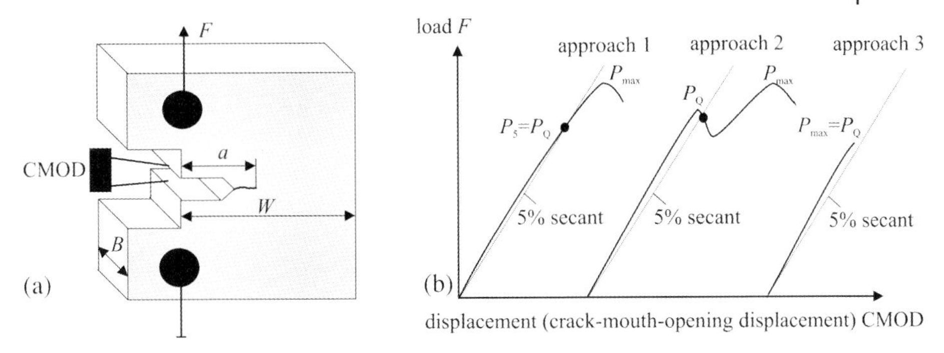

Fig. 3.12 (a) Specimen geometry of CT specimens according to ASTM E399; (b) evaluation procedure to determine P_Q (for details see text).

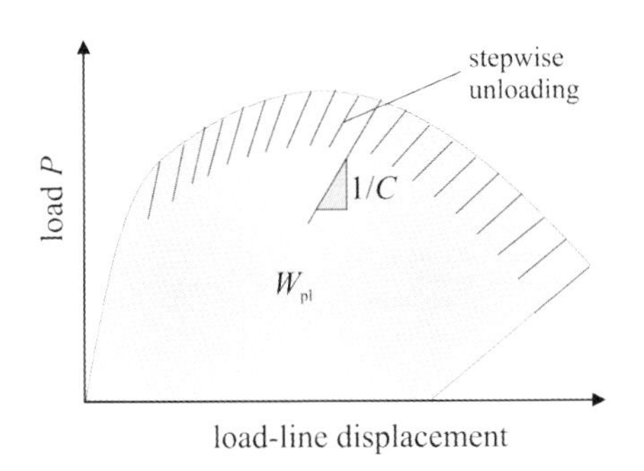

Fig. 3.13 Stepwise unloading procedure for relating a/W to the compliance C to determine the J integral (according to ASTM E813; the plastic area under the load-displacement curve W_{pl} can be used for estimation, cf. Eq.. 3.7 [56]).

The preliminary value for the fracture toughness K_{Ic} follows from P_Q and a valid material parameter when Eq. (3.5) and the condition $P_{max} \leq 1.1 P_Q$ are fulfilled:

$$K_{Ic} = \frac{P_Q}{B\sqrt{W}} f\left(\frac{a}{W}\right) \tag{3.6}$$

The geometrical functions $f(a/W)$ can be obtained for many common geometries in tables available in ASTM E399-90 or in [65]. The strict geometrical limits (prevailing plane strain conditions at the crack tip) for the validity of fracture-toughness data determined according to ASTM E399-90 restrict its application to only relatively brittle or large specimens; for very ductile materials the application of the J integral is often the more suitable choice.

The critical value of the J integral, J_{Ic}, as a criterion for unstable elastic–plastic crack propagation can be determined according to ASTM E813. This requires the indirect evaluation of the crack length a by determining the compliance by stepwise unloading of a notched specimen, as shown in Fig. 3.13. The J integral can be derived also from the load displacement curve according to

$$J = J_{el} + J_{pl} = \frac{K^2\left(1-\nu^2\right)}{E} + \frac{\eta W_{pl}}{B b_0} \tag{3.7}$$

with W_{pl} equal to the plastic work, i.e., the area below the load–displacement curve, η is a dimensionless constant (SENB specimen: $\eta = 2$) and b_0 is the original length of the uncracked ligament.

3.2.2
In Situ Observation of the Crack Length

Alternatively or in addition to fracture mechanics experiments, crack propagation in engineering materials can be monitored directly by microscopy or indirectly by the replica technique.

Several of the results discussed in Chapter 6 were obtained by an optical microscope attached to a servohydraulic testing machine (Fig. 3.14) in such a way that it can be moved in three axes to find and to follow cracks at the specimen surface (traveling microscope). This approach can be automated, using image-analysis software and step motors to locate and focus the optical microscope. Such an approach can also be used to automatically measure the crack length vs. the number of cycles N at high temperatures, when using quartz lamps or inductive heaters and a microscope with a long focal length. Such an approach has been used (e.g., [109]) to evaluate thermomechanical fatigue experiments.

In a quite simple way, the propagation behavior of fatigue cracks can be studied by means of the replica technique [90, 94, 103]. For this purpose, a polymer film (e.g., cellulose acetate) is wetted with acetone and pressed against the surface of a cyclically loaded specimen at maximum tensile load. The capillary attraction of the cracks draws in the semi-dissolved polymer film. When the polymer film is dried, the specimen surface with slip traces and the crack pattern is reproduced and can be evaluated by optical microscopy or SEM (cf. Section 6.4.2). Bowles and Schijve

Fig. 3.14 Optical microscope attached to a servohydraulic testing machine to observe crack propagation.

[110] developed a modification of the replica approach that allows a quantitative representation of the three-dimensional shape of fatigue cracks. For this purpose, a polymer resin is vacuum-infiltrated into the crack and hardened. Afterwards, the metal in the vicinity of the crack is etched away, leaving behind the polymeric replica of the original crack that reveals the microstructural features of the crack faces and, hence, important information on crack-propagation mechanisms.

In several more recent studies, computer tomography was used to describe deformation and damage mechanisms in a three-dimensional manner. Beside macroscopic approaches that are useful, e.g., for analysis of the deformation behavior of cellular materials [111], high-energy brilliant synchrotron radiation has been used to study fatigue-crack-propagation mechanisms, reaching a resolution of between 0.5 and 1 μm [112–114]. Interactions with grain boundaries were made visible by wetting grain boundaries in aluminum alloys, using liquid gallium [112, 115]. Local crystallographic orientations in three dimensions can be obtained by the Bragg reflection of the synchrotron radiation on respective lattice planes (3DXRD) [116, 117] in a way similar to electron backscattered diffraction (EBSD) in an SEM instrument, as described in more detail in Section 3.3.2.

Very helpful for understanding interactions between alloy microstructure and crack-propagation mechanisms is the use of *in situ* techniques in electron microscopy like miniature mechanical testing machines (cf. [118]). Because of the analytical capabilities (EBSD, energy-dispersive X-ray spectroscopy, EDS) and the high resolution of modern SEM instruments, the crack-propagation rate can be correlated with local microstructural features and with the evaluation of fracture-mechanics parameters on the microscale, e.g., the range of crack-opening displacement (ΔCOD) [119]. Most common are electromechanically driven deformation stages, since they can easily be operated within small vacuum chambers [120–124]. However, *in situ* fatigue studies in an SEM instrument have been carried out also by means of servohydraulic testing machines having the advantage that the cycling frequency is not limited to below $f = 1$ Hz as is the case for conventional electromechanically driven systems (see e.g. [121]). To study fatigue-crack-propagation mechanisms *in situ* in the SEM instrument, a piezoactuator-operated system has recently been developed that allows for cycling frequencies above $f = 50$ Hz [125]. Figure 3.15 shows a commercially available *in situ* deformation stage and the prototype of a piezoactuator-driven system.

Probably the main problem with using miniature testing systems is achieving perfect alignment of the specimens, which is required for a well-defined stress–strain state within the gauge length. Fulfilling this requirement becomes particularly difficult for experiments in a transmission electron microscope carried out to observe *in situ* dislocation motion as the most basic fatigue mechanism. For such experiments the gauge length must be electron transparent, and hence must be of the order of 150 nm thick. Due to the tricky specimen preparation (which has become easier thanks to FIB precision milling) and problems in interpretation of the results, only a few studies have reported on *in situ* deformation in a TEM instrument (e.g., [126–129]).

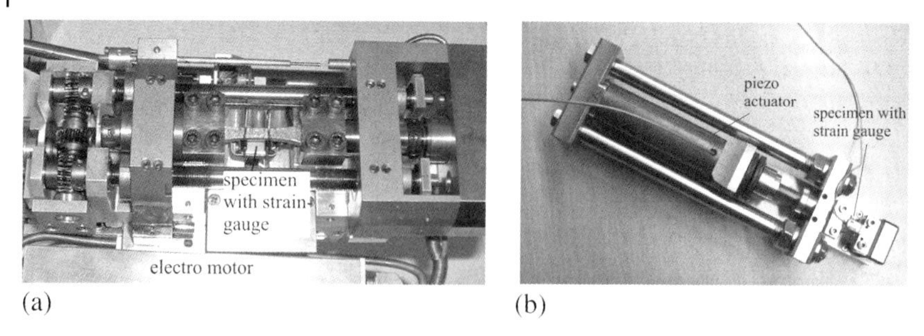

Fig. 3.15 Miniature testing machines for *in situ* deformation experiments in a scanning electron microscope: (a) electromechanical concept (Kammrath & Weiss, TU Dresden); (b) piezoactuator-driven prototype (Univ. Siegen).

3.3
Methods of Microstructural Analysis and Quantitative Characterization of Grain and Phase Boundaries

Neglecting the very special *in situ* techniques mentioned above, fatigue and fracture research is typically separated into mechanical testing followed by microstructural analysis (metallography) to reveal a deeper understanding of the failure mechanisms.

In addition to standard metallographical methods, like optical microscopy, microhardness testing, etc., analytical SEM has become a very powerful tool [130]. Beside its capability in fracture surface imaging without elaborate specimen preparation, the fast development of automated EBSD allows the quick collection of structural information – crystallographic orientations – of the surface grains. Hence, EBSD provides a convenient way to evaluate microtexture, grain and phase size and distribution with high resolution but, in contrast to TEM, for representatively large areas.

Owing to its significance to the research work discussed in Chapters 5 and 6, the fundamental concept of the EBSD technique and the corresponding evaluation process is briefly explained in the following.

3.3.1
Analytical SEM: Topography Contrast to Study Fracture Surfaces

For fracture surface examination at magnifications higher than about ×20, SEM is essentially the only common imaging technique available, thanks to the high depth of focus that can be used down to the resolution limit of modern SEM instruments being of the order of magnitude of 1 nm (field emission gun SEM). In many cases, fracture surface analysis yields important information helping to identify failure causes, e.g., nontolerable overloads, material defects, wrong heat treatment or welding processes. As an example, small amounts of hydrogen (c_H = 1–4 wt.ppm) in high-strength steels cause a change from ductile transgranular to brittle inter-

Fig. 3.16 Hydrogen-induced failure of (a) high-strength spring steel (intercrystalline fracture) and (b) plain carbon steel (cleavage fracture).

granular fracture (see Fig. 3.16a and Section 6.5) or in the case of plain carbon steel cause cleavage (see Fig. 3.16b); i.e., transition from plastic slip to brittle separation of (001) planes in the same way as occurs below the brittle–ductile transition temperature (BDTT).

Fatigue-crack propagation can be identified in many cases by the presence of fatigue striations within the fracture surface, particularly in the case of ductile alloys. As discussed in more detail in Section 6.2.2, the spacing between the striations corresponds to one fatigue cycle. Hence, the correlation between crack depth, da/dN and the local stress intensity range ΔK or the load history can be estimated. This is shown in Fig. 3.17 showing fatigue striations in the fracture surface of two copper specimens loaded by constant strain amplitude fatigue (Fig. 3.17a) and an incremental step test (Fig. 3.17b).

Often lateral crack propagation becomes visible by beach marks, due to environmental attack during crack arrest, and by crack-course lines, due to microstructure changes along the crack front (see Fig. 3.18). Then the location of crack initiation can be identified, allowing one to attribute failure to, for example, a large nonmetallic inclusion to be analyzed by EDS [130], or a notch as a geometrical stress con-

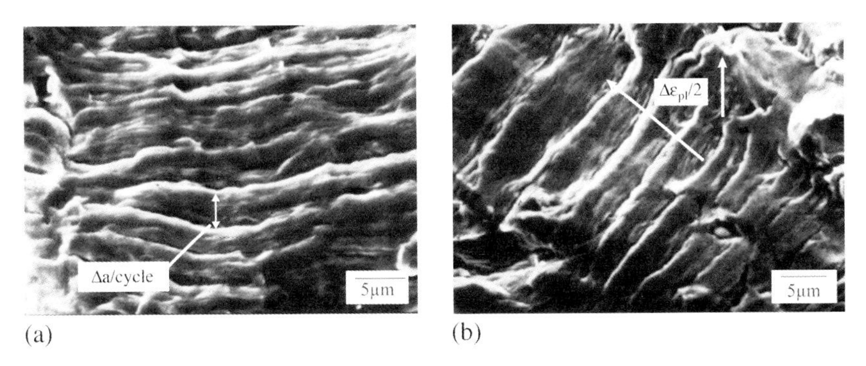

Fig. 3.17 Fatigue striations in the fracture surface of cyclically deformed copper (a) at constant plastic-strain amplitude and (b) during an incremental step test (IST).

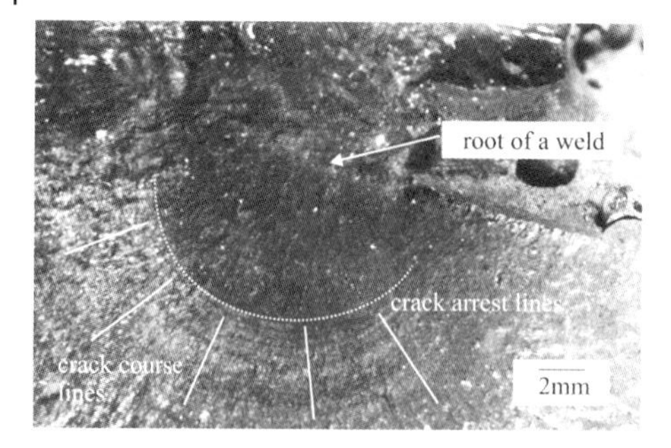

Fig. 3.18 Macroscopic view of a fracture surface showing crack arrest and crack course lines (see markers; welded agitator made of austenitic stainless steel).

centrator. This is described in more detail in Chapter 5. Fracture surface analysis generally needs much experience but often provides many important details about the load spectrum a component experienced during service life. Useful interpretation support can be found, e.g., in [131, 132]. An alternative way to quantitatively evaluate fracture surfaces is by stereographic analysis of two images, taken from the same location but slightly different tilt angles ($\Delta\alpha = 3$–$8°$). Such a technique was applied by Bichler and Pippan [133] to study crack-propagation mechanisms by comparing the three-dimensional topography of two opposing fracture surfaces.

3.3.2
SEM Imaging by Backscattered Electrons and EBSD

While rough surface topologies are studied by means of secondary electron contrast, flat polished surfaces yield electron contrast mainly by backscattered electrons (BSEs). Backscattering of primary electrons occurs due to elastic deflection in the Coulomb field of the positively charged nuclei of the near-surface atoms. Hence, the gain of BSEs depends on the atomic weight; i.e., high-density specimen areas appear white (high BSE signal) and low-density areas appear dark (low BSE signal). Therefore, BSE imaging allows the localization of nonmetallic inclusions, like Al_2O_3, SiC, TiN, etc., as crack-nucleation sites. In addition to the atomic weight, the gain of BSEs depends on the inclination angle of incident primary electrons with respect to the orientation of the lattice planes of the surface grains as shown schematically in Fig. 3.19a. Low-index lattice planes lying parallel to the incident primary electron beam act as "channels" and therefore resulting in a low BSE intensity. Figure 3.19b shows an example of the electron channeling contrast (ECC) revealing the grain and subgrain distribution for the nickel-based superalloy Waspalloy.

ECC can also be used to make characteristic dislocation patterns visible since BSE gain also depends on the local defect density. For pure copper or nickel the typical ladder structure of persistent slip bands (cf. Chapters 4 and 5) and the vein structure of areas with lower dislocation densities were shown (see Fig. 3.20).

The electron channeling technique can also be used to quantitatively evaluate local crystallographic orientations. By tilting the electron beam interacting with the local set of lattice planes ("rocking-beam" technique) the BSE gain depends on the

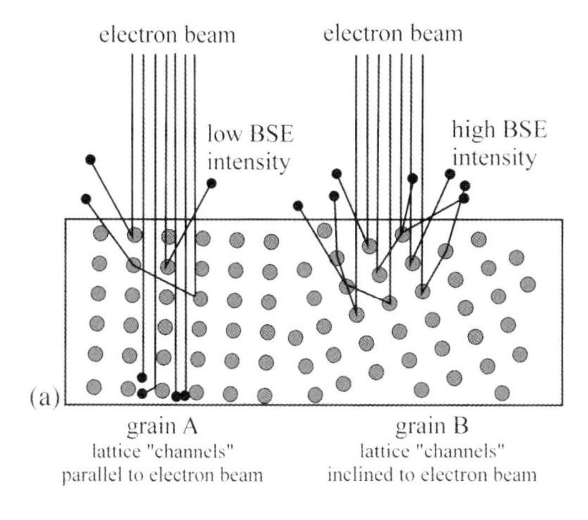

Fig. 3.19 Electron channeling contrast; (a) schematic representation, (b) revealing the microstructure of a polished specimen of the nickel-based superalloy Waspalloy.

incidence angle of the primary beam resulting in intensity minima and maxima, the geometric position of which can be correlated with the local crystallographic orientation with respect to the specimen surface [135].

More conveniently, the crystallographic orientation distribution of surface grains can be determined by means of EBSD, a technique that has experienced a tremendous development during the last 15 years and has become a standard

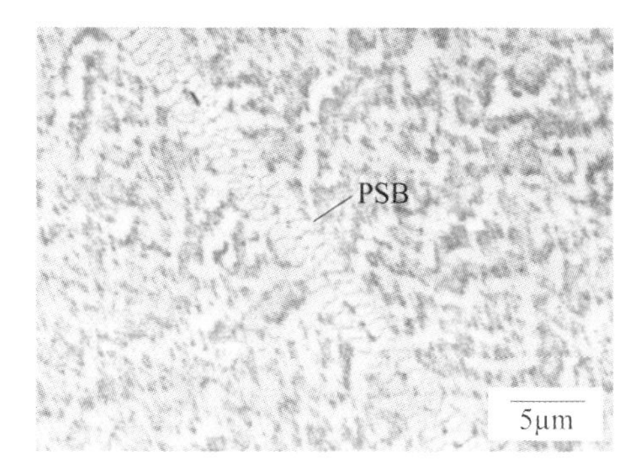

Fig. 3.20 Characteristic arrangement of dislocations, persistent slip bands within a vein structure (compare Section 4.4) made visible by electron channeling contrast (after [134]).

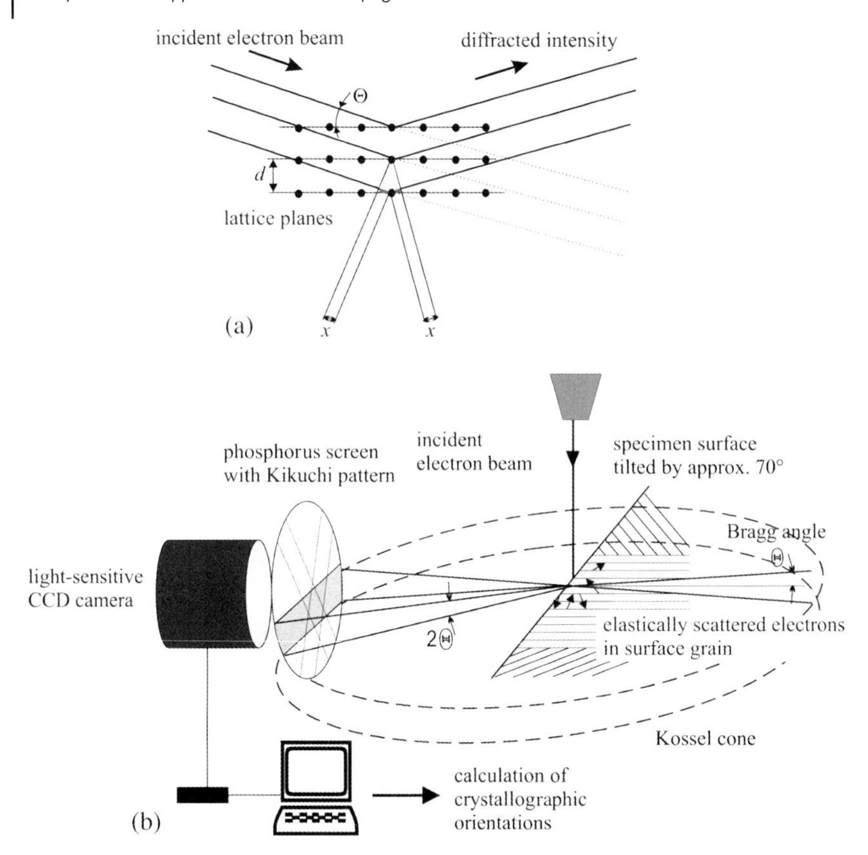

Fig. 3.21 (a) Bragg diffraction at lattice planes of a crystalline solid; (b) schematic representation of the working principle of the EBSD technique.

method to be used in combination with SEM [136, 137]. EBSD is extremely helpful in evaluating and interpreting mechanical failure [138]. Analogous to electron diffraction techniques used with TEM, the EBSD technique is based on diffraction of electrons by the lattice planes, according to the Bragg condition (Fig. 3.21a):

$$2d\sin\Theta = n\lambda \tag{3.8}$$

As shown schematically in Fig. 3.21b, elastically scattered electrons coming from spatially all directions in the interaction volume of the electron beam are reflected by the lattice planes. If the incident angle of the electrons is equal to the Bragg angle Θ, they experience an intensity increase in the corresponding leaving direction, since the phase shift of the electron waves reflected by neighboring lattice planes of spacing d is an integer (n) multiple of the wavelength λ. The Bragg condition is fulfilled for electrons reaching the lattice planes from the top and from the bottom, hence resulting in two interference cones, the so-called Kossel cones. These cones have large opening angles ($\alpha = 180° - \Theta$) due to the small wavelength of elastically scattered electrons λ [Å] $\approx (150/U\ [\mathrm{V}])^{0.5}$, with U being the electron acceleration voltage [102]. When placing a phosphorus screen into the interference cones, the intersections with the two cones above and below the reflecting set of lattice planes appear as two quasi-parallel lines. Since the same effect is present for essentially all sets of lattice planes, a pattern of lines, the so-called Kikuchi lines, is formed. The course of each set of lines is parallel to the reflecting set of lattice planes. Intersection points represent the direction of common zone axis. The width of the lines increases with increasing lattice plane indexes; i.e., low-index planes, e.g., {110} planes, result in narrow bands, since due to the low *d* spacing the Bragg angle Θ is very small. If the position of the phosphorus screen relative to the specimen position is known, the Kikuchi pattern can be used to determine locally the crystallographic orientation with respect to the specimen coordinate system (see Section 3.3.3). A very helpful compilation of EBSD Kikuchi patterns for various crystal structures can be found in [139].

Figure 3.22 shows an example of an indexed Kikuchi pattern obtained by EBSD for the fcc nickel-based superalloy Waspalloy (Fig. 3.22a) and the technical realization of the EBSD technique in a SEM instrument (Fig. 3.22b). To extract the crystallography information from the overall gain of BSEs, it is required to tilt the specimen surface by about 60–80°; i.e., about 10–30° with respect to the electron beam direction. Then, the main BSE intensity leaves the specimen downwards by an angle of about 20–60° with respect to the electron beam (glancing incidence). Only BSEs that are diffracted by the lattice planes according to Bragg's law leave the specimen in the direction of the phosphorus screen. The glancing incidence restricts the electron interaction volume to a small surface area, and therefore strong absorption of diffracted electrons is avoided. The phosphorus screen located in front of the specimen surface by a distance of about 30 mm (to ensure a high spatial angle) is monitored by a very light-sensitive CCD (charge coupled device) or SIT (silicon intensified target) camera through a quartz window or a vacuum-tight bellows. The video image is analyzed by image-manipulation software. High-quality Kikuchi patterns for fast evaluation (up to 50 patterns per second; see Section

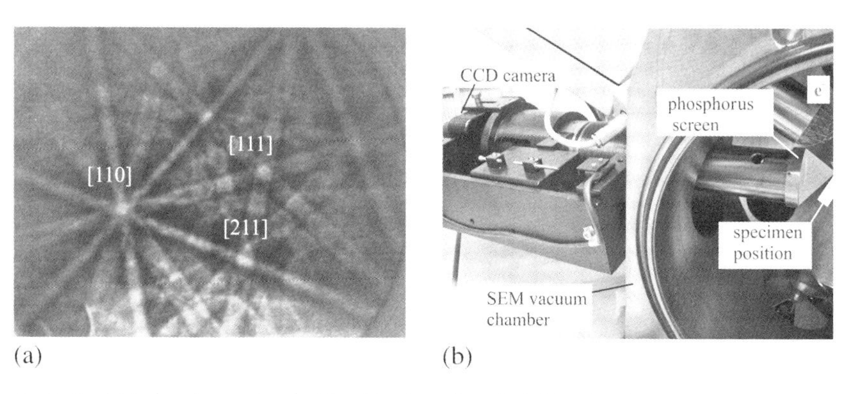

Fig. 3.22 (a) Indexed example of a Kikuchi pattern for a fcc lattice structure (nickel-based superalloy Waspalloy); (b) commercially available realization of the EBSD technique (TSL) in a scanning electron microscope (Philips SEM XL30).

3.3.3) are obtained by extracting the pure Kikuchi-line information from the image by means of background subtraction (removing the intensities caused by nondiffracted BSEs).

The small information depth of approximately 100 nm due to the glancing incidence is the reason for the great demands regarding the metallographic surface finish. Standard metallography, applying grinding with SiC paper followed by polishing on conventional polishing wheels, is usually not sufficient, since high plastic distortions during grinding prohibit homogeneous diffraction and the appearance of Kikuchi lines (see [140]). The easiest way to avoid plastic distortions within the surface layer is the application of electropolishing, a treatment that has been successfully used to study the initiation and propagation behavior of microstructurally short fatigue cracks (see Chapters 5 and 6). Alternatively, specimens can be mechanically prepared, applying an appropriate polishing technique by which the occurrence of plastic distortions within the surface layer is avoided. This requires a stepwise grinding procedure, where during each step the plastic zone resulting from the preceding step is removed. Final surface finishing should be carried out by means of several hours of vibro-polishing (typically 3–8 h) using 0.05-μm SiC or Al_2O_3 suspension.

It is worth mentioning that the large spatial angle between specimen and phosphorus screen restricts EBSD analysis methods to the evaluation of crystallographic orientations. By increasing the distance between the phosphorus screen and specimen the spatial angle is decreased with the consequence that, on the one hand, the section of the Kikuchi pattern becomes smaller, but, on the other hand, the Kikuchi band width and hence the sensitivity to small changes in the lattice constant due to elastic strains increases. This was shown, for example, by Wilkinson [141] for SiGe layered structures.

The Bragg reflection of the characteristic X-ray radiation (normally used only for EDS analysis) on the lattice planes yields the so-called Kossel lines, which are intersections through Kossel cones, which have a much lower opening angle than the EBSD cones, due to the much higher wavelength of X-rays as compared to electrons. These Kossel lines are of substantially higher sensitivity with respect to lattice spacing and further structural information, and can be used, e.g., for high-resolution local residual stress measurements [142].

3.3.3
Evaluation of Kikuchi Patterns: Automated EBSD

The Kikuchi pattern allows the exact determination of the position of the local crystal coordinate system with respect to the specimen coordinate system, the latter typically defined by the specimen normal vector (N) and two perpendicular directions lying in the surface plane, usually denoted as rolling direction (RD) and transverse direction (TD), referring to the historic background of texture research for rolled materials (cf. Fig. 3.25). This correlation requires a careful calibration of the system geometry; i.e., the relative positions of specimen surface and phosphorus screen in combination with the image analysis software. Such a calibration can be achieved by placing a <001>-oriented Si single crystal in such a way into the system that a {001} plane is oriented parallel to the connecting line from specimen to phosphorus screen. By marking, e.g., the [114] and the [001] zone axis on the screen, the system can be calibrated. Of course, a calibration can be done on the specimen to be studied itself, when the crystal structure is known.

Using EBSD systems of the first generation, it was necessary to evaluate manually the Kikuchi pattern with respect to the angles between the bands in combination with crystallography data sets in look-up tables. The local crystallographic orientation was determined by generating a calculated Kikuchi pattern that fits to the measured one.

The increasing interest in the EBSD technique as a unique method for measuring local crystallographic orientations of relatively large samples (as compared to TEM), led to a rapid technical development. Modern EBSD systems control the electron beam, placing it automatically point by point over a predefined area. Background subtraction for the Kikuchi pattern and indexing is done automatically, the latter using the Hough transformation [143], which transfers each band in two-dimensional coordinates allowing a fast evaluation. In combination with fast CCD cameras more than 50 measurements per second are possible, allowing the generation of highly resolved crystallographic orientation mappings consisting of hundred thousands of individual data points.

The EBSD technique and the respective evaluation procedures are subject of many overview articles, e.g., [144–146], and textbooks, e.g., [136, 137].

3.3.4
Orientation Analysis Using TEM and X-Ray Diffraction

In a similar way to the EBSD approach, local crystallographic orientations can be determined using TEM, either by selected area diffraction (SAD) or convergent beam electron diffraction (CBED) [102, 147], as shown schematically in Fig. 3.23. While CBED results in the generation of Kossel cones, representing Kikuchi bands on the imaging phosphorus screen, SAD is based on the Bragg reflection of virtually parallel electrons and yields a point pattern on the imaging phosphorus screen. According to Fig. 3.23, this pattern represents the specimen lattice by means of the reciprocal lattice. Evaluation of the angle relationships and distances with respect to the central [001] reflex of the incident electron beam ($\vec{g}$ vectors; see e.g. [102] for details) allows the determination of the local crystal structure and crystallographic orientation. The main disadvantages in using TEM is the elaborate way of specimen preparation and the small electron-transparent area available for analysis, restricting orientation measurements in most cases to only a few grains/subgrains.

Probably the oldest method to determine crystallographic orientations is the diffraction of X-rays. Again, the basic physical principle is described by the Bragg relationship (Eq. 3.8), differing from electron diffraction techniques only in the substantially higher wavelength of the X-rays. Since precision-focusing and exact positioning of X-rays is not possible in the same way as for electrons in an electron microscope, X-ray techniques are predominantly employed for phase, macrotex-

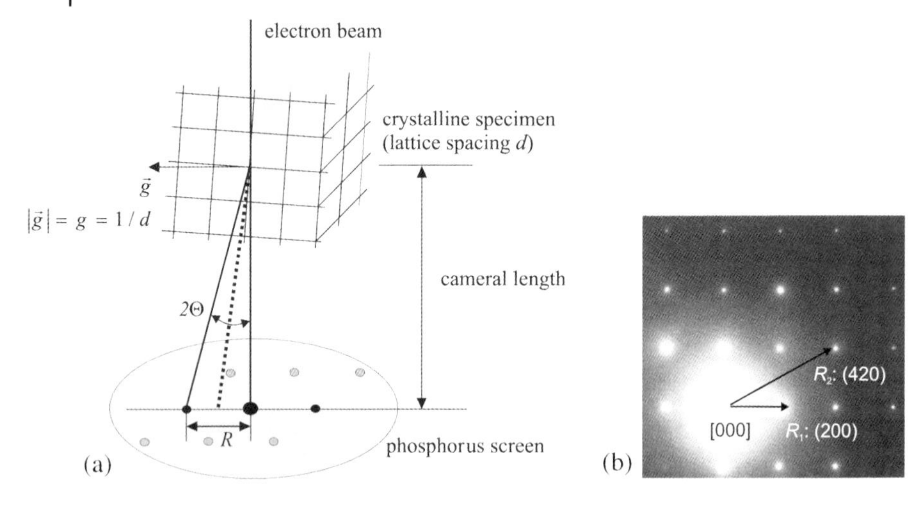

Fig. 3.23 Origin (a) and example (b) of SAD pattern in a transmission electron microscope.

ture and residual stress analysis, as well as for single-crystal orientation measurements (see e.g. [148]). For the latter kind of orientation measurements the simple setup of a Laue camera can be used. Figure 3.24 shows schematically the functional principle of a backscattering Laue camera (Fig. 3.24a), which has been applied, e.g., to produce diffusion-bonded bicrystals for crack-propagation studies (see Section 6.5), and a Laue X-ray photograph of a single-crystalline specimen with the <001> directions oriented perpendicular and 20° rotated with respect to the image plane (Fig. 3.24b). The evaluation of the point diagrams is possible either manually by using a Beranger diagram in combination with a Wulf net [148] or by computer software.

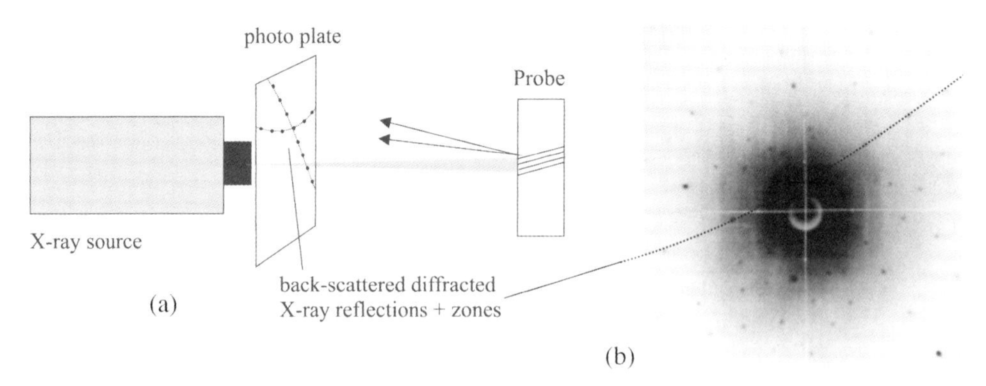

Fig. 3.24 Schematic representation of a backscattering Laue camera; (b) example of a Laue pattern of an alloy 718 single crystal.

3.3.5 Mathematical and Graphical Description of Crystallographic Orientation Relationships

The electron and X-ray diffraction techniques described above allow the experimental determination of the crystallographic orientation of individual grains and subgrains as well as the lateral distribution. It is common to represent the results by means of three-dimensional rotations of the crystal coordinate systems; i.e., for cubic systems spanned by the <001> axes, $\vec{x}_1$, $\vec{x}_2$, $\vec{x}_3$, with respect to the specimen coordinate system. The specimen coordinate system is defined by the normal vector perpendicular to the specimen surface (z direction) $\vec{r}_{\mathrm{N}}$ and two vectors lying within the surface typically related to the loading or rolling direction (x direction) $\vec{r}_{\mathrm{RD}}$ and the respective transverse direction (y direction) $\vec{r}_{\mathrm{TD}}$. The relationship between the two coordinate systems, as shown in Fig. 3.25, is given by the orientation matrix (rotation matrix) $\underline{\underline{M}}$ containing the respective direction cosine:

$$\begin{array}{c} \\ \underline{\underline{M}} = \begin{array}{c}[100]\\ [010]\\ [001]\end{array} \end{array} \begin{array}{c} \begin{array}{ccc} x & y & z \end{array} \\ \begin{bmatrix} \cos\alpha_1 & \cos\beta_1 & \cos\gamma_1 \\ \cos\alpha_2 & \cos\beta_2 & \cos\gamma_2 \\ \cos\alpha_3 & \cos\beta_3 & \cos\gamma_3 \end{bmatrix} \end{array} = \begin{bmatrix} a_{11} & a_{12} & a_{13} \\ a_{21} & a_{22} & a_{23} \\ a_{31} & a_{32} & a_{33} \end{bmatrix} \tag{3.9}$$

The unit vectors of the crystal coordinate system ($\vec{e}_x$) can be transferred into the specimen coordinate system ($\vec{e}_r$) by the following relationship:

$$\vec{e}_x = \underline{\underline{M}}\vec{e}_r \tag{3.10}$$

Equations (3.9) and (3.10) can be applied to polycrystals by extracting the misorientation matrices $\underline{\underline{M}}_i$ for the individual crystallites i holding the crystal coordinate system fixed. The data can be further processed by computer and graphically represented by colored orientation distribution mappings or (inverse) pole figures of, e.g., the <001> axes, as is shown in Fig. 3.25 (cf. [136]).

Beside the misorientation matrix, crystallographic relationships are frequently described by misorientation angles Θ referring to minimum rotations around misorientation axes $[uvw]$ to bring neighboring crystallites into agreement (see Fig. 3.26a). Taking into account that (1) the misorientation matrix is symmetrical and (2) the cross products of all column and row vectors are zero; i.e., the vectors are perpendicular to each other, the relationship between the elements of the misorientation matrix a_{ij}, the misorientation angle Θ and the misorientation axis $[uvw]$ can be given by the following equations:

$$\cos\Theta = \frac{(a_{11} + a_{22} + a_{33} - 1)}{2} \tag{3.11}$$

$$u : v : w = \sqrt{a_{11} + 1} : \sqrt{a_{22} + 1} : \sqrt{a_{33} + 1} \tag{3.12}$$

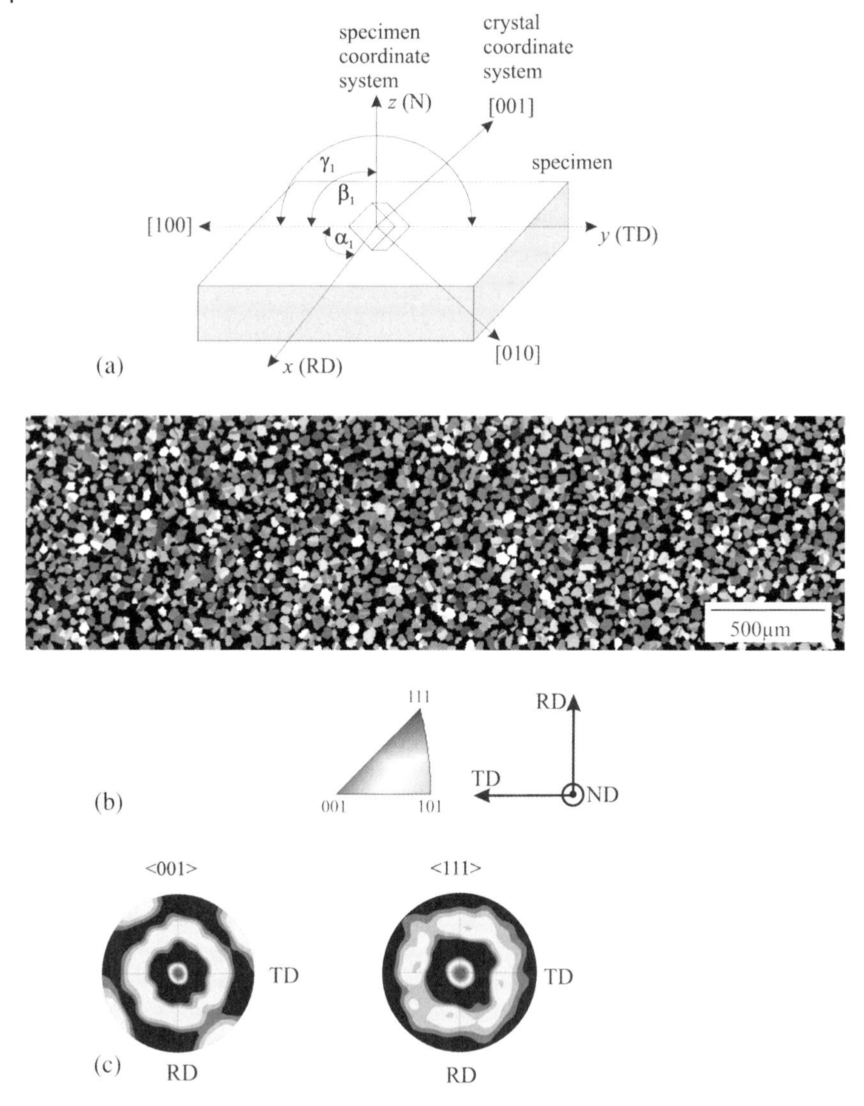

Fig. 3.25 (a) Relationship between the specimen and crystal coordinate systems and graphical representation of orientation data as (b) crystallographic orientation mapping (austenite phase in duplex steel) and (c) corresponding pole figures (according to [317]).

In the opposite way, one can obtain the elements of the misorientation matrix by knowing the misorientation angle Θ and direction [*uvw*] (cf. [137]). Due to the symmetry of cubic crystals (body-centered cubic and face-centered cubic), the misorientation angle can be considered as a rotation of {001} planes of the neighboring grains with the normal vectors N_1 and N_2, both perpendicular to the misorientation angle [*uvw*]. The misorientation angle can be calculated using the vector product:

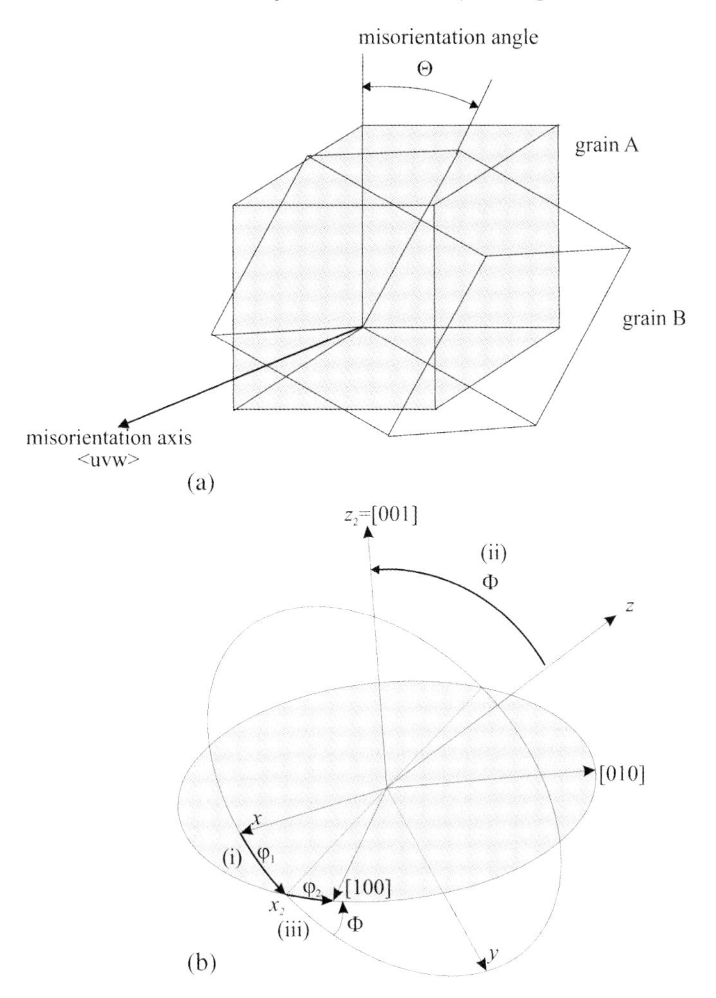

Fig. 3.26 Description of crystallographic misorientations by means of (a) the misorientation angle axis scheme and (b) Euler angles (for details see text).

$$\Theta = \arccos\left[\frac{\vec{N}_1\vec{N}_2}{|\vec{N}_1||\vec{N}_2|}\right] \tag{3.13}$$

A further, very common way to describe the transformation of coordinate systems is the application of Euler angles. Three rotations are required to bring the axes of the <001> crystal coordinate system into agreement with a reference coordinate system spanned by x, y, z (Fig. 3.26b): (1) rotation around the z axis by the angle φ_1, (2) rotation around the new x_2 axis by the angle Φ and (3) rotation around the z_2 axis (= [001] crystal coordinate) by the angle φ_2 [137]. Mathematically, the rotations (1) to (3) are described by the product of three rotation matrices $\underline{\underline{m}}_1$, $\underline{\underline{m}}_2$ and $\underline{\underline{m}}_3$ that corresponds to the misorientation matrix

$$\underline{\underline{M}} = \overbrace{\begin{bmatrix} \cos\varphi_1 & \sin\varphi_1 & 0 \\ -\sin\varphi_1 & \cos\varphi_1 & 0 \\ 0 & 0 & 1 \end{bmatrix}}^{\underline{\underline{m}}_1} \cdot \overbrace{\begin{bmatrix} 1 & 0 & 0 \\ 0 & \cos\Phi & \sin\Phi \\ 0 & -\sin\Phi & \cos\Phi \end{bmatrix}}^{\underline{\underline{m}}_2} \cdot \overbrace{\begin{bmatrix} \cos\varphi_2 & \sin\varphi_2 & 0 \\ -\sin\varphi_2 & \cos\varphi_2 & 0 \\ 0 & 0 & 1 \end{bmatrix}}^{\underline{\underline{m}}_3} \tag{3.14}$$

It is important to mention that the misorientation formalism refers only to three of five degrees of freedom of grain boundary geometries. The spatial position of the boundary plane requires two further degrees of freedom, given by the grain-boundary normal vector $\vec{N}_{GB}$. But even with five degrees of freedom the curved nature and the atomistic structure of the boundaries remain unconsidered. The latter are of great significance for many material properties, e.g., grain-boundary diffusivity (cf. Section 6.5).

The experimental evaluation of the complete sets of five degrees of freedom for a polycrystalline material is difficult and elaborate. The combination of EBSD measurements and stepwise polishing is a possible way for three-dimensional microstructure reconstruction (cf. [149, 150] and Section 6.4.1). Alternatively, highly focused high-energy X-ray or synchrotron radiation can be used to scan the crystallographic orientations of surface grains in three dimensions, and hence the orientation of the grain-boundary plane can be determined (see [116]). The orientation relationship of the surface grains together with the surface trace of the grain-boundary planes can be used in a statistical approach developed by Saylor et al. [151] together with a grain-boundary-energy minimization method to obtain the five-parameter grain-boundary data based on planar sections.

Among the grain boundaries with a continuous distribution of misorientation angles there exist a set of so-called special grain boundaries having a certain fraction of coincident lattice sites (coincident-site-lattice grain boundaries, CSL GBs) [152]. The CSL scheme assumes the lattices of neighboring grains to be interpenetrating (see Fig. 3.27). The reciprocal value of the fraction of coincident lattice sites

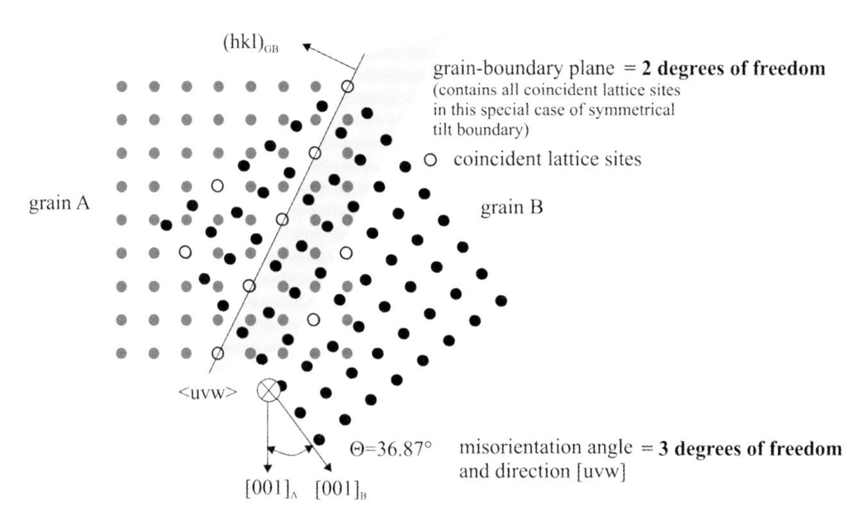

Fig. 3.27 Schematic representation of a symmetrical Σ5 tilt boundary.

of the two interpenetrating lattices is accounted for by the parameter Σ. In the case of a twin boundary for instance with a misorientation angle of $\Theta = 60°$ (misorientation axis: [111]) every third lattice site of the two neighboring grains is a coincident one, and hence such a grain boundary is termed a Σ3 grain boundary. Figure 3.27 shows a Σ5 boundary revealing the central problem of the CSL scheme: only in the special case of symmetrical tilt boundaries is the fraction of coincident lattice sites valid for the actual grain-boundary plane.

Nevertheless, it is well established in the literature to define grain boundaries with Σ values between Σ1 and Σ29 as special grain boundaries (see [153] and Section 6.5) assigning them outstanding properties with respect to a polycrystalline material's mechanical and corrosion behavior. The exact coincidence misorientation angle can only be determined for the formation of twins having a physically well-defined misorientation relationship. In the case of all other kinds of grain boundaries, a certain CSL parameter is assumed when coincidence can be established by geometrically necessary dislocations. According to this requirement, which is commonly used to distinguish low-angle grain boundaries (Σ1, $\Theta < \Theta_0 = 15°$, established by aligned edge dislocations), Brandon [154] proposed a criterion for the tolerance angle $\Delta\Theta$ around the exact CSL misorientation, which can be accommodated by geometrically necessary dislocations:

$$\Delta\Theta = \Theta_0 \frac{1}{\sqrt{\Sigma}} \tag{3.15}$$

A detailed overview of the CSL formalism and respective data are given in the textbooks by Randle [137, 152]. Table 3.1 gives a brief overview of CSL misorientation angles and axes.

Table 3.1 Selection of CSL misorientation angles (Θ) and axes [uvw] (after [137]).

uvw	Σ	Θ	uvw	Σ	Θ	uvw	Σ	Θ	uvw	Σ	Θ
100	5	36.9	110	3	70.5	111	3	60	210	3	131.8
	13a	22.6		9	38.9		7	38.2		5	180
	17a	28.1		11	50.5		13b	27.8		7	73.4
	25a	16.3		17b	86.6		19b	46.6		9	96.4
	29b	43.6		19a	26.5		21a	21.8		15	48.2
				27b	31.6					21b	58.4
										23	163.0
										27a	35.4
										29a	112.3

uvw	Σ	Θ	uvw	Σ	Θ	uvw	Σ	Θ	uvw	Σ	Θ
211	3	180	221	5	143.1	310	5	180	320	7	149.0
	5	101.5		9	90		7	115.4		11	100.5
	7	135.6		9	180		11	144.9		13a	180
	11	63.0		13b	112.2		13b	76.7		17b	122.0
	15	78.5		17b	61.9		19a	93.0		19b	71.6
	21b	44.4		25b	73.7		23	55.6		29a	84.1
	25b	156.9		29a	46.4						
	29a	149.6									

3.3.6
Microstructure Characterization by TEM

For the understanding of the most fundamental mechanisms during mechanical deformation of materials, TEM is probably the experimental key technique, even though specimen preparation is often difficult, and the extremely small portion of electron-transparent material represents the mechanically loaded component or specimen in a very limited way. However, only TEM made the direct observation of dislocation arrangements possible for the first time (see [17]). As shown schematically in Fig. 3.28a, a dislocation line is a local lattice distortion leading to a slight inclination of the lattice planes. If the position of the distorted lattice planes with respect to the incident electron beam corresponds to the Bragg condition (under two-beam condition) the dislocation line reflects intensity, and hence appears as a dark line (Fig. 3.28b, see [102] for details). Even more detailed information about the arrangement of single dislocations and other lattice defects down to the atomic length scale can be obtained by high-resolution TEM (HRTEM).

The simplest way to prepare TEM specimens from metals and alloys is electrolytic thinning. For this purpose the area to be studied has to be cut out from the component or specimen and ground by SiC paper (1200 grit) down to a thickness of approximately 80–100 μm. From these thin sheets small discs are punched out of 3 mm diameter to fit in a standard TEM specimen holder (Fig. 3.29a). To obtain an electron-transparent area the disc is provided with a hole having flanks that are thin enough to let electrons through (depending on the electron energy, the maximum thickness for electron transparency is of the order of 1 μm). The hole is generated either by electrolytic jet polishing (Fig. 3.29b) or Ar^+ ion bombardment (ion thinning, see Fig. 3.29c). The ion-thinning method requires a further intermediate preparation step, dimple grinding, and allows thinning of a predefined area, e.g., interfaces or crack tips. This kind of precision thinning was greatly improved by invention of the FIB technique, allowing micromachining of TEM specimens from

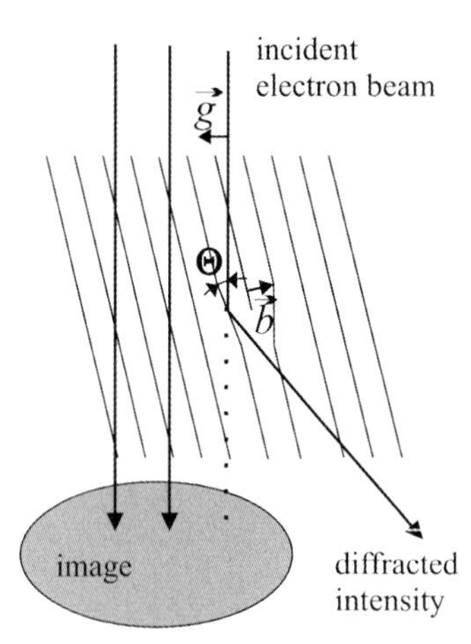

Fig. 3.28 Schematic representation of the contrast generation for imaging of dislocations in TEM.

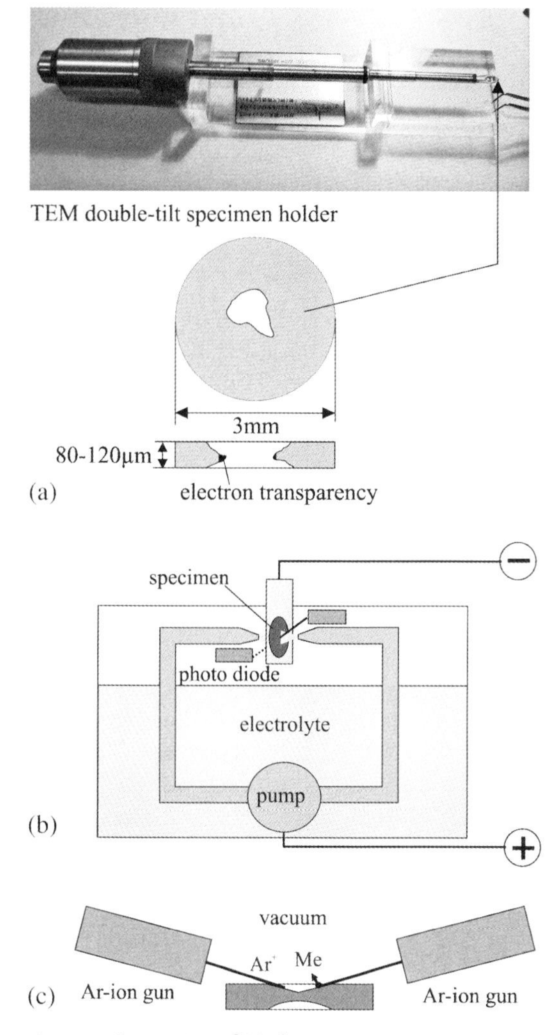

Fig. 3.29 Preparation of (a) electron-transparent TEM specimens by (b) electrolytic jet polishing and (c) ion polishing (schematic).

relatively large components or specimens by means of a Ga ion beam during simultaneous SEM observation (dual-beam FIB).

Detailed and comprehensive descriptions of imaging, diffraction and analyses techniques in combination with TEM can be found in numerous review publications. A good introductory and complete overview is given by the textbooks of von Heimendahl [102] and Wiliams and Carter [147].

3.3.7
Further Methods to Characterize Mechanical Damage Mechanisms in Materials

Even though analytical electron microscopy is the most important tool to study microstructural changes in materials caused by mechanical deformation, there are a great variety of special techniques to correlate plastic deformation, crack initiation and crack propagation with microstructural features and the fundamental physical mechanisms.

As already mentioned in Section 3.3.4, *X-ray diffraction* yields important information on phase composition and macroscopic textures of materials (see [148] for details). The core of conventional X-ray diffractometers is a kinematic setup, consisting of an X-ray source and an X-ray detector, mounted on a step-motor-adjustable goniometer having a movable specimen stage in its center (see Fig. 3.30). This arrangement allows continuous changing of the geometry in such a way that the angles Θ between X-ray source and specimen surface as well as between specimen and detector are always the same. Hence, maxima in the intensity vs. 2Θ scan correspond to lattice planes fulfilling the Bragg condition (Eq. 3.8). Since the lattice spacing d is proportional to Θ, X-ray diffraction allows the measurement of lattice constants and residual stresses, the latter based on the fact that elastic lattice strains cause a shift in the 2Θ peaks from the equilibrium position that can be evaluated by rotating the specimen around its surface normal vector ($\sin^2 \psi$ method [148]).

In the following sections it is shown that even in the case of a perfectly polished specimen, mechanical loading leads to the appearance of surface roughness in the form of pronounced slip bands that may act as crack-initiation sites. The evolution of such kind of surface topography can be quantitatively analyzed by means of *atomic force microscopy* (AFM), a technique that allows surface roughness measurements down to the atomic length scale (see [155] for details). Conventional AFM systems use a sharp silicon tip (normally made by FIB milling) applied to a cantilever that moves across the surface to be analyzed. Interatomic forces move the tip,

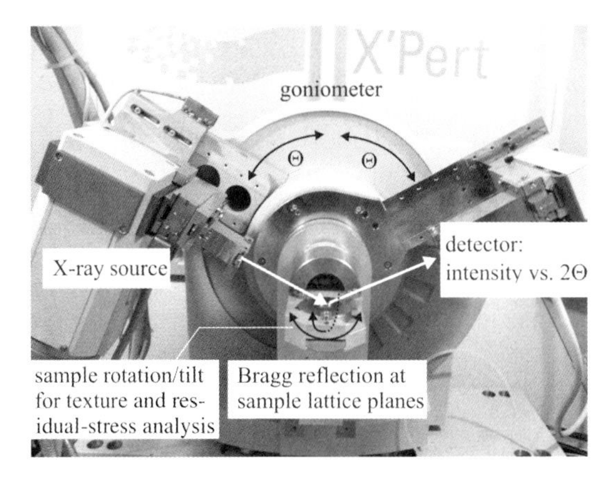

Fig. 3.30 Working principle of an X-ray diffractometer for phase and texture analysis as well as for residual-stress measurements.

leading to deflection of the cantilever. These deflections are measured and evaluated by laser light reflected by the cantilever towards a photodiode.

A nondestructive way to measure deformation-induced defects, i.e., dislocations, is to use the phenomenon of *positron annihilation*. The physical principle, explained, e.g., in [156], is based on positron radiation interacting with the specimen surface. In general, the positrons annihilate with free electrons in the metal surface, but the survival period depends on the local concentration of lattice defects, like dislocations or vacancies. By precise measurement of the positron survival times (in picoseconds) or evaluation of the line shape of the Doppler broadening of the annihilation radiation (so-called *S* parameter, see [156]), the plastic-deformation state of mechanically deformed specimens can be analyzed. Even though in several studies the application of positron annihilation to predict the residual fatigue life was reported [156–158], a clear correlation between the fatigue-damage state (e.g., damage parameter) and the positron annihilation signal is not possible yet.

A further nondestructive method, being highly sensitive to strain and microstructural changes due to fatigue and/or creep damage in ferromagnetic materials, is the *magnetic Barkhausen emission* (see e.g. [159]). This emission is based on discrete jumps of domain walls under the influence of a magnetic field with stepwise changing flux density.

As a relatively new approach for nondestructive analysis of fatigue damage evolution the *electrochemical fatigue sensor* (EFS) is applied to technical components. This technique uses the area to be analyzed as anodes of a two-electrode cell, which is operated by a constant voltage in the passivating regime of electrochemical corrosion. During fatigue loading, the cell current is permanently recorded as a measure of damage. It was shown that crack-initiation events lead to superimposed peaks in the current vs. time course [160].

In particular the new groups of engineering materials require the development of new tailored testing methods. For instance, inhomogeneous materials, e.g., cellular metals or carbon-fiber-reinforced polymers, are characterized by a technique originally developed for medical applications: *computer tomography* (CT). In a similar way to human bodies the three-dimensional-structures of such materials or defects can be obtained. As already mentioned in Section 3.2.2, high-energy synchrotron radiation is used to study small defects and cracks at the micrometer length scale (μCT).

As a further example, new nanosized materials require completely different mechanical testing setups. The material properties of carbon nanotubes cannot be studied by conventional tensile tests. As an alternative approach, electrostatic testing methods were developed that bring the structures in vibration within electron microscopes. By applying solid-state mechanics these vibrations can be used to derive the mechanical properties of nanostructures (see e.g. [161]).

3.4
Reproducibility of Experimentally Studying the Mechanical Behavior of Materials

Mechanical damage of materials depends on a variety of physical mechanisms acting in combination. These mechanisms influence each other and depend strongly on the environmental and material conditions, e.g., temperature, partial pressures of gaseous species, humidity, chemical composition of the material, heat-treatment conditions, grain size, surface treatment, etc. These relationships and interactions have to be accounted for, when carrying out any tests to derive material properties. Only then can one avoid misinterpreting different experimental results, which fall within a certain scatter band, as some kind of physical mechanisms. Anybody, who works in the field of experimental materials science and technology knows that two mechanical experiments carried out in exactly the same way never yield the same results. Reasons for the substantial scattering of experimental materials data are the stochastic distribution of microstructural features, like grain size, microtexture, defects and chemical composition, as well as unavoidable inaccuracies in measuring, differences in specimen preparation and environmental conditions.

Figure 3.31 gives for the example of fatigue testing an overview of avoidable and unavoidable factors affecting the reproducibility of experimental studies.

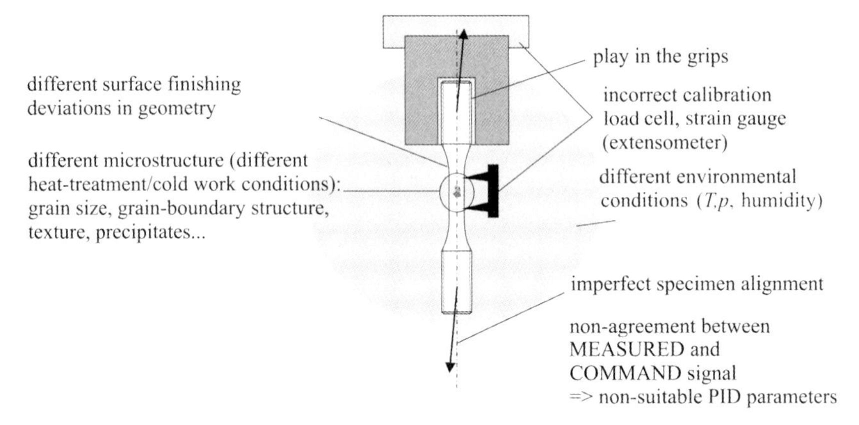

Fig. 3.31 Factors affecting the reproducibility of results of fatigue testing and crack-propagation measurements.

4
Physical Metallurgy of the Deformation Behavior of Metals and Alloys

The complete set of phenomena that are discussed in the following sections can be attributed to the deformation behavior of a material under the influence of a mechanical load. Depending on the mechanical stress level and the microstructure of a material in a volume element of the specimen or a component, deformation can be either elastic and fully reversible, or elastic plastic, resulting in nonreversible changes in the material's microstructure. In the preceding chapters, these two general deformation mechanisms were described by means of a continuum-mechanics isotropic approach, which is based on the isotropic elastic constants, Young's modulus E, shear modulus G, compression modulus K, Poisson ratio ν and the macroscopic yield strength σ_Y. This is schematically shown in Fig. 4.1a and b. Correlation between the physical mechanisms of deformation and the material's microstructure, being substantial for a sound understanding of the crack initiation and propagation behavior, requires the introduction of direction-dependent material laws, accounting for the anisotropy of elastic deformation (Fig. 4.1c) and dislocation motion along slip planes (Fig. 4.1d). This kind of elastic plastic deformation in polycrystals depends strongly on the efficiency of grain boundaries as barriers for slip transmission, which is subject of Chapters 5 and 6.

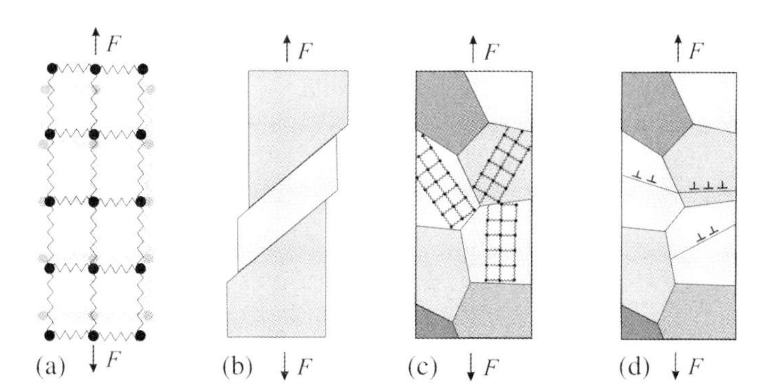

Fig. 4.1 Simplified schematic representation of isotropic (a) elastic and (b) plastic deformation as well as microstructure-dependent anisotropic (c) elastic and (d) plastic deformation.

Fatigue Crack Propagation in Metals and Alloys: Microstructural Aspects and Modelling Concepts. Ulrich Krupp

ISBN: 978-3-527-31537-6

The following discussion of the metallurgical principles of the deformation behavior of metals and alloys should be understood as a link between the phenomenological approaches introduced in Chapter 2 and the mechanism-based approaches to crack propagation that are the subject of Chapters 5 to 7. As regards a comprehensive overview, the present chapter might appear to be incomplete in several aspects. For more detailed information on the subject the reader is referred to the textbooks (1) on physical metallurgy of, e.g., Gottstein [162] (in German), Haasen [163] and Haasen and Cahn [164], and (2) on the theory of dislocations of, e.g., Hull and Bacon [165] and Nabarro [166].

4.1 Elastic Deformation

When applying a small mechanical force to a cylindrical tensile specimen, it will exhibit an extension in the axial loading direction and a contraction in the radial direction. After unloading, the specimen will return to its original shape. The extent of this kind of elastic deformation varies greatly among different materials and depends on the properties of the active metallic bonding. In a simplified way, a metallic material consists of metal cations in a lake of free electrons acting as the "glue" that holds the structure together (long-range attractive forces, see Fig. 4.2a). Once overlapping, the metal cations experience a sudden increase in repulsion (short-range repulsive forces, see Fig. 4.2a). Consequently, elastic deformation can be considered as a deviation from the equilibrium state, which is described by the atomic-distance potential U vs. the atomic spacing r. This relationship can be expressed in the form of a Taylor series [85]:

$$U(r) = U(r_0) + \left(\frac{dU}{dr}\right)_{r=r_0}(r - r_0) + \frac{1}{2}\left(\frac{d^2U}{dr^2}\right)_{r=r_0}(r - r_0)^2 + \frac{1}{3!}\left(\frac{d^3U}{dr^3}\right)_{r=r_0}(r - r_0)^3 + \cdots \tag{4.1}$$

For small deviations $(r - r_0)$ from the equilibrium atomic spacing r_0, Eq. (4.1) can be cut off after the quadratic term or, for slightly higher deviations, after the cubic term. The atomic-distance potential $U(r_0)$ for the equilibrium spacing r_0 corresponds to the bonding energy E_b, and hence the first derivative dU/dr at r_0 is essentially equal to zero, i.e., there are no interatomic forces active at equilibrium. Taking this into account, the second derivative d^2U/dr^2 should represent the strength of the increase in the force when r deviates slightly from r_0 (elastic deformation). Together with the actual elastic strain $(r - r_0)/r_0$ one obtains Hooke's law:

$$\sigma \propto \frac{F}{r_0^2} = \left(\frac{d^2U}{dr^2}\right)_{r=r_0}\left(\frac{r - r_0}{r_0}\right) \Rightarrow \sigma = E\overbrace{\varepsilon}^{(r-r_0)/r_0} \quad \text{or} \quad \sigma = E\varepsilon + k\varepsilon^2 \tag{4.2}$$

with E being the isotropic Young's modulus and k the elasticity constant representing nonlinear elastic behavior, as has been observed in earlier work, e.g., by

Sommer et al. [167] for the roller bearing steel 100Cr6 (SAE 52100, see Fig. 4.2b) and Floer [90] for the β-titanium alloy LCB. According to the second derivative of the atomic-distance potential d^2U/dr^2 (cf. Fig. 4.2), the elastic stiffness can be considered to be of a constant value for small displacements in the vicinity of the equilibrium spacing r_0.

The description of a material's elastic behavior by assuming one single elastic constant, the Young's modulus according to Eq. (4.2), is only justified when the specimen or the component consists of a sufficiently high number of randomly oriented crystallites, or when the elastic properties are independent of the crystallographic direction (elastic isotropy). Otherwise, when taking anisotropic elastic properties of the grains into account (Fig. 4.3a), Hooke's material law has to be used in its more general form, linking the stress tensor σ_{ij}, acting on a small volume element as shown in Fig. 4.3b, with the strain tensor ε_{kl} by means of the elasticity/stiffness tensor C_{ijkl} in the following way:

$$\sigma_{ij} = C_{ijkl}\ \varepsilon_{kl} \qquad (4.3)$$

For the sake of clarity, the stresses σ_{ij}, $i \neq j$ are mostly denoted as shear stresses τ_{ij} in the following.

In its general form the stiffness tensor C_{ijkl} is a tensor of fourth order with $3^4=81$ elements. To avoid any rotation of the small volume element in Fig. 4.3b the shear stresses acting in opposite directions must be essentially of the same value: $\tau_{xy}=\tau_{yx}$, $\tau_{xz}=\tau_{zx}$, $\tau_{yz}=\tau_{zy}$. Hence the stress as well as the strain tensor are symmetric, reducing the number of elements to 36 (cf. Eq. (4.4), stress and strain tensor can be written as vectors). Furthermore, reversibility of elastic deformation re-

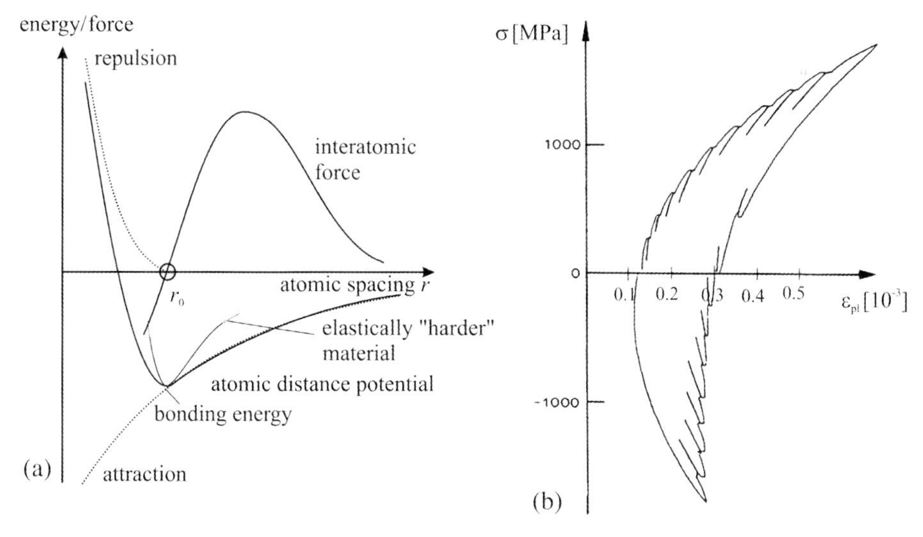

Fig. 4.2 (a) Schematic representation of the relationship between atomic-distance potential U and atomic spacing r; (b) stress–strain diagram revealing nonlinear elastic behavior of the roller-bearing steel 100Cr6 (after [167]).

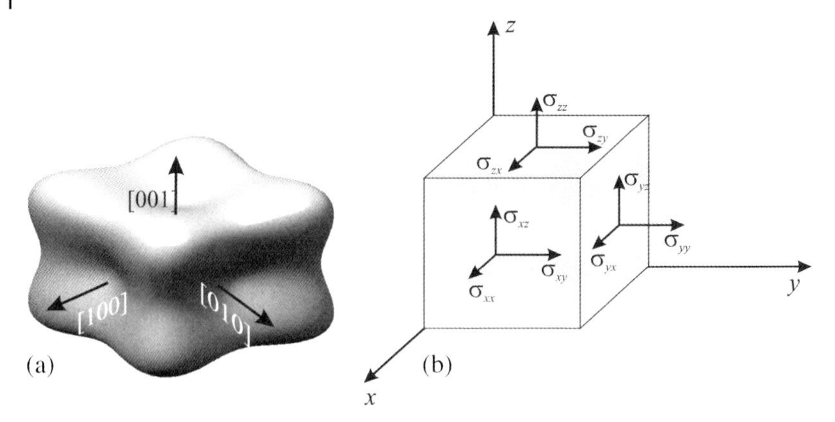

Fig. 4.3 (a) Three-dimensional distribution of Young's modulus (stiffness) for an anisotropic material (pure iron, $A = 2.4$); (b) volume element with the components of the three-dimensional stress tensor (for details see text).

quires $c_{ij} = c_{ji}$, and reduces the number of elements to 21. Finally, the symmetry of cubic lattice structures, face-centered cubic (fcc) and body-centered cubic (bcc), results in only three independent elastic constants c_{11}, c_{12} and c_{44} (hexagonal double-packed (hdp) lattice structure has five independent constants c_{11}, c_{12}, c_{13}, c_{33} und c_{44}). Equation (4.4a) represents Hooke's law in its general form, while Eq. (4.4b) accounts for the simplifications due to crystal symmetry, i.e., the <001> directions of the crystal coordinate system are coincident with the local specimen coordinate system in Fig. 4.3b:

$$\begin{bmatrix} \sigma_{xx} \\ \sigma_{yy} \\ \sigma_{zz} \\ \tau_{yz} \\ \tau_{zx} \\ \tau_{xy} \end{bmatrix} = \begin{bmatrix} c_{11} & c_{12} & c_{13} & c_{14} & c_{15} & c_{16} \\ c_{21} & c_{22} & c_{23} & c_{24} & c_{25} & c_{26} \\ c_{31} & c_{32} & c_{33} & c_{34} & c_{35} & c_{36} \\ c_{41} & c_{42} & c_{43} & c_{44} & c_{45} & c_{46} \\ c_{51} & c_{52} & c_{53} & c_{54} & c_{55} & c_{56} \\ c_{61} & c_{62} & c_{63} & c_{64} & c_{65} & c_{66} \end{bmatrix} \cdot \begin{bmatrix} \varepsilon_{xx} \\ \varepsilon_{yy} \\ \varepsilon_{zz} \\ \gamma_{yz} \\ \gamma_{zx} \\ \gamma_{xy} \end{bmatrix} \tag{4.4a}$$

$$\begin{bmatrix} \sigma_{xx} \\ \sigma_{yy} \\ \sigma_{zz} \\ \tau_{yz} \\ \tau_{zx} \\ \tau_{xy} \end{bmatrix} = \begin{bmatrix} c_{11} & c_{12} & c_{12} & 0 & 0 & 0 \\ c_{12} & c_{11} & c_{12} & 0 & 0 & 0 \\ c_{12} & c_{12} & c_{11} & 0 & 0 & 0 \\ 0 & 0 & 0 & c_{44} & 0 & 0 \\ 0 & 0 & 0 & 0 & c_{44} & 0 \\ 0 & 0 & 0 & 0 & 0 & c_{44} \end{bmatrix} \cdot \begin{bmatrix} \varepsilon_{xx} \\ \varepsilon_{yy} \\ \varepsilon_{zz} \\ \gamma_{yz} \\ \gamma_{zx} \\ \gamma_{xy} \end{bmatrix} \tag{4.4b}$$

As shown in Table 4.1 (data from [168]), the relative orientation of the crystallites with respect to the loading direction is of substantial significance for a great variety of metals and alloys. Due to the anisotropy of the elastic properties, even small

Table 4.1 Elastic anisotropy of several metallic materials (after [168]).

Metal	Anisotropy factor $A = \frac{G_{(011)}}{G_{(001)}} = \frac{2c_{44}}{c_{11} - c_{12}}$	Young's modulus in [111] direction, $E_{[111]}$ (GPa)	Young's modulus in [001] direction, $E_{[001]}$ (GPa)	Ratio $E_{[111]}/E_{[001]}$
Aluminum	1.219	76.1	63.7	1.19
Copper	3.203	191.1	66.7	2.87
Gold	2.857	116.7	42.9	2.72
Iron	2.512	272.7	125.0	2.18
Tungsten	1.000	384.6	384.6	1.00

values of remote mechanical stresses may cause local stress concentrations and crack initiation at the grain boundaries. This is discussed in detail in Section 5.3. Only for metals exhibiting an anisotropy factor of $A \approx 1$ (ratio between the shear modulus in the (001) plane and the (011) plane, Eq. 5.5) or structures with a very high number of grains within the loaded cross section is the application of isotropic elastic constants, Young's modulus $E = \sigma_{xx} / \varepsilon_{xx}$, shear modulus $G = \tau_{xy} / \gamma_{xy}$, the Poisson ratio $\nu = -\varepsilon_{yy}/\varepsilon_{xx}$ and the compression modulus $K = -Vdp/dV$ meaningful. Only two of these constants are independent, according to the relationships

$$G = \frac{E}{2(1+\nu)}, \quad K = \frac{E}{3(1-2\nu)}, \quad \frac{E}{G} = \frac{9}{3+G/K} \tag{4.5}$$

Any given applied stress tensor σ_{ij} can be transferred into the stress tensor σ'_{ij} in an arbitrary crystal coordinate system of individual grains, by means of the rotation matrix $\underline{\underline{M}}$ (see Section 3.3.4) and the respective transposed matrix $\underline{\underline{M}}^{\mathrm{T}}$:

$$\sigma'_{ij} = \underline{\underline{M}}^{\mathrm{T}} \, \sigma_{ij} \, \underline{\underline{M}} \tag{4.6}$$

It is always possible to define a coordinate system with a stress tensor, where the shear stresses vanish and where the stress state is represented only by normal stresses, the principal stresses σ_1, σ_2 and σ_3 (cf. Section 2.3.2):

$$\underline{\sigma} = \begin{bmatrix} \sigma_1 & 0 & 0 \\ 0 & \sigma_2 & 0 \\ 0 & 0 & \sigma_3 \end{bmatrix} \tag{4.7}$$

In this case the maximum shear stresses are acting in a plane inclined 45° with respect to the mean stresses and can be calculated for $\sigma_1 \geq \sigma_2 \geq \sigma_3$ as

$$\tau_{\max} = \frac{1}{2}(\sigma_1 - \sigma_3) \tag{4.8}$$

4.2
Plastic Deformation by Dislocation Motion

Following the analysis of elastic deformation as a small deflection from the equilibrium atomic spacing r_0 in Fig. 4.2a one can conclude that, once a critical distance r for the cohesive force F_c is exceeded, the atomic bonds are going to break. Thus, one would expect an immediate separation of the specimen or the component. However, metallic materials experience irreversible plastic deformation far below the theoretical value of the separation strength. Experimental analysis of the mechanical behavior of metallic materials has shown that this kind of irreversible deformation can be attributed to slip of crystallographic lattice planes. According to Fig. 4.4a slip of lattice planes is only possible if the energy maxima of the atom sites can be overcome by the resolved shear stress. Assuming a sinusoidal course of the stress required for permanent slip along the slip plane [169], the theoretical critical shear stress τ_{th} for initializing of crystallographic slip is given as

$$\tau_{th} = \frac{Gb}{2\pi d} \tag{4.9}$$

with b being the absolute value of the Burgers vector, G the material's shear modulus and d the respective lattice spacing. The actual critical shear stress to activate slip in real materials is three to four orders of magnitude smaller than the theoretical value predicted by Eq. (4.9) [162, 169]. An explanation for the considerably lower value of the actual critical shear stress is provided by the dislocation theory. Instead of simultaneous motion within the whole slip plane, the integration of a dislocation as a line defect allows the consecutive motion of an inserted half plane, as shown schematically in Fig. 4.4b (motion of an edge dislocation). When comparing the force required to move a rug either as a whole (Fig. 4.4a) or by running waves (Fig. 4.4b), it becomes obvious that dislocation motion requires much smaller shear stress than rigid motion of the whole plane. Assuming a sinusoidal stress course similar to that for rigid slip, Peierls [170] and Nabarro [171] derived the so-called Nabarro–Peierls stress τ_P, accounting for the resistance of a slip band to dislocation motion:

$$\tau_P = \frac{2G}{1-\nu} \exp\left(-\frac{2\pi W}{b}\right) \tag{4.10}$$

where ν is the Poisson ratio and W the area within which the lattice is distorted by the dislocation.

Generally, one distinguishes between two different extreme kinds of dislocations: edge and screw dislocations. As already mentioned, the term edge dislocation refers to a two-dimensional lattice defect in the form of an additional inserted half plane as shown schematically in Fig. 4.5a. In the case of a screw dislocation, two halves of a crystal are shifted against each other starting from the dislocation line, as shown schematically in Fig. 4.5b. The geometry of dislocations is defined by their Burgers vector $\vec{b}$ and dislocation line vector $\vec{s}$ (Fig. 4.5). Dislocation lines

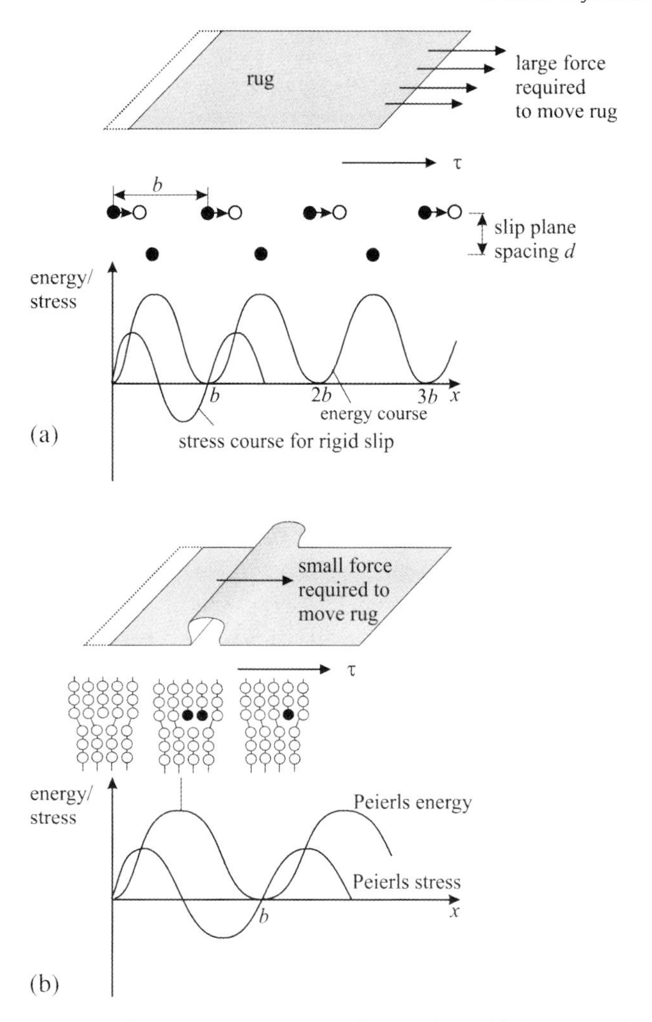

Fig. 4.4 Schematic representation and rug analogy of deformation of a crystalline structure by (a) rigid slip and (b) dislocation motion (here, edge dislocations).

cannot end within the crystal. Assuming the surface of the crystal section in Fig. 4.5b is covered by an additional lattice plane, it becomes obvious that the screw dislocation merges into an edge dislocation with the dislocation line vector shown as a dashed line in Fig. 4.5b. This corresponds to the structure of dislocation rings, which are introduced later in this section.

By analogy with pure elastic deformation discussed in the preceding section, the introduction of dislocations yields a change in the atomic spacing from its equilibrium value r_0 resulting in an elastic stress field around the dislocation cores. This

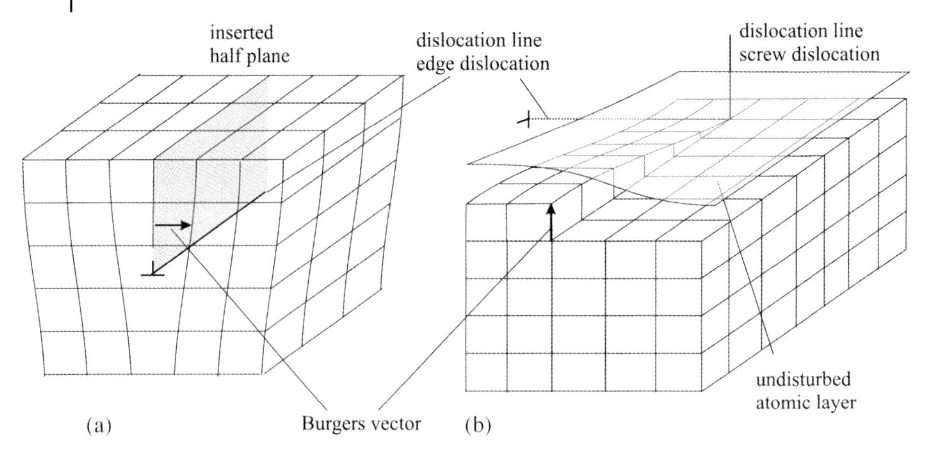

Fig. 4.5 Simplified representation of (a) an edge dislocation and (b) a screw dislocation, the latter in combination with an edge dislocation (see dashed line).

can be described mathematically in an easy way by assuming a cylindrical body placed around a screw dislocation according to Fig. 4.6a. By means of the shear strain $\gamma_{\Theta z}$ depending only on the distance r from the dislocation core the shear stress $\tau_{\Theta z}$ in the z direction can be determined:

$$\gamma_{\Theta z} = \frac{b}{2\pi r} \quad \text{and} \quad \tau_{\Theta z} = G\,\gamma_{\Theta z} = \frac{Gb}{2\pi r} \tag{4.11}$$

Analysis of the stress field around an edge dislocation (see Fig. 4.6b) yields the local lattice expansion below and compression above the slip plane. Hence, there are normal stress and strain contributions in x and y directions. For the normal stresses one obtains [165]

$$\sigma_{xx} = -\frac{Gb}{2\pi(1-\nu)}\frac{y\left(3x^2+y^2\right)}{\left(x^2+y^2\right)^2}, \quad \sigma_{yy} = \frac{Gb}{2\pi(1-\nu)}\frac{y\left(x^2-y^2\right)}{\left(x^2+y^2\right)^2}, \quad \sigma_{zz} = \nu\left(\sigma_{xx}+\sigma_{yy}\right) \tag{4.12}$$

The respective shear stresses are given as

$$\tau_{xy} = \tau_{yx} = \frac{Gb}{2\pi(1-\nu)}\frac{x\left(x^2-y^2\right)}{\left(x^2+y^2\right)^2}, \quad \tau_{xz} = \tau_{zx} = \tau_{yz} = \tau_{zy} = 0 \tag{4.13}$$

In addition to the residual stress field caused by the individual dislocations and the applied stress, the overall stress state of a crystalline body is determined by mutual interaction forces between the dislocations. A shear stress acting on a slip plane imposes a force F (or as a vector $\vec{F}$) per unit length onto a dislocation according to the Peach–Koehler equation:

$$F = \tau \cdot b \text{ or in its general form } \vec{F} = \left(\sigma_{ij}\cdot\vec{b}\right)\times\vec{s} \tag{4.14}$$

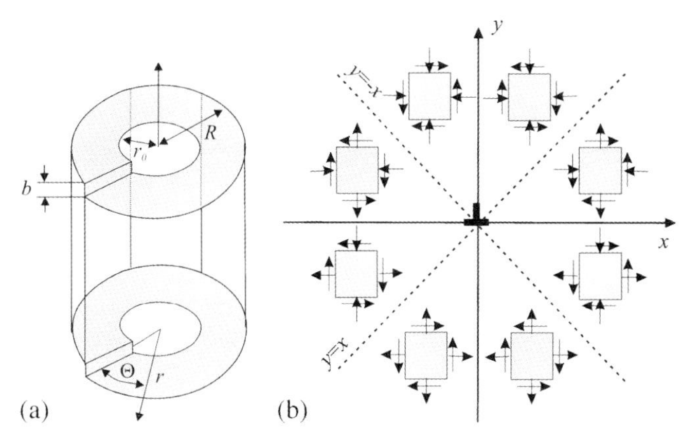

Fig. 4.6 (a) Elastic distortion of a screw dislocation; (b) elastic stress field around an edge dislocation (directions of the normal and shear stresses, according to [172]).

By inserting Eqs. (4.12) and (4.13) into Eq. (4.14) one obtains the forces between two edge dislocations on parallel slip planes, according to the geometrical relationship shown in Fig. 4.7a in the following form:

$$\begin{bmatrix} F_x \\ F_y \end{bmatrix} = \frac{Gb^2}{2\pi r(1-v)} \begin{bmatrix} \cos\Theta(1-2\sin^2\Theta) \\ \sin\Theta(1+2\cos^2\Theta) \end{bmatrix} = \frac{Gb^2}{2\pi r(1-v)} \begin{bmatrix} \dfrac{x(x^2-y^2)}{(x^2+y^2)^2} \\ \dfrac{y(3x^2+y^2)}{(x^2+y^2)^2} \end{bmatrix} \tag{4.15}$$

Hence, dislocations arranged mutually under an angle of $\Theta = 90°$ or $\Theta = 45°$ are in a force-free metastable rest position. Most stable are $\Theta = 90°$ arrangements, which are, for example, responsible for the formation of low-angle grain boundaries during recovery and early recrystallization processes. The $\Theta = 45°$ arrangement of two dislocations of opposite sign (Fig. 4.7b) is termed a dislocation dipole.

In terms of plastic deformation, the maximum of F in Eq. (4.15) $F_{x,\max}$ refers to the applied shear stress on the respective slip plane that is required to move dislocations. Accounting for mutual interactions with dislocations on parallel slip planes of a dislocation density of ρ_p and the requirement to overcome the maximum force for $\cos\Theta(1 - 2\sin^2\Theta) = 1$, the shear stress allowing dislocations to pass τ_{pass} can be given as the following [162]:

$$\tau_{\text{pass}} = \alpha_1 Gb\sqrt{\rho_p} \tag{4.16}$$

Here, α_1 is a geometry factor and $d = 1/\sqrt{\rho_p}$ the mean distance between dislocations on parallel slip planes. However, besides the interactions with parallel dislocations, the resistance to slip is influenced by cutting processes with dislocations on non-parallel planes. These cutting processes lead generally to an energy increase, resulting in the formation of either (1) kinks, lying within the slip plane, or (2) jogs,

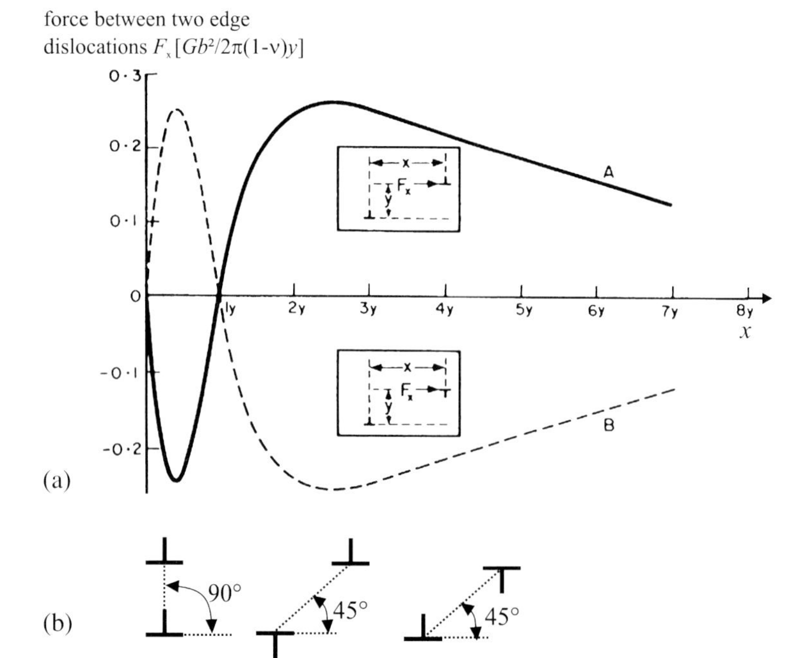

Fig. 4.7 (a) Forces between two edge dislocations; (b) force-free dipole arrangements.

when the dislocation jumps into a different slip plane ([165], see Fig. 4.8). The interactions with nonparallel dislocations, so-called forest dislocations of a density of ρ_f can be expressed in a similar way as is done in Eq. (4.16) for parallel dislocations. By superimposing the shear stress τ_{pass}, allowing dislocations to pass, and the additional shear stress τ_{cut}, required to cut the forest dislocations, one obtains the value for the critical resolved shear stress τ_0 to activate plastic slip on a slip plane:

$$\tau_0 = \alpha G b \sqrt{\rho} \tag{4.17}$$

with the geometry factor $\alpha \approx 0.5$ and the overall dislocation density $\rho = \rho_p + \rho_f$.

In Eq. (4.17) the influence of the temperature on the shear rate $\dot{\gamma} = \rho b v$ (Orowan equation, with v being the dislocation velocity) is neglected. Thermal activation stimulates lattice vibrations, and hence it facilitates dislocation glide, which can be represented by the following Arrhenius-type expression for the shear rate:

$$\dot{\gamma} = \dot{\gamma}_0 \exp\left(-\frac{\Delta G(\tau_a)}{kT}\right) \tag{4.18}$$

where $\Delta G(\tau_a)$ is the activation energy for (screw) dislocations to overcome the Peierls energy maxima at an applied shear stress τ_a. Figure 4.8b shows schematically how a screw dislocation overcomes sectionally the Peierls energy maxima, a

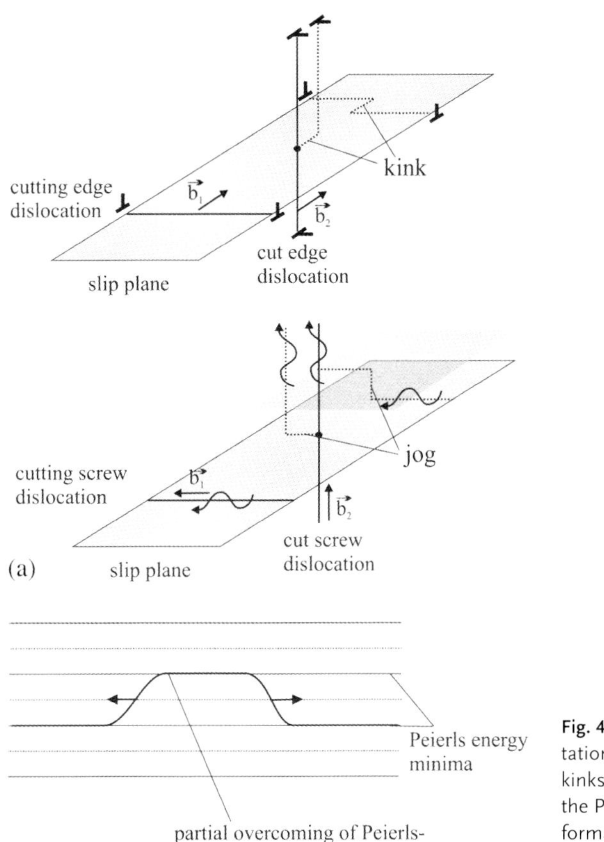

Fig. 4.8 Schematic representation of (a) the formation of kinks and jogs; (b) overcoming the Peierls energy maxima by formation of a double kink (after [162]).

process that depends strongly on the temperature and is described by a critical shear stress τ_0, which decreases substantially with increasing temperature:

$$\tau_0 = \tau_P - AT^2\left(\ln\frac{\dot{\gamma}}{\dot{\gamma}_0}\right)^2 \tag{4.19}$$

In Eq. (4.19) A is a constant and τ_p the Nabarro–Peierls stress (Eq. 4.10). In fcc and hdp metals the Nabarro–Peierls stress is considerably smaller than the critical shear stress according to Eq. (4.17) (Fig. 4.9a). In these cases, dislocation motion is determined mainly by elastic interactions, i.e., τ_{pass} (Eq. 4.16) has to be overcome having only a slight temperature dependence due to the shear modulus G. Furthermore, cutting processes by forest dislocations yield a further temperature-dependent contribution.

In general, the critical shear stress to activate dislocation motion along a slip band can be separated into an athermal and a thermal contribution (see Fig. 4.9b). In the case of bcc metals, where the critical shear stress can even be dominated by

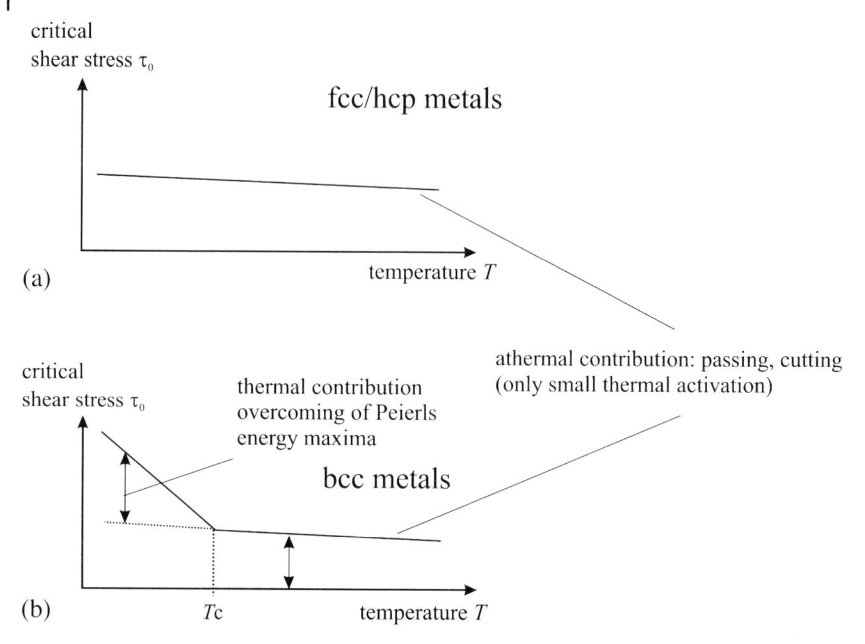

Fig. 4.9 Critical shear stress vs. temperature, shown schematically for (a) fcc/hcp metals and (b) bcc metals.

the Nabarro–Peierls stress, a pronounced temperature dependence of plastic deformation can be observed for low temperatures (see Fig. 4.9b) and in a similar way for high plastic strain rates (according to Christ [85] for $\dot{\varepsilon}_{pl} \geq 10^{-2}\ s^{-1}$), further limiting thermally activated dislocation motion. This increases the critical temperature T_c in Fig. 4.9a to values higher than room temperature. It should be mentioned that this kind of strain-rate effect is not observed for all kinds of bcc alloys, e.g., bcc β-titanium alloy Timetal LCB (low-cost beta, cf. Chapters 5 and 6) does not show any strain-rate dependence of the yield strength [90]. This can probably be attributed to the high concentration of alloying elements, which leads to an additional increase in τ_{pass} (hindering of dislocation motion by solid solution strengthening).

During sustained plastic deformation, the value of the critical shear stress does not remain constant. Increasing slip leads to an increase in the critical shear stress. While in the beginning of the deformation process dislocations can move unhindered through the material's crystallites (stage A in Fig. 4.10a), an increasing dislocation density makes this kind of free movement increasingly difficult, finally resulting in the activation of secondary slip systems. If dislocations from such a secondary system encounter dislocations of the primary system they form very immobile arrangements, the so-called Lomer–Cottrell locks, blocking the movement of subsequent dislocations. This leads to a substantial increase in the critical shear stress and the overall dislocation density (stage B in Fig. 4.10).

Generation of new dislocations can occur at the grain boundaries, by glide multiplication or by means of a Frank–Read source [165]. The mode of operation of a

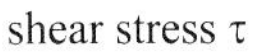

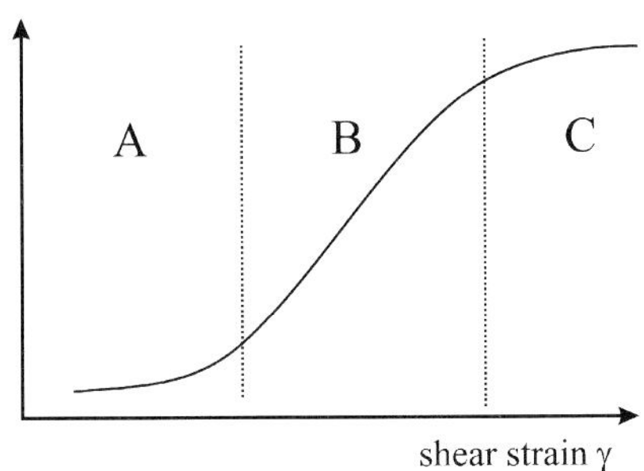

Fig. 4.10 Schematic representation of a strain-hardening curve for fcc single crystals exhibiting three regimes A, B and C.

Frank–Read source can be described according to Fig. 4.11 [172]: under the influence of mechanical stress the dislocation line, blocked at the points D and D′, will bow out, developing a curvature with decreasing radius R according to (cf. Eq. 4.17)

$$\tau_0 = \frac{\alpha G b}{R} \qquad (4.20)$$

until reaching a minimum value of R (Fig. 4.11b). At this point the dislocation line becomes unstable, and hence the line further expands, while τ decreases (Fig. 4.11c). Finally, the dislocation-line segments M and N in Fig. 4.11c annihilate and form a new dislocation ring (Fig. 4.11d).

Returning to Fig. 4.10, it is shown that when further deforming a material, the strong increase in strain hardening during stage B slows down (stage C in Fig. 4.10). This can be attributed to an increase in the activation of cross slip and annihilation events. In the case of fcc metals, screw dislocations move on close-packed {111} planes in <110> direction by dissociation into two Shockley partial dislocations, separated by a stacking fault (see Fig. 4.12), according to the scheme $1/2<110> \to 1/6<211> + 1/6<12\bar{1}>$. Cross slip, however, is only possible by complete dislocations. Hence, the separation length of the partial dislocations needs to

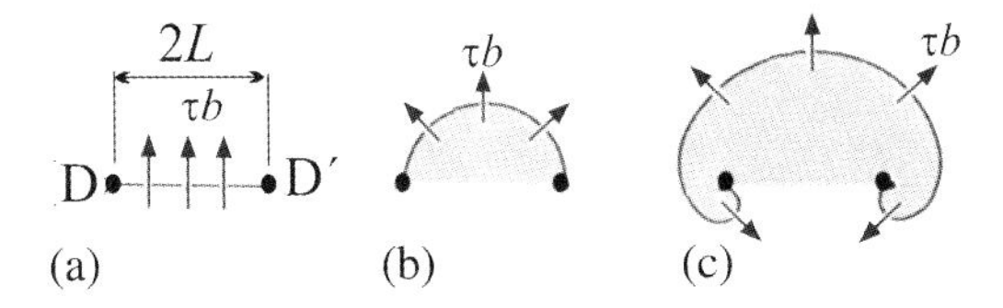

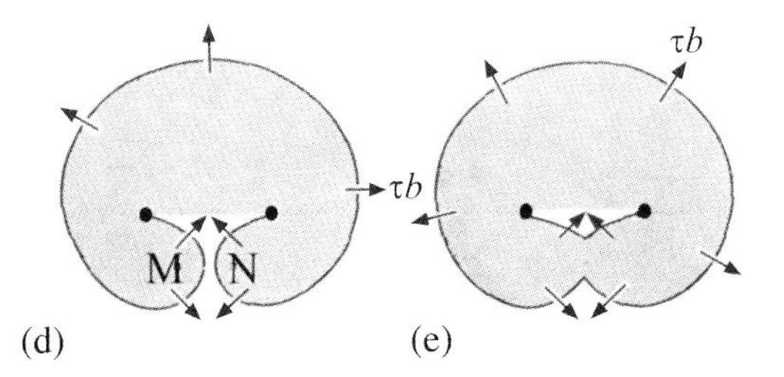

Fig. 4.11 Dislocation multiplication by a Frank–Read source: (a) initial state, (b) bowing out of the dislocation line, (c) instable growth of the dislocation line, (d) formation of a dislocation ring and (e) reaching the initial state once again.

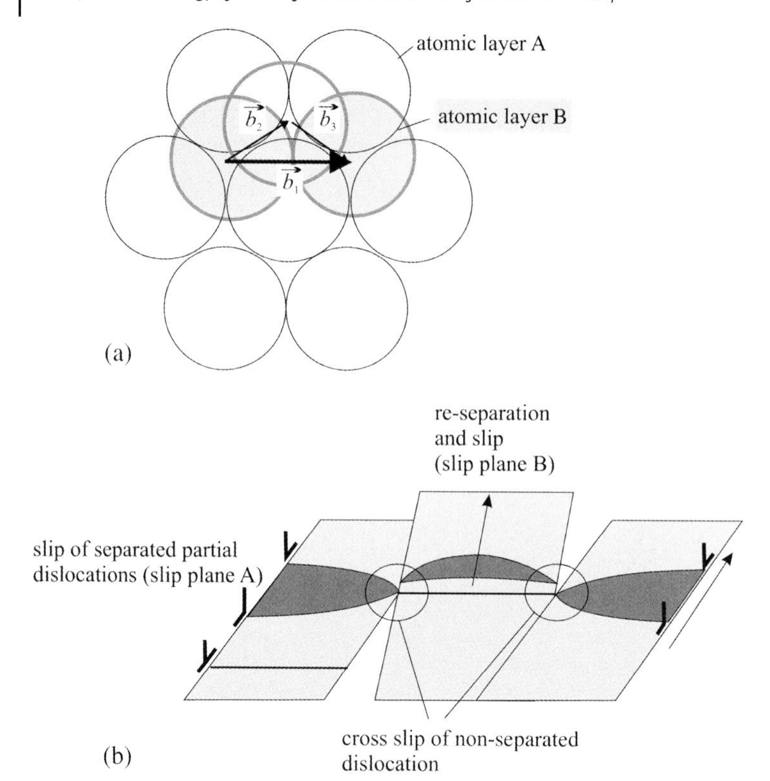

Fig. 4.12 (a) Mechanism of formation of Shockley partial dislocations ($\vec{b}_1 = \vec{b}_2 = \vec{b}_3$); (b) cross slip of nonseparated dislocations.

narrow down according to Fig. 4.12. For slip on the alternate slip plane the dislocation may dissociate into partial dislocations again. This process depends on the stacking fault energy (SFE): Plastic deformation of materials having low SFE involves wide expansion of the partial dislocation separation, and hence makes cross slip more difficult (cf. [85]). In the case of alloys, additional factors have to be taken into consideration: Substitutionally or interstitially dissolved atoms contribute to the friction stress in such a way that recombination of the partial dislocations is hindered and planar slip is promoted [173]. In general, a higher resistance to cross slip, which can also be attributed to short-range order effects in alloys, leads to a change in the plastic deformation behavior from wavy slip, e.g., in the case of pure copper, to planar slip, e.g., in the case of brass.

Summarizing, for pure single-crystalline materials the following contributions add up to the critical shear stress τ_0 to activate plastic slip on a slip plane:

- Nabarro–Peierls stress: τ_P, resistance to the movement of a single dislocation on a slip plane.
- Passing stress: τ_{pass}, interactions with the elastic stress field of dislocations on parallel slip planes.

- Cutting stress: τ_{cut}, increase of the critical shear stress due to cutting processes with nonparallel forest dislocations.
- Strengthening mechanisms: formation of barriers by locking and multiplication of dislocations.

The sum of these contributions will be used in Chapter 7, devoted to modeling by means of the term friction stress τ_{fr}.

It should be noted that for most technical materials the friction stress depends substantially on the chemical composition [162]. Alloying elements having larger or smaller atom diameters yield to an additional elastic stress field and an increase in τ_{fr} (lattice parameter effect of solid-solution strengthening). Furthermore, these elements may contribute to τ_{fr} by altering the shear modulus G (dielastic interaction of solid-solution strengthening). A considerable strengthening effect can be obtained by precipitation hardening, in the case of small coherent particles by cutting, in the case of larger particles by bowing the dislocation line between the particles (Orowan mechanism). For a detailed introduction of strengthening mechanisms the reader is referred to relevant textbooks, either on the mechanisms, e.g., Hull and Bacon [165], or the materials, e.g., [174] for lightweight alloys.

Additional deformation mechanisms have to be considered when loading specimens or components at elevated temperatures above approximately 0.4 times the homologous melting temperature of the respective material (see also Sections 4.4.2 and 7.3). Due to diffusion of vacancies towards dislocation cores, the dislocations become able to leave their slip planes and to climb. This kind of pure thermally activated deformation mechanism has been termed dislocation climb. The respective deformation rate $\dot{\varepsilon}_{st} = (d\varepsilon/dt)_{\text{steady state}}$ can be described by means of Norton's creep law, where thermal activation is accounted for by the activation energy Q being of the same order of magnitude as the activation energy for a material's self-diffusion. The stress dependence is represented by the Norton exponent n, typically having values of $n = 5$:

$$\dot{\varepsilon}_{st} = A\left(\frac{\sigma_a}{G}\right)^n \exp\left(\frac{-Q}{kT}\right) \tag{4.21}$$

At very high temperatures and at the same time low mechanical stresses, dislocation creep becomes negligible and the deformation process is governed by solid-state diffusion, either along grain boundaries (Coble creep, moderate temperatures) or bulk diffusion (Nabarro–Herring creep, high temperatures). A comprehensive and very detailed description of creep deformation and damage processes is given by Evans [175].

Which deformation mechanism is the prevailing one for given loading conditions, i.e., given temperature and applied stress level, can be assessed by the so-called deformation-mechanisms maps of Frost and Ashby [176]. As an example, Fig. 4.13 shows a deformation-mechanism map for aluminum.

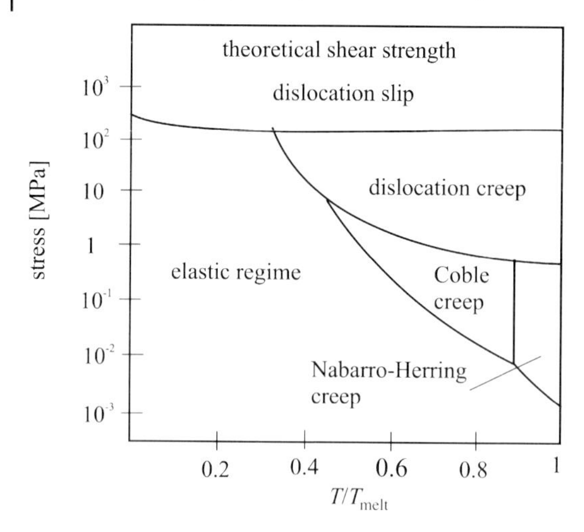

Fig. 4.13 Deformation-mechanism map for aluminum according to Frost and Ashby [176].

4.3 Activation of Slip Planes in Single- and Polycrystalline Materials

While in the preceding section plastic deformation was considered as dislocation motion along a nonspecified slip plane in a single-crystalline material, in the following the significance of the crystal structure, the geometrical arrangement of possible slip planes and the plastic deformation in a polycrystal is discussed.

In general, metallic materials can be distinguished by means of their lattice structure, with the bcc, the fcc and the hexagonal close-packed (hcp) being the most important ones. The fcc lattice exhibits twelve slip systems of type {111}<110> (four {111} planes, each with three <110> directions; see Fig. 4.14a). The slip planes are close-packed planes with an interplane spacing of $d = a/\sqrt{h^2 + k^2 + l^2}$. Due to the high packing density of 0.78, the critical shear stress on these slip planes is comparatively small (of the order of $\tau_0 = 0.35$–0.7 MPa [177]), being responsible for the superior ductility of fcc materials.

In contrast to this, bcc materials have close-packed directions (<111> directions) but no close-packed planes. Slip occurs always in <111> direction, either on six independent {110} planes (twelve slip systems of type {110}<111>), or on twelve independent {112} slip planes and 24 independent {123} slip planes. Even though bcc materials exhibit 48 slip systems (see Fig. 4.14b) the lower packing density of 0.68 yields a critical shear stress of the order of τ_0 = 35–70 MPa [177], being substantially higher than that of fcc metals. Finally, hexagonal metals exhibit three independent {0001}<1120>-type slip systems (see Fig. 4.14c). Depending on the aspect ratio of the unit cell axes a/c, pyramidal and prismatic slip becomes important (see dashed lines in Fig. 4.14c).

Alternatively to plastic slip, crystals can change their shape by spontaneous formation of mechanical twins. Twinning does not involve structural changes by dislocation motion. Deformation is solely based on a mirror-symmetrical shear trans-

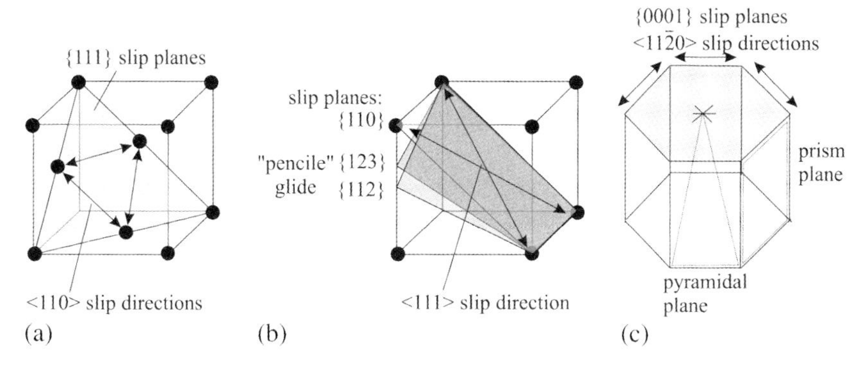

Fig. 4.14 Slip systems in (a) fcc, (b) bcc and (c) hexagonal crystal systems.

formation of a part of the crystal along the twinning plane (see Fig. 4.15a). The geometry of twinning deformation is shown in Fig. 4.15b for the fcc lattice with {111} being the twinning plane and <112> being the twinning direction (bcc: {112}<111>; hcp: $\{10\bar{1}2\} < 10\bar{1}1 >$).

Which slip system is going to be activated in a single-crystalline specimen under the influence of uniaxial stress σ_a depends on the magnitude of the resolved shear stress τ_a on the various slip systems that can be calculated by means of Schmid's law. Hence, the dependence of τ_a on the angle λ between slip direction and stress axis and the angle φ between slip plane normal and stress axis according to Fig. 4.16a can be written as

$$\tau_a = \sigma_a(\cos\varphi \cdot \cos\lambda) = \sigma_a M_S \qquad (4.22)$$

with M_S being the Schmid factor, having a maximum value of $M_S = 0.5$ (for $\lambda = \varphi = 45°$, maximum shear stress). For a given applied stress state the slip system with the highest Schmid factor will be the first one to be activated.

As soon as plastic deformation sets in, slip along the primary slip system leads to a gradual shift of the crystal with respect to the loading axis (see Fig. 4.16b). Since in the case of a single crystal this shift is suppressed by the grips (in the case of a polycrstal it is suppressed by the surrounding grains), a rotation of the slip sys-

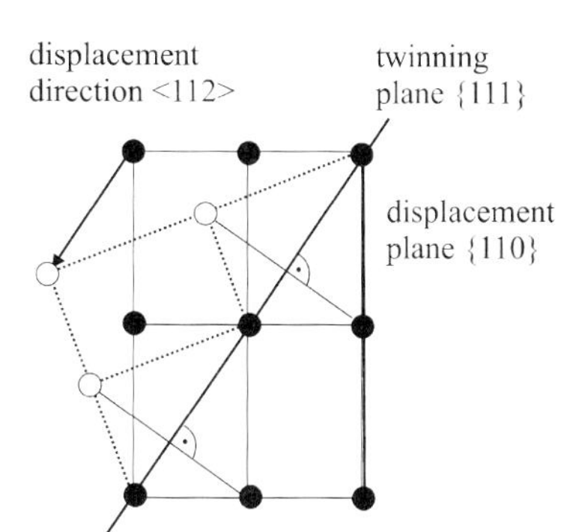

Fig. 4.15 Geometry of twinning in fcc metals.

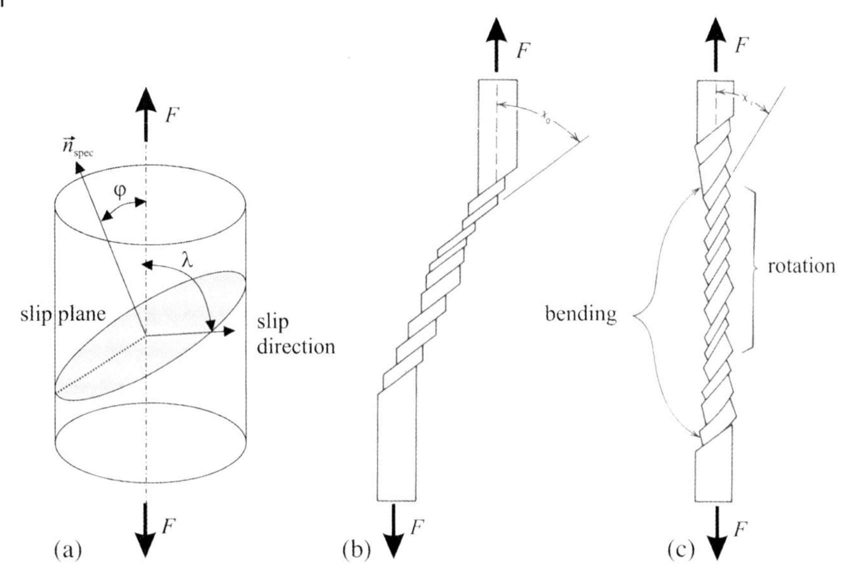

Fig. 4.16 Determination of the Schmid factor for (a) single-crystalline tensile specimen and (b) corresponding change in shape. (c) Required crystal rotations to maintain the shape of the deformed crystal.

tem is required (Fig. 4.16c). This involves a change in the Schmid factor, and hence the activation of a second conjugate slip system (see [168, 178]).

In a polycrystalline specimen or component, initial plastic deformation occurs in those grains exhibiting the highest Schmid factors. The resulting change in shape needs to be accommodated by initially elastic deformation of the surrounding grains, leading to stress concentration preferentially along the grain boundaries. Further deformation leads to plastic deformation of grains having lower Schmid factors. Only when all grains are plastified is the macroscopic yield strength σ_Y reached [162].

It should be noted that dislocation motion along slip bands cannot easily proceed across the grain boundaries because of a discontinuity in the crystal lattice that is tilted at the grain boundary by the misorientation angle Θ (see Section 3.3.5). Along three-dimensional grain boundaries, the lattice planes touch each other only pointwise, since the misorientation between the neighboring lattices is not only given by the tilt angle Φ but also by the twist angle ξ (see Section 6.2). Hence, in the case of high-angle grain boundaries, slip transmission is an indirect process, which is according to Gemperlova et al. [179] substantially influenced by the dislocation structure of the respective grain boundaries themselves. In a very simplified way, the interaction between gliding dislocations and the grain boundaries can be considered as dislocation pileup ahead of the boundaries as shown schematically in Fig. 4.17. By n piled-up dislocations the stress acting on the grain boundaries increases from $\tau_1 = M_{S1}\sigma_a$ to $\tau_{max} = n\tau_1$. Due to the dislocation pileup, the shear stress acting on a dislocation source in the neighboring grain experiences an increase by

$\beta(r_0)\,\tau_{max}$, where the function $\beta(x)$ correlates the decrease of the pileup stress with the distance r_0 from the grain boundary. Hence, the condition for plastic deformation in the neighboring grain can be expressed by the critical shear stress τ_c that needs to be exceeded to activate dislocation emission at r_0 (cf. Fig. 4.17):

$$\tau_c = M_{S2}\sigma_a + \beta(r_0)\cdot\overbrace{\frac{\pi(1-\nu)}{Gb}\frac{d}{2}\tau_1}^{n}\cdot\tau_1 \tag{4.23}$$

Under the assumption that $M_{S2}\sigma_a$ can be neglected, Eq. (4.23) yields the general condition for slip transmission from an initially plastified grain into the neighboring grain:

$$\text{(a) } \tau_1^2 d = \text{const} \Rightarrow \text{(b) } \tau_Y = \tau_0 + \frac{k'}{\sqrt{d}} \tag{4.24}$$

Equation (4.24b) corresponds to the well-known Hall–Petch relationship [180, 181] with the material-specific constant k' that represents in a simple way the relationship between the yield strength, here written for shear as τ_Y, and the grain size d. As the basic idea of the approach mentioned above, the yield strength is assumed to be reached as soon as a dislocation pileup in any arbitrary grains due to a shear stress τ_1 causes the activation of a dislocation source in one of the neighboring grains. For very large grains ($d \rightarrow \infty$) at least the critical shear strength τ_0 (= friction stress τ_{fr}) for dislocation motion in a single crystal needs to be exceeded (cf. Section 4.2 and Schmid's law, Eq. 4.22).

If more than one or all grains of a polycrystal are plastified, deformation by only activating primary slip systems is not possible, because the changes in shape of the crystallites would not be compatible. To maintain the compatibility of all the grains, the activation of five independent slip systems is required. Therefore, an infinitesimal strain $d\varepsilon$ can be expressed by infinitesimal shear strains $d\gamma$ on five independent slip planes as follows:

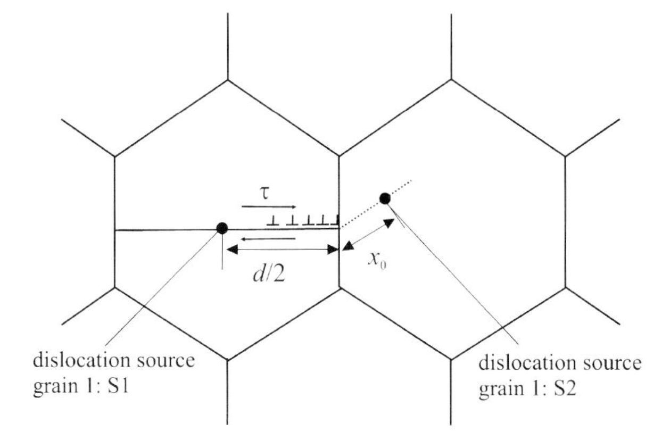

Fig. 4.17 Dislocation pileup and dislocation source activation to explain the Hall–Petch relationship (for details see text and Eqs. (4.23) and (4.24)).

$$d\varepsilon = \frac{1}{M_T} \overbrace{\sum_{s=1}^{5} d\gamma_s}^{\Gamma} \qquad (4.25)$$

under the condition that the sum of these shear strains Γ is minimal; M_T is the so-called Taylor factor, having a value of $M_T = 3.06$ for a random distribution of crystallographic orientations of the grains. On the other hand, averaging over the Schmid factors M_S of a random orientation distribution yields the so-called Sachs factor $M_{Sa} = 2.24$, which should be considered for cyclic deformation at low strains. The reciprocal values of the Sachs and the Taylor factors give the mean Schmid factor (cf. Section 7.2.2).

4.4 Special Features of the Cyclic Deformation of Metallic Materials

The general nature of dislocation motion during cyclic deformation is similar to the one during monotonic deformation, as addressed in the preceding section. But even the phenomenological features of cyclic deformation as summarized in Section 2 allow one to draw the conclusion that there must be fundamental differences between monotonic and cyclic deformation. The following section provides a brief look at the most essential differences while more detailed information on the fatigue behavior of metals and alloys can be found elsewhere, e.g., in the textbooks of Suresh [36] (also nonmetals), Christ [85], Schijve [182], Bilý [183] and Ellyin [184]. As an important feature relating the fundamental mechanisms of dislocation motion with the macroscopic fatigue behavior, the glide character and the formation of characteristic dislocation structures have been identified by means of transmission electron microscopy. Figure 4.18a gives a simplified but general overview of these characteristic dislocation structures depending on the dominant glide character (planar or wavy) and the plastic-strain amplitude $\Delta\varepsilon_{pl}/2$ [185, 186].

Comparing the cumulative plastic deformation during fatigue loading at constant plastic-shear-strain amplitude $\Delta\gamma_{pl}/2 = \text{const}$

$$\Gamma = 4\sum_{i=1}^{N} \frac{\Delta\gamma_{pl,i}}{2} \qquad (4.26)$$

with the plastic-shear strain γ_{pl} during a monotonic tensile test according to Fig. 4.18b shows that the level of cyclic strain hardening is two orders of magnitude smaller, provided the test was started with an annealed specimen having a low dislocation density.

Certainly, the maximum value of the plastic-shear strain of the cyclic tests is far below that of the monotonic tests. That is because rotation of the slip planes is not required due to the absence of large changes in shape during cyclic loading (see Fig. 4.16c), and hence activation of secondary slip planes in combination with a steep increase in work hardening is negligible (see Fig. 4.10a, stage B). Generation

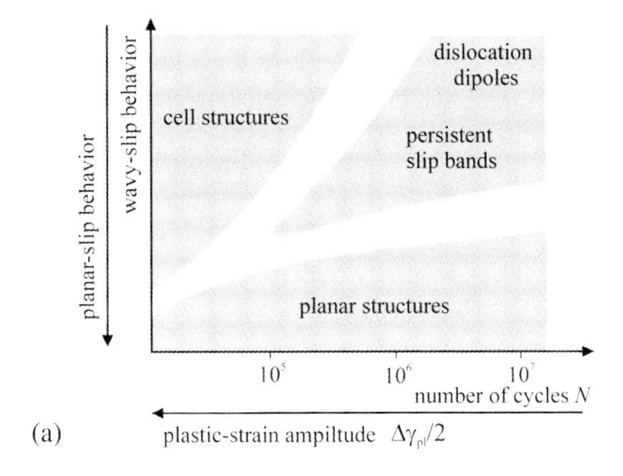

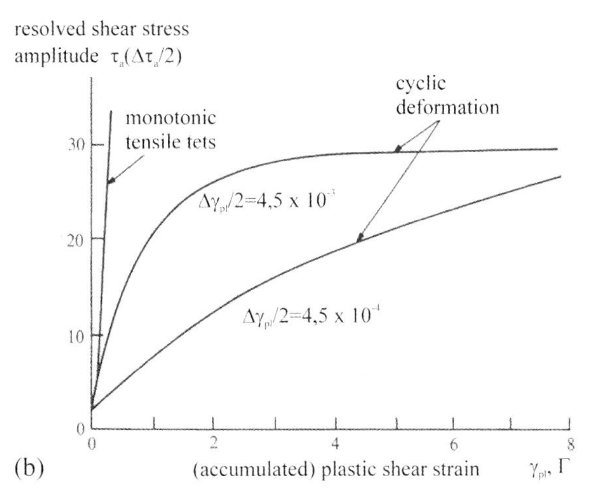

Fig. 4.18 (a) Dislocation arrangement in cyclically deformed fcc metals depending on the respective glide characteristics and the loading conditions (after [185, 186]). (b) Resolved shear stress vs. cumulative plastic shear in a copper single crystal oriented for single slip (stress axis parallel to $[\bar{1}23]$, slip system: $(111)[\bar{1}01]$) (after [188]).

and movement of dislocations are often restricted to the primary slip planes. According to Mughrabi [187], cyclic deformation is determined by approximately the same number of positive and negative dislocations. That means, in contrast to monotonic deformation, no long-range internal stresses are induced. Cyclic hardening can be attributed to interactions of mutual attracting edge dislocations and dislocation locking by dipole formation. These dipoles form characteristic bundle/vein structures at low plastic shear strain amplitudes $\Delta\gamma_{pl}/2$ as shown for stage A in Fig. 4.19 (see also Fig. 3.20). With increasing number of cycles, both an increase in the dislocation density within the veins as well as an increase in the fraction of veins per volume of deformed material of up to 50% can be observed [36].

A further feature of cyclic plasticity becomes evident when the plastic shear strain amplitude $\Delta\gamma_{pl}/2$ overcomes stage A in Fig. 4.19a. In the subsequent stage B the resolved shear stress τ_a in cyclic saturation remains constant (plateau stress) when the strain amplitude $\Delta\gamma_{pl}/2$ is increased. This effect can be attributed to the appearance of persistent slip bands (PSBs, cf. Section 5.2.3). Within these PSBs, plastic deformation can be a factor of approximately 100 higher than within the bundle/vein-structured matrix areas, e.g., in copper $(\Delta\gamma_{pl}/2)_{matrix} = 6 \times 10^{-5}$, $(\Delta\gamma_{pl}/2)_{PSB} = 7.5 \times 10^{-3}$ [85]. According to the rule of mixing of Winter [189], an increase in the plastic shear strain amplitude is mainly due to an increasing volume fraction f_{PSB} of PSBs:

$$\frac{\Delta\gamma_{pl}}{2} = f_{PSB}\left(\frac{\Delta\gamma_{pl}}{2}\right)_{PSB} + (1 - f_{PSB})\left(\frac{\Delta\gamma_{pl}}{2}\right)_{Matrix} \tag{4.27}$$

It should be noted that Winter's rule of mixing is a simplification; in reality ongoing plastic deformation yields to strengthening effects even within the PSBs. The existence of a plateau stress τ_p (see Fig. 4.19, for copper $\tau_p = 28$ MPa) can be attributed to the PSB formation stress that decreases while the plastic shear strain amplitude $\Delta\gamma_{pl}/2$ increases [85, 190]. Contrary to the low dislocation mobility within the bundle/vein structures, dominated by reversible movement of dislocation rings between the dislocation-rich veins (so-called flip-flop mechanism [36]), edge dislocations within the walls of the PSBs can easily bow out into the channels causing the high deformability of the PSB structures.

The saturation state within the PSBs is determined by a dynamic equilibrium between dislocation multiplication by dislocations, bowing out into the channels of the PSBs, and the annihilation of dislocations on slip planes where the distance has fallen below the critical annihilation distance [194]. This kind of annihilation yields to the generation of vacancies, and hence to an increase of the PSB volume, finally resulting in the formation of protrusions on the specimen/component surface. The implication of these protrusions for the initiation of fatigue microcracks is explained in more detail in Section 5.2.3. Thompson et al. [20] had shown already in 1956 that slip traces (surface protrusions) on fatigued copper specimens, which were removed by electrolytic polishing, appear again at the same positions during further cycling. They were the first who termed these kinds of slip traces "persistent".

Plastic-shear-strain amplitude exceeding stage B in Fig. 4.19 again yields a further increase of the saturation shear stress. This can be attributed to the set-in of cross slip on secondary slip bands and the contribution of strengthening mechanisms, e.g., Cottrell–Lomer locks (see Section 4.2). During this transition the bundle/vein-PSB structure rearranges itself towards a labyrinth or cell structure as shown in Fig. 4.19 (stage C). It is worth mentioning at this point that deformation experiments carried out by Wang et al. [193] on copper single crystals, which were oriented for multiple slip, did not show a pronounced plateau behavior. Instead, they found labyrinth structures even for small values of the shear-strain amplitude $\Delta\gamma_{pl}/2$.

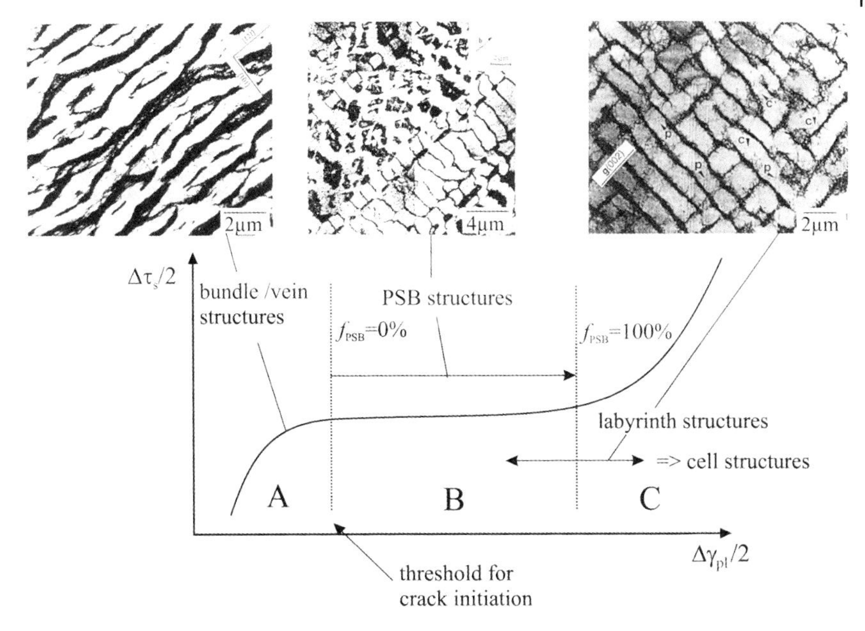

Fig. 4.19 Cyclic stress–strain curve (amplitude of the resolved saturation shear stress $\Delta\tau_s/2$ vs. the plastic-shear amplitude $\Delta\gamma_{pl}/2$) for single-crystalline copper with the respective dislocation structures (TEM micrographs taken from [191]). (A) Bundle-vein structure after cyclic deformation [192]; (B) development of the PSB structure from dissolving veins [193]; (C) labyrinth structures.

In several studies, which are summarized by Suresh [36], it was shown that PSBs occur also within polycrystalline materials, particularly within the surface grains. For higher, technically relevant plastic-shear-strain amplitudes, the shape compatibility requirements promote multiple slip, leading to the formation of cell structures instead of PSBs. An empirical quantification of such cell structures is possible according to Plumtree and Pawlus [195], who correlated the average cell size λ with the saturation stress amplitude $\Delta\sigma_s/2$, the friction stress σ_{fr} (cf. Section 4.2) and a constant B depending on the stacking fault energy as follows:

$$\frac{\frac{\Delta\sigma_s}{2} - \sigma_{fr}}{E} = B\frac{b}{\lambda} \tag{4.28}$$

Up to this point, the discussion of the cyclic deformation behavior has been restricted to fcc metals. It should also be mentioned that the cyclic plastic deformation behavior of bcc metals exhibits some characteristic features (for details see [162, 196]). The main reason for these characteristic features is the fact that in the bcc lattice screw dislocation cores cannot dissociate into partial dislocations under the formation of a stacking fault, leading to a high value of the Nabarro–Peierls stress (cf. Section 4.2). Hence, the plastic deformability of bcc metals depends substantially on the mobility of screw dislocations with the consequence of a strong

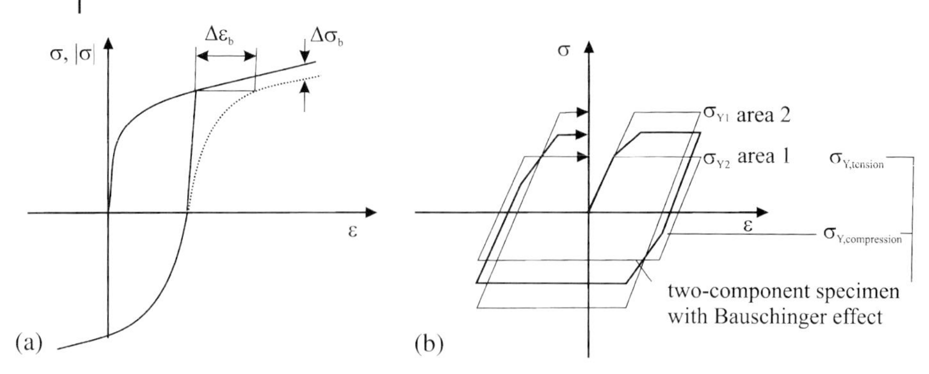

Fig. 4.20 (a) Schematic representation of the stress–strain curve for fully reversed loading; (b) explanation of the Bauschinger effect by means of a simple two-component model.

temperature and strain-rate dependence as well as a pronounced asymmetry of dislocation glide in tension and compression. Fatigue experiments with high strain rates result in strong cyclic hardening and the formation of a pronounced cell structure [36]. However, small amounts of impurities may shift the deformation behavior of bcc metals towards the behavior of fcc metals. Mughrabi et al. [197] (for pure iron single crystals doped with 30 ppm carbon) and Pohl et al. [198] (for plain carbon steel) found PSBs that appeared very similar to those observed in fcc copper and nickel.

A further important aspect of cyclic deformation of metals and alloys is the occurrence of the Bauschinger effect [6]. As shown schematically in Fig. 4.20a, the yield strength exhibits a substantial decrease after load reversal. This can be quantified according to Fig. 4.20a by means of the Bauschinger stress $\Delta\sigma_b$ and the Bauschinger strain $\Delta\varepsilon_b$. Even though a quantitative description of the Bauschinger effect would require the analysis of the influence of the dislocation structure induced during the initial tensile deformation on the following load reversal into compression, a simplified qualitative explanation is possible under the assumption that polycrstalline metallic materials deform like composite structures with locally different yield strengths. Following a simple two-component Masing model (cf. [38]) with two areas of the same size and exhibiting elastic ideal plastic behavior with yield strengths of σ_{Y1} and σ_{Y2} (thin lines in Fig. 4.20b), respectively, plastic deformation of the composite structure sets in, once the yield strength of the weaker area 1 is exceeded (yield strength in tension: $\sigma_{Ytension}$). After reaching maximum tension and subsequent load reversal, the weaker area 1 will be plastified, when the compressive stress exceeds $\sigma_{max} - 2\sigma_{1Y}$, determining the yield strength σ_{Ycompr} of the composite structure in compression. Hence, it is obvious that the initial yield strength is higher than after subsequent load reversal, i.e., $\sigma_{Ytension} > |\sigma_{Ycompr}|$.

5
Initiation of Microcracks

5.1
Crack Initiation: Definition and Significance

The definition of crack initiation depends on the length scale being considered. Basically, one may define crack initiation as the local overcoming of interatomic cohesive forces in combination with the formation of new surfaces. Fundamental studies on brittle cracking and the analysis of basic crack-growth mechanisms justify high-resolution experimental methods and the application of atomistic modeling; however, the scope of this section is a mesoscopic correlation of the metallurgical principles of elastic and plastic deformation with the occurrence of cracks of a technically relevant size, i.e., cracks of the same length scale as that of the significant microstructural features, e.g., the grain size, or the mean diameter of precipitates or pores. The further discussion in Chapter 6 considers cracks as already being present, and is focused on the interactions between crack-propagation rates and the local microstructural features. This propagation eventually causes material failure by fracture.

During fatigue loading, crack initiation usually occurs at the specimen surface (see, e.g., Lukas [199]), and is generally caused by stress concentrators, giving rise to local plastic deformation or cracking and detachment of brittle precipitates. Local stress–strain concentration can be attributed to a variety of microstructural inhomogeneities (see, e.g., Lindley and Nix [200]). The most prominent of these are:

- surface roughness (due to the manufacturing process) or notches;
- surface protrusions due to the formation of pronounced (persistent) slip bands [201];
- nonmetallic inclusions, that might be broken as a consequence of the rolling process [202];
- second phases, precipitates or pores; and
- grain and phase boundaries, because of the elastic and plastic anisotropy of the microstructure of polycrystalline materials.

In particular for high-cycle fatigue (HCF) loading conditions and high-strength materials, up to 90% of a technical component's fatigue life can be determined by the phase of crack initiation and the propagation of microstructurally short cracks

Fatigue Crack Propagation in Metals and Alloys: Microstructural Aspects and Modelling Concepts. Ulrich Krupp

ISBN: 978-3-527-31537-6

[32, 33]. This supports the significance of quantitative analyses of the physical mechanisms being responsible for local stress concentration, plastic deformation and, eventually, the initiation and propagation of short fatigue cracks. The early propagation mechanisms of these cracks are strongly related to the local microstructual features. Only when the sizes of these features are small as compared to the crack length is prediction of crack propagation rates da/dN possible on the basis of linear-elastic fracture mechanics, i.e., the Paris law for long fatigue cracks (cf. Chapter 2). However, in the case of HCF loading, most of the fatigue life is already passed, when short cracks have been grown to long fatigue cracks. Figure 5.1 shows schematically for intermediate- and long-life fatigue the development of fatigue failure during the fatigue life of metals and alloys (cf. Polak [203]).

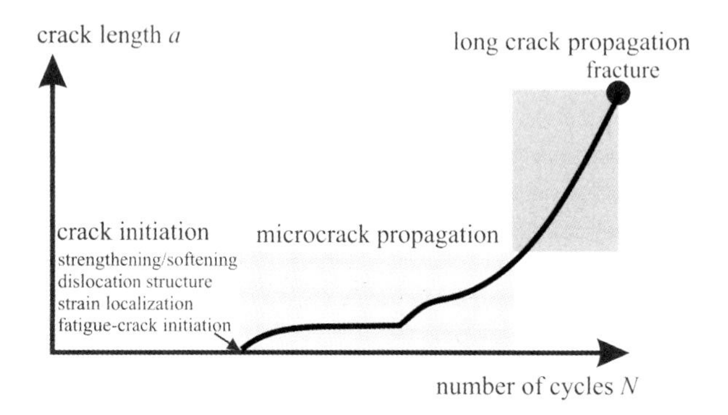

Fig. 5.1 Development of fatigue damage in polycrystalline metals and alloys.

5.1.1
Influence of Notches, Surface Treatment and Residual Stresses

In technical components, the typical initiation sites for fatigue cracks are notches, either given by engineering design or due to surface roughness caused by machining. The influence of such notches of various radii on the initiation and propagation or nonpropagation of microstructurally short cracks was studied by Vallellano et al. [204, 205] by means of a micromechanical model (cf. Chapter 7 for micromechanical modeling). Andrews and Sehitoglu [206] showed by experimental studies on 4340 steel that the influence of surface roughness becomes the key factor when the mean surface roughness R_a exceeds the diameter of intrinsic material defects or the grain size. In addition to the mean value, also the standard deviation of the surface roughness, which can reach the same order of magnitude as the mean value of the surface roughness itself, was accounted for in a model of the stress distribution. Following this model, the strong scattering of fatigue life data being a typical feature of HCF-loaded specimens can be attributed to the distribution of metallurgical defects as well as to a high standard deviation of surface roughness data.

In order to reduce the risk of promoting crack initiation at the surface, many high-stressed components are surface treated by means of, for example, shot peening, tensile pre-stressing, roller hardening or laser-shock peening (see, e.g., [207] for details). The compressive residual stresses in the surface layer, resulting from such kinds of surface treatments, reduce the crack-driving effect of applied remote tensile stresses in the surface layer. Almer et al. [208] used X-ray diffraction in combination with finite-element (FE) calculations to study fatigue-crack initiation in the presence of compressive residual stresses in 1080 steel. They found a gradual reduction of the residual stresses due to cyclic softening. The important effect of the cyclic-loading-induced reduction of residual stresses depends strongly on the surface or pre-straining treatment and determines the delay in the initiation of fatigue cracks. Frequently, it has been observed that introduction of compressive residual stresses in combination with HCF loading leads to crack initiation underneath the surface layer, an effect which can be attributed to a tensile stress peak just below the surface layer of compressive-residual stresses [209]. Such cracks – an example is shown in Fig. 5.2b – are difficult to detect by means of conventional nondestructive testing, like optical microscopy, scanning electron microscopy or the replica technique (cf. Section 3.2).

Table 5.1 summarizes the general relationship between surface treatment condition, crack initiation and crack-propagation behavior.

Certainly, crack initiation can be caused by surface modification during service, e.g., corrosion or erosion by particles. Peters et al. [210, 211] studied the influence of foreign object damage (FOD), caused by high-velocity particles (e.g., occurring during the operation of aeroengines), on the behavior of short fatigue cracks in the titanium alloy Ti-6Al-4V, a material that is commonly used for compressor blades in gas turbines. It was shown that FOD promotes the initiation of fatigue cracks. On the other hand, further crack propagation is hindered by the compressive residual stress field in the vicinity of the FOD sites, in agreement with the statements in Table 5.1.

Table 5.1 Significance of surface properties for the fatigue life of metallic materials (according to [207].

	Fatigue crack initiation	Fatigue crack propagation
Surface roughness	Accelerating	Negligible
Cold work (dislocation density)	Retarding	Accelerating
Residual stresses	Negligible	Retarding

5.2 Influence of Microstructual Factors on the Initiation of Fatigue Cracks

5.2.1 Crack Initiation at the Surface: General Remarks

Not only in the case of technically rough surfaces, but also for finely polished specimens, crack initiation is predominantly found at the surface. Exceptions to this rule are (1) cracks originating at particularly large precipitates or pores within the bulk, or (2) crack initiation during fatigue loading at very low remote stress levels. In the latter case, recently termed very high-cycle fatigue (VHCF) or ultrahigh-cycle fatigue (UHCF) (see [212]), fatigue damage accumulation requires billions of cycles, and the effect of plane stress at the surface promoting slip band formation and crack initiation is negligible compared to the stress concentration at inner inclusions or material defects.

For most applications the surface layer exhibits the highest mechanical stress levels in combination with missing strain constraint normal to the surface. This kind of plane-stress condition eases the initiation of microplasticity at the surface, mostly promoted by additional stress concentrators, e.g., high-angle grain boundaries or nonmetallic inclusions. This is the case even at low remote stress amplitudes [199]. However, the distribution of mechanical stresses is not homogeneous throughout the material. It depends on shape, size and crystallographic orientation of the grains, as well as on the distribution of secondary phases, nonmetallic precipitates and pores (cf. [182]).

During crack initiation under HCF-loading conditions at the surface, activation of single slip systems prevails [213], while crack propagation into the bulk tends to require multiple slip due to the strong constraint by the surrounding grains. This is in particular the case for ductile face-centered cubic (fcc) materials, leading to the formation of fatigue striations that can be attributed to alternate activation of different slip systems. This is discussed in more detail in Section 6.2.2 on fatigue crack propagation.

A key point of fatigue crack initiation at the surface is the phenomenon of slip irreversibility. Reversal of plastic deformation, e.g., from tension to compression, is only partially reversible. Consequently, each cycle causes a small increment of fatigue damage. Adsorption of oxygen from the atmosphere at the fresh metal surface of the emerging slip bands is probably the main reasons for slip irreversibility, material separation and crack initiation.

5.2.2
Crack Initiation at Inclusions and Pores

Pores in metals and alloys are as microcavities mostly a consequence of material shrinkage during the casting and solidification process, and can be basically considered as existing microcracks. This is in particular the case when their size falls within the dimension of the most characteristic microstructural feature of the material, e.g., the grain size or the phase distribution in two- or multi-phase materials. An example is the fatigue damage process of magnesium alloys, where cracks typically emanate from pores formed during dendritic solidification [103].

For crack initiation at nonmetallic inclusions or secondary phases one can distinguish between two general mechanisms:

- Different elastic properties or different coefficients of thermal expansion (CTE) of the parent material and the precipitates may raise the mechanical stress in the vicinity of the interface during mechanical or thermal loading and, eventually, cause interfacial crack initiation. Of course, the level of this stress increase depends strongly on the size of the inclusions. Normally, small nonmetallic inclusions (<5 μm) are considered as harmless for materials of low purity and components without special requirements on the surface finish. They might be harmful for high-cleanliness materials, which are used, e.g., for structural aircraft components. Depending on the structure relationship between substrate and precipitates – coherent (e.g., γ' precipitates in nickel-based superalloys), semi-coherent or noncoherent – the respective interfaces are the weakest sites of a material if they are noncoherent or the particles themselves if the interface is semicoherent. For example, in the case of sulfur-containing steels, failure often occurs as a consequence of the separation of the interface between bulk and MnS precipitates [36].

- Especially large, nonmetallic precipitates, e.g., large Al_2O_3 inclusions in steels, tend to brittle cracking during cold working or during service. These cracked precipitates can be considered already as small cracks in the same way as pores, exhibiting an even higher stress intensity at the crack tip.

Of course, precipitates are often unavoidable impurities as are most of the nonmetallic inclusions in steels, but essential elements of the microstructure. For instance, small precipitates are used for the stabilization of grain boundaries (to avoid unwanted grain growth, e.g., during high-temperature service of superalloys), particle strengthening, stiffening or to improve the wear resistance. Except for the latter, where typically large SiC particles are used (see Fig. 5.2a), most of these precipitates are nanoscale particles, having a negligible effect on local stress concentration. Hence, a metallurgical compromise has often to be found between reinforcement by particles or dispersoids and the introduction of new crack initiation sites. An example of crack initiation in a SiC-reinforced aluminum alloy Al8019+12.5%SiC is shown in Fig. 5.2a; as another example, subsurface fatigue crack initiation in a laser-shock-peened high-strength titanium alloy is shown in Fig. 5.2b.

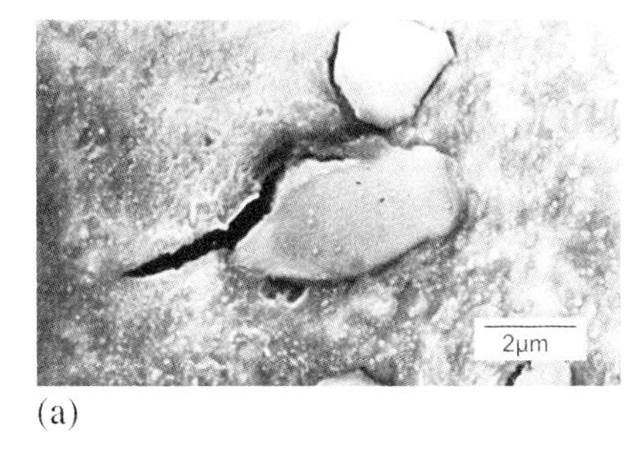

(a)

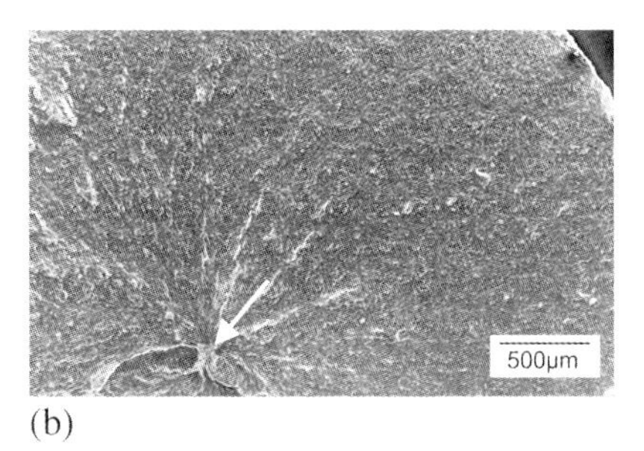

(b)

Fig. 5.2 Fatigue-crack initiation (a) at a SiC particle in a SiC-reinforced Al 8019 alloy (after [82]) and (b) in the bulk of a surface-treated (laser-shock-peened) specimen of the β-titanium alloy LCB in the technical $\alpha+\beta$ heat treatment condition.

In the case of the higher strength steels, crack initiation at nonmetallic inclusions such as Al_2O_3, MnS and SiO_2 is the most frequent cause for failure. In the case of aluminum alloys, crack initiation occurs predominantly at coarse intermetallic precipitates, e.g., the S phase (Al_2CuMg). Studies of the short-crack behavior in the aluminum alloy Al 2219-T815 carried out by Morris [214, 215] revealed that, for most cases, fatigue crack initiation occurred at intermetallic precipitates and only occasionally at grain boundaries.

5.2.3
Crack Initiation at Persistent Slip Bands

Even in the case of idealized materials, having no defects, no precipitates and no grain boundaries, and being perfectly polished, cyclic deformation will lead sooner or later to crack initiation, provided some infinitesimal amount of localized irreversible plastic slip is present. Dislocations that are transported towards the surface generate a slip step, where fresh metal surface is exposed to oxygen from the environment (Fig. 5.3a). Absorption of oxygen impedes complete rewelding of the slip step, which is moved back into the bulk during load reversal (Fig. 5.3b). As a consequence, local decohesion along the slip band generates crack-initiation sites (Fig. 5.3c). This mechanism can be indirectly proven by fatigue experiments in vacuum, as was done, for example, in the early work of Thompson et al. [20], typically resulting in substantially longer fatigue life than for corresponding experiments in air.

Beside the environmental effect, cyclic slip irreversibility can be attributed to mutual interactions between dislocations, which are subject of Section 4.2. The dynamic equilibrium between dislocation multiplication by the Frank–Read mechanism and the annihilation of screw and edge dislocations in persistent slip bands (PSBs) leads to the generation of vacancies, which eventually cause macroscopic roughness along the intersection lines between PSBs and originally polished specimen surface

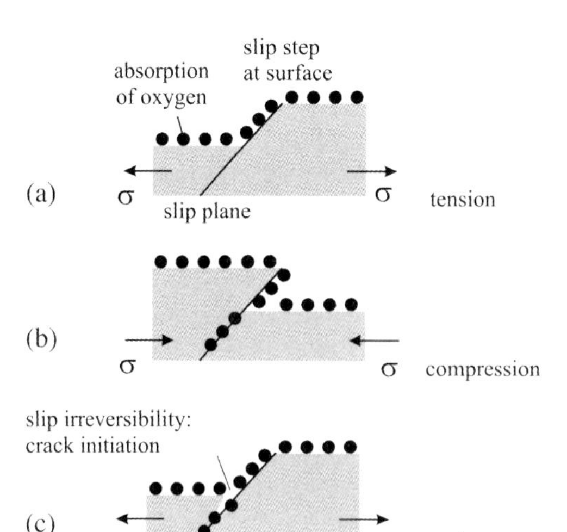

Fig. 5.3 Crack initiation due to environmentally assisted slip irreversibility: (a) exposure of fresh metal surface; (b) absorption of oxygen; (c) local decohesion.

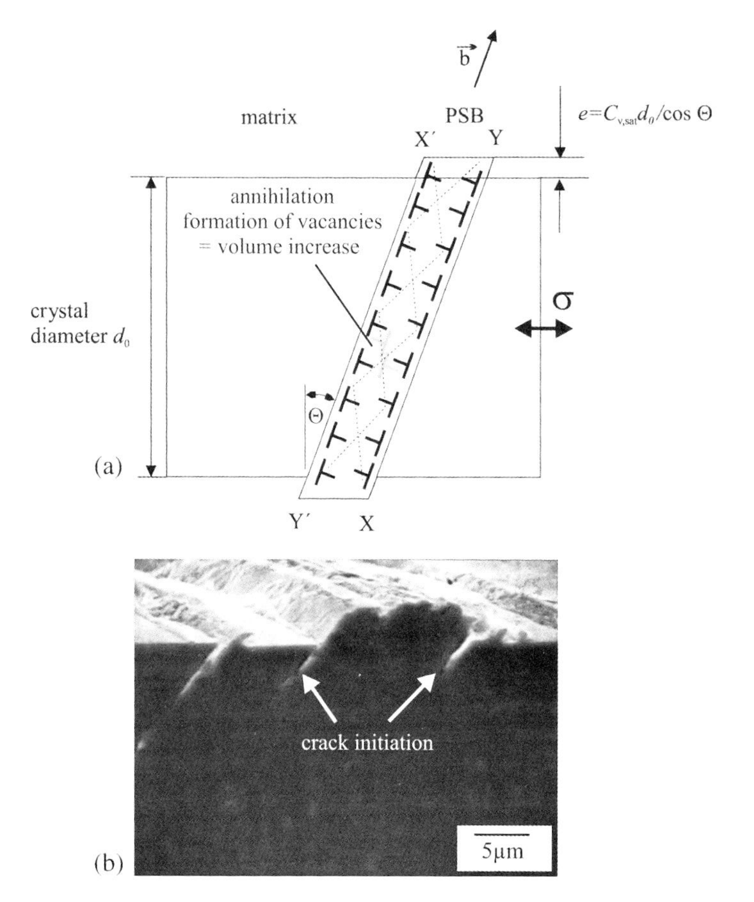

Fig. 5.4 (a) Schematic representation of the mechanism of extrusions/protrusions at a persistent slip band according to the EGM model [217]. (b) Crack initiation at the PSB–matrix interface in cyclically deformed copper ($\Delta\varepsilon/2 = 0.002$, 60 000 cycles); scanning electron micrograph of a cross section obtained by means of the sharp corner sectioning technique (after [219]).

[199, 216]. The physical mechanism leading to the formation of this kind of surface roughness (mean value: $R_a \approx 1$ μm) consisting of extrusions and intrusions (notch radius $r \approx 0.1$ μm [203]) can be explained by means of a model, published by Essmann, Gösele and Mughrabi (EGM model [217]), in the following way (see Fig. 5.4).

It is known that two dislocations of opposite sign annihilate when the distance between them falls below a certain value y, being $y_s = 50$ nm in the case of screw dislocations and $y_e = 1.6$ nm in the case of edge dislocations [36]. Within the PSBs, mainly consisting of edge dislocations, annihilation leads either to the formation of vacancies or interstitials, depending on whether the corresponding annihilating lattice half planes overlap (interstitial dipoles) or are distant by a length of $y < y_{e/s}$, respectively. This mechanism becomes possible because of the generation and movement of dislocations on various individual slip planes within the PSB during the alternating tension and compression periods. Annihilation of vacancy-type dislocation dipoles results in a change of the operating slip plane, i.e., the effective slip planes become slightly inclined with respect to the direction of the PSB. This is represented by the line XX′ and YY′ for load reversal in Fig. 5.4a. The volume, newly created by the vacancies, manifests itself by the formation of extrusions, or intrusions in the case of predominant interstitial-type annihilation. This process is active until a saturation concentration $C_{Va,sat}$ is reached.

Those dislocations that are not going to be annihilated are mobile along the paths of the dashed lines, represented schematically in Fig. 5.4a, either moving to the free surface or to the bulk–PSB interface. Hence, rows of interface dislocations directed towards the inner PSB are formed. Assuming all of these dislocations emanate at the free surface, a protrusion of height e according to Fig. 5.4a would be the consequence. Due to their mutual repulsion, the increase in the interface dislocation density gives rise to the establishment of internal compressive stresses σ_{iPSB} within the PSB. These compressive stresses are superimposed by the applied remote stress σ_a leading to stress peaks in the tensile phase at the points X and X′ in Fig. 5.4a. Ma and Laird [218, 219] showed for cyclically deformed copper single crystals that, indeed, these stress peaks cause crack initiation at the flanks of the PSB at the specimen surface (see Fig. 5.4b). This was supported by applying an interferometric method, which revealed an inhomogeneous strain distribution within the PSBs with higher displacements at the PSB–matrix interface [219]. The harmful effect of the surface protrusions became evident from fatigue experiments by Thompson et al. [20] and Basinski et al. [220], where the surface roughness and surface cracks developed as a consequence of PSB formation were periodically removed by polishing. This procedure led to an enormous increase in the fatigue life of the respective specimens as compared to untreated specimens. However, this effect should not be attributed to the removal of PSB steps only, but also to strain delocalization, which has been reported by Witmer et al. [221]. PSBs formed later in fatigue life exhibit a much smaller local strain than those formed early in fatigue life. Consequently, this kind of delocalized strain is much more reversible, giving rise to extended fatigue life.

The mechanisms active within PSBs were first studied for single-crystalline copper and nickel. However, the impact of PSB formation on the crack initiation in technical polycrystalline materials is similar. Again, the formation of protrusions visible as slip bands at the surface is the key factor [199]. Investigations on polycrystalline copper by Figueroa and Laird [222] and Kim and Laird [223] revealed that for moderate strain amplitudes transcrystalline cracks are initiated at the intersection points between PSBs and grain boundaries. Sauzay and Gilormini [224] analyzed the slip range in surface PSBs at low remote stress amplitudes and used an energy balance to account for slip irreversibility as the origin for crack initiation. At higher strain amplitudes and a correspondingly higher volume fraction of PSBs, the high number of slip steps, reaching heights up to 1 μm along the grain boundaries, promotes intercrystalline crack initiation [223, 225]. This was supported by transmission electron microscopy studies of Huang and No [226], who found protrusions along the PSB–grain boundary interfaces. However, for high strain amplitudes they report crack initiation at grain boundaries without PSB interactions, but accompanied by pronounced formation of dislocation cell structures. The appearance of cell structures is an indication of uniform strain within the respective grains, causing relative displacements from grain to grain and, eventually, the formation of transgranular cracks emanating from the grain boundary.

Although the nature of plastic deformation in body-centered cubic (bcc) materials is different from that in fcc materials (cf. Chapter 4), the significance of slip bands is quite similar. Instead of the PSB structure, introduced in the section above, the slip bands in many bcc alloys can be attributed to planar dislocation arrangements, where slip irreversibility is due to dislocation motion on individual parallel slip planes. PSB structures in bcc alloys have been reported only in a few papers, e.g., by Pohl et al. [198]. In general, bcc alloys may behave as fcc alloys, when the mobility of edge and screw dislocations is made similar by alloying [197]. Studies on a β-titanium alloy in a solution-heat-treated bcc structure [90, 227] have shown that intercrystalline crack initiation is governed by the same slip-step mechanism that was observed by Kim and Laird [223, 225] in polycrystalline copper. An example is discussed in Section 5.4.1 and shown in Fig. 5.18.

5.3 Crack Initiation by Elastic Anisotropy

5.3.1 Definition and Significance of Elastic Anisotropy

The elastic response of a crystalline material to mechanical loading depends on the direction of the loading with respect to the crystal orientation and on the symmetry of the atomic structure. Since most technical materials are polycrystalline, the direction dependence of the elastic material properties vanishes, and one can introduce the quasi-isotropic Young's modulus E. However, single-crystalline materials often exhibit a strong anisotropy of the elastic properties (cf. Table 4.1). This is technically used, for example, for thermally and mechanically high-loaded gas turbine blades (first and second stage), which are made of single-crystalline nickel-based superalloys. The orientation is adjusted in such a way that the elastically softest [001] direction lies parallel to the direction of the maximum thermal expansion (radial direction). Hence, for a given strain the elastic regime is increased and plastic deformation causing damage is avoided. Figure 5.5 represents the elastic anisotropy as the Young's modulus for the various crystallographic directions in the fcc lattice of the single-crystalline nickel-based superalloy PWA 1480 (according to [228, 229]).

When the mechanical behavior of a quasi-isotropic polycrystal on the length scale of the grain size is examined closely, the elastic anisotropy is seen to be a key factor determining local deviations from the remote stress, the latter being macro-

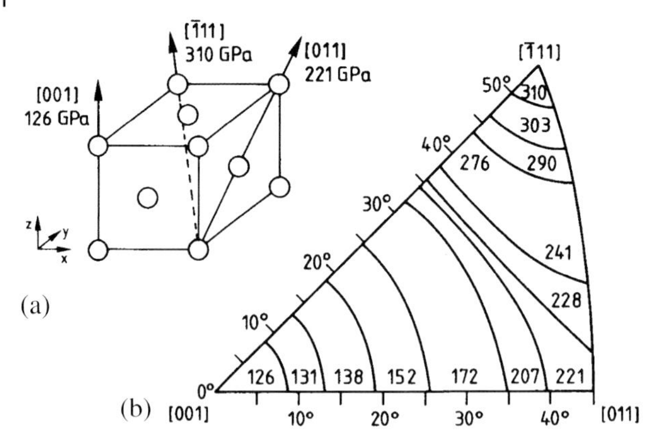

Fig. 5.5 Relationship between the Young's modulus and the crystallographic orientation for the single-crystalline nickel-based superalloy PWA 1480 at room temperature (after [228]).

scopically in the elastic regime. Differences in crystallographic orientation of the grains cause stress peaks at the grain boundaries and triple lines that may exceed locally the yield strength. This is shown schematically in Fig. 5.6. This kind of local plastic deformation may eventually cause fatigue crack initiation.

Mathematically, the elastic anisotropy can be described by considering the general material law (cf. Section 4.1) for the relationship between the strain vector ε_{ij} and the stress vector σ_{kl}:

$$\varepsilon_{ij} = S_{ijkl}\sigma_{kl} \tag{5.1}$$

While in Eq. (4.3) (Hooke's law) the stress is related to the strain by means of the stiffness tensor C_{ijkl}, here the strain is related to the stress by means of the compliance tensor S_{ijkl}, which can be written in a simplified form with only three independent components (cf. Section 4.1)

$$S_{ijkl} = \begin{bmatrix} s_{11} & s_{12} & s_{12} & 0 & 0 & 0 \\ s_{12} & s_{11} & s_{12} & 0 & 0 & 0 \\ s_{12} & s_{12} & s_{11} & 0 & 0 & 0 \\ 0 & 0 & 0 & s_{44} & 0 & 0 \\ 0 & 0 & 0 & 0 & s_{44} & 0 \\ 0 & 0 & 0 & 0 & 0 & s_{44} \end{bmatrix} \tag{5.2}$$

assuming cubic crystal symmetry.

The elastic anisotropy factor A is defined as the ratio between the shear modulus in the (011) plane:

$$G_{(011)} = \frac{E_{[001]}}{2(1+\nu)} = \frac{1}{2(s_{11} - s_{12})} \tag{5.3}$$

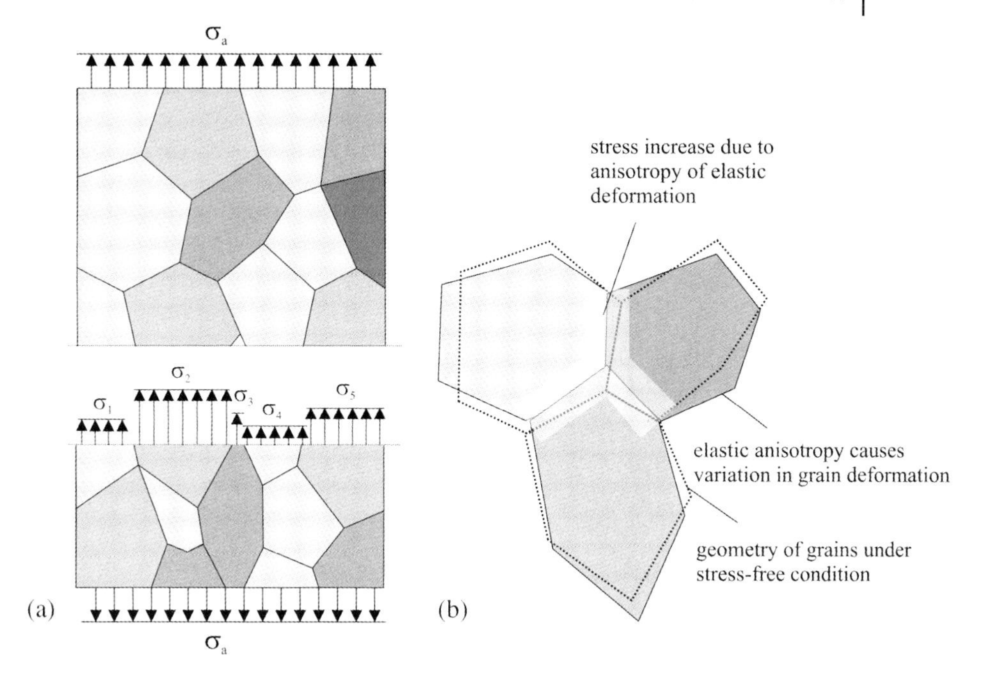

Fig. 5.6 Schematic representation of the influence of crystallographically anisotropic material properties: (a) inhomogeneous stress distribution (simplified for normal stresses only); (b) resulting anisotropy of elastic deformation.

with $E_{[001]}$ being the Young's modulus in [001] direction, and the shear modulus in the (001) plane:

$$G_{(001)} = \frac{1}{s_{44}} \tag{5.4}$$

Hence the anisotropy factor A can be given as

$$A = \frac{G_{(001)}}{G_{(011)}} = \frac{2(s_{11} - s_{12})}{s_{44}} \tag{5.5}$$

5.3.2
Determination of Elastic Constants and Estimation of the Elastic Anisotropy

For many pure metals the elastic constants for the major crystallographic orientations are known and listed in tables, e.g., in [230]. However, for most technical alloys this is not the case, and an experimental evaluation of the elastic behavior is required. When single-crystalline specimens are available, such an evaluation can be done in a straight forward way, either by mechanical deformation in different crystallographic orientations in the elastic regime or by means of ultrasonic measurements. The latter uses the relationship between the velocity of sound (transversal waves c_T and longitudinal waves c_L) measured by the time interval between two successive back-face echoes in a specimen of density ρ and dynamic Young's modulus E according to [231]

$$E = \rho \frac{3c_L^2 c_T^2 - 4c_T^4}{c_L^2 - c_T^2} \tag{5.6}$$

In the following section an approach is introduced that allows the estimation of elastic single-crystal constants from polycrystalline specimens [232] using a combination of crystallographic orientation mapping by electron backscattered diffraction (EBSD; see Section 3.3.2) and local strain measurements using an interferometric strain/displacement gauge (ISDG; see Section 3.1.3).

From the local displacement, parallel to the loading axis and within an individual grain p, to be measured by ISDG (see Fig. 5.7), one can obtain the component s_{11}^p of the grain's compliance tensor S_{ijkl}^p, referring to the specimen coordinate system and assuming that the stress active in the grain interior can be approximated by the remote stress σ_a. According to Section 3.3.5, the specimen coordinate system and the crystal coordinate system are correlated by means of the rotation matrix. Hence, the component s_{11}^p of the grain's compliance tensor can be expressed as a function of the unknown components of the compliance tensor S_{ijkl} in the <001> crystal coordinate system (cf. Eq. 5.2):

$$s_{11}^p = s_{11} - (2s_{11} - 2s_{12} - s_{44}) \cdot (l_{1p}^2 l_{2p}^2 + l_{2p}^2 l_{3p}^2 + l_{3p}^2 l_{1p}^2) = s_{11} - (2s_{11} - 2s_{12} - s_{44}) \cdot \Gamma_p \tag{5.7}$$

with l_{1p}, l_{2p} and l_{3p} being the direction cosines between the <001> crystal axes and the loading axis (see Fig. 5.7) summarized in the orientation function Γ_p of the grain p.

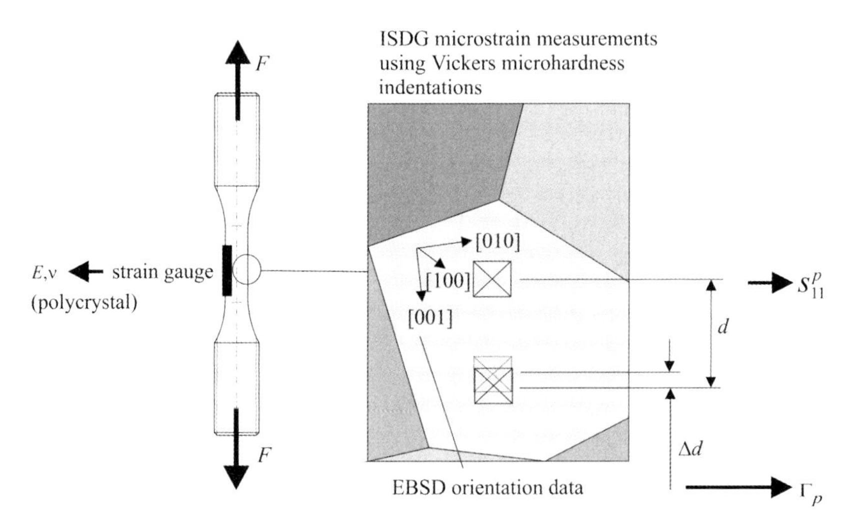

Fig. 5.7 Arrangement for global and local strain measurements using strain gauges and the ISDG technique.

Equation (5.7) contains three unknown components s_{11}, s_{12} and s_{44} of the single-crystal compliance tensor S_{ijk} and the measured values for s_{11}^p and Γ_p. When at least two measurements are obtained in two individual grains ($p = 1$; $p = 2$), a part of Eq. (5.7) can be eliminated and one obtains for the single-crystal constant s_{11}

$$s_{11} = \left(\frac{\Gamma_2}{s_{11}^1} - \frac{\Gamma_1}{s_{11}^2} \right) \cdot \frac{1}{\Gamma_2 - \Gamma_1} \tag{5.8}$$

For a complete description of the elastic anisotropy the constants s_{12} und s_{44} are also required. These can be estimated by the concepts of Voigt and Reuss [233–236], that are based on the borderline cases that the microstructure is composed either of parallel arranged grains (Voigt, Fig. 5.8a) or sequentially arranged grains (Reuss, Fig. 5.8b). Thus, the quasi-isotropic elastic constants can be estimated on the basis of the single-crystalline components of the compliance tensor as follows:

$$E_{\text{Reuss}} = \frac{5}{3s_{11} + 2s_{12} + s_{44}}, \qquad \nu_{\text{Reuss}} = \frac{-2s_{11} - 8s_{12} + s_{44}}{6s_{11} + 4s_{12} + 2s_{44}} \tag{5.9}$$

$$E_{\text{Voigt}} = \frac{5 - 3B}{(5 - B)s_{11} - 2Bs_{12}}, \qquad \nu_{\text{Voigt}} = \frac{-5s_{12} + B(s_{11} + 2s_{12})}{5s_{11} + (B - 2)(s_{11} + 2s_{12})} \tag{5.10}$$

The constant parameter B in Eq. (5.10) includes the elastic anisotropy A according to $B = 1 - A$. Using Eq. (5.5) and Eqs. (5.7) to (5.10) allows the components s_{11}, s_{12} and s_{44} of the single-crystal compliance tensor S_{ijkl} to be estimated, provided the quasi-isotropic constants Young's modulus E and the Poisson ratio ν are known from measurements.

The above-mentioned method was applied to determine the elastic anisotropy A (Eq. 5.5) of the metastable β-titanium alloy Timetal LCB (low-cost β) used as a model alloy in the solution-heat-treated condition. This alloy was also subject to funda-

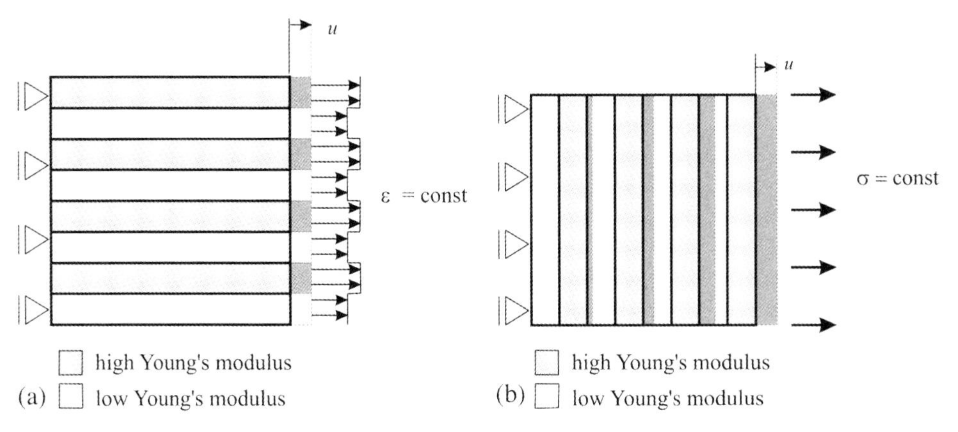

Fig. 5.8 Elastic properties of composites: concepts of (a) Voigt for grains arranged in parallel and (b) Reuss for grains arranged sequentially.

Table 5.2 Evaluation of intercrystalline crack initiation in the β-titanium alloy LCB (solution heat treated) after 60% of fatigue life ($\Delta\sigma/2 = 600$ MPa, $R = -1$).

	Fraction of all 596 evaluated grain boundaries (%)	**Fraction of all 41 intercrystalline cracked grain boundaries (%)**
Low-angle grain boundaries	15	0
High-angle grain boundaries	85	100
Special $\Sigma1$ to $\Sigma49$ grain boundaries	26	35
$\Sigma3$ grain boundaries	9	20
Random nonspecial grain boundaries	74	65

mental studies of microcrack propagation, which are discussed in Chapter 6. By choosing relatively cheap alloying elements (FeMo pre-alloying), LCB is foreseen for high-strength applications in the aircraft and automotive industries. Its chemical composition and the standard heat treatment parameters are given in Table 5.2. Figure 5.9a shows its microstructure in the technical thermomechanically processed and aged condition consisting of bcc β grains, primary hcp α precipitates at the β grain boundaries and secondary α needles within the grain interior. To obtain the solution-heat-treated condition (Fig. 5.9b) the LCB material was annealed above the β transus temperature. It should be mentioned that this kind of heat treatment leads to an enormous increase in the grain size. For more details of titanium alloys see [237] and of metastable titanium alloys and Timetal LCB see [238–243].

Coarse-grained LCB specimens were provided with several Vickers microhardness indentations within individual grains for ISDG microstrain measurements and strain gauges to measure the quasi-isotropic Young's modulus ($E = 85$ GPa) and the Poisson ratio ($\nu = 0.365$) [227]. To compensate for the large scattering of experimental data and the error due to the assumption of a constant stress to be of

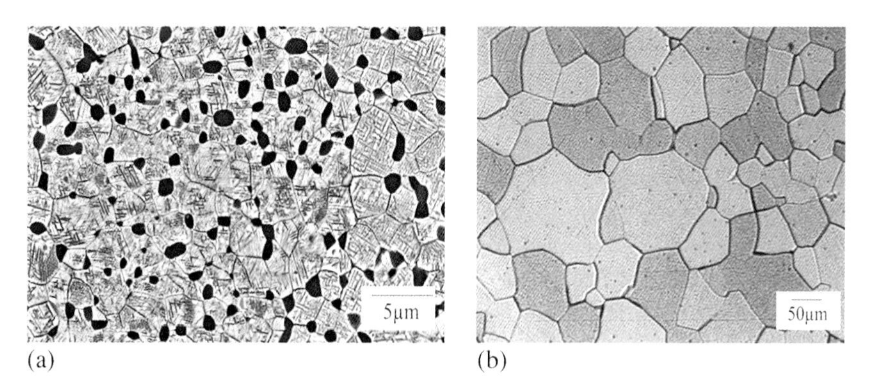

Fig. 5.9 Microstructure of the β-titanium alloy LCB: (a) after technical heat treatment – thermomechanical processing + aging (primary α precipitates at the grain boundaries (black) and secondary α needles in the grain interior); (b) solution-heat-treated condition (single-phase β).

the value of the remote stress within the grains, the estimated single-crystal data were processed by means of the FE method. This approach allowed the local strain to be calculated by comparing the calculated strain values based on the Voigt and Reuss estimation introduced above with the actually measured ones. Hence, it was possible to derive more reliable single-crystal compliance values by an iterative optimization procedure [244]. Finally, for the titanium alloy LCB an anisotropy factor of $A \approx 0.7$ was obtained, and used for the FE calculations described in Section 5.3.3 to calculate the elastic anisotropy stresses. The anisotropy of the elastic properties of bcc titanium was shown in earlier work to be represented by anisotropy factors of $A = 1.7$ in the case of Ti-40V [245] and $A = 2.25$ in the case of Ti-10Cr [246], i.e., the <001> directions are elastically weaker than the <111> directions (cf. Table 4.1). Probably, the differences in those anisotropy factors can be attributed to variations in the chemical compositions and the respective physical bonding conditions.

It should be mentioned that the results of the approach described here should be used only for a qualitative analysis of the relationship between stress concentration, due to elastic anisotropy, and crack initiation. The assessment of quantitative data would require precise measurements of the elastic constants on single-crystalline specimens, a real three-dimensional model of the specimen's polycrystalline microstructure and consideration of crystal plasticity in the FE model. Such kinds of three-dimensional simulations were carried out by Vehoff et al. [247], and were the subject of an interdisciplinary project by Anagnostou et al. [248].

5.3.3
FE Calculations of Elastic Anisotropy Stresses to Predict Crack Initiation Sites

The values of the stress peaks in the vicinity of the grain boundaries caused by elastic anisotropy depend substantially on the shapes and the mutual geometric position of the respective grains. This greatly complicates an analytical calculation for arbitrary polycrystalline microstructures, even if applied to simplified geometries such as twin boundaries or bicrystals (see Section 5.3.4 and [91, 127, 249–253]).

For the example of cyclically deformed specimens of the β-titanium alloy LCB in the single-phase bcc condition, areas of the microstructure containing crack initiation sites were analyzed by means of the FE method, using the anisotropy factor $A=0.7$. For this purpose, scanning electron micrographs of the respective damaged microstructure areas were meshed by 40·48·8 anisotropic hexagonal 8-knod finite elements (hex8 elements) using the FE preprocessing software PATRAN. Since the micrographs provide only the surface of actually three-dimensional grains, the grain boundaries were considered to be oriented orthogonal to the surface, represented by a thickness of the microstructure area of eight elements. The anisotropic fine-meshed microstructure area is embedded within an isotropic area, which is fixed at the one end and loaded by the applied remote stress of $\sigma_a = 600$ MPa. The deformation behavior of the isotropic area is determined by the measured quasi-elastic Young's modulus of single-phase bcc LCB ($E = 84.5$ GPa, $\nu = 0.365$). The whole configuration is represented in Fig. 5.10. Finally, the stress distribution was calculated by the FE software ABAQUS.

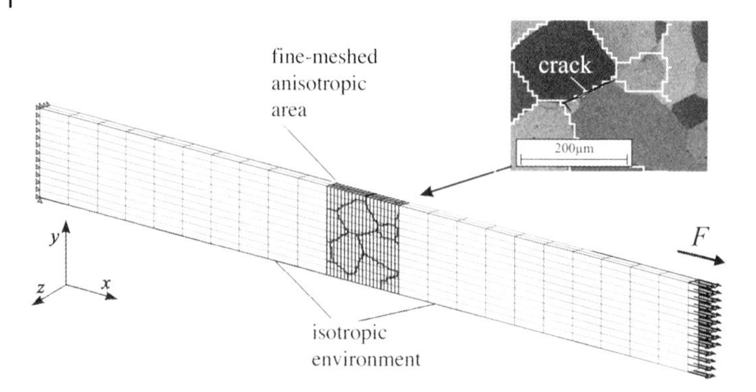

Fig. 5.10 Finite element mesh to represent real microstructure areas with microcracks, obtained by scanning electron microscopy examinations, embedded in an isotropic environment.

As fatigue experiments carried out on the β-titanium alloy revealed, more than 80% of all cracks initiated at the grain boundaries [254]. An example of such a kind of crack initiation during cyclic deformation at $\Delta\sigma/2 = 600$ MPa at a stress ratio of $R = -1$ is shown in Fig. 5.11a. Several cracks initiated along the grain boundary between grains A and B and propagated as slip-band cracks into both grains. The principal normal stress distribution, calculated by the anisotropic FE model for the microstructure area shown in Fig. 5.11a, showed increased stress levels along those grain boundaries that indeed acted as crack initiation sites (Fig. 5.11b).

In a quite similar way, intercrystalline crack initiation can be attributed to elastic anisotropy stresses. An example is shown in Fig. 5.12, again for the titanium alloy LCB in the single-phase bcc condition. Using the anisotropy factor of $A = 0.7$, maximum stress values were obtained at those grain boundaries that experienced actual crack initiation.

Of course, the susceptibility to crack initiation does not only depend solely on the absolute value of the stresses obtained by the FEM approach; the direction of the maximum shear stresses with respect to the slip planes of the grains involved is of even greater significance. Taking this into account requires extension of the FE microstructure model by crystal plasticity. Such a procedure has been reported by Anagnostou et al. [248], and is also the subject of ongoing work on an austenitic–ferritic duplex steel by Köster [255]. For several examples he could predict the appearance of slip bands as a result of elastic–plastic anisotropy.

Experimental support for the hypothesis that elastic anisotropy is the key factor for the occurrence of high shear stresses on a material's slip planes at low remote stress amplitude was obtained by the experimental evaluation of active slip systems and their Schmid factors with respect to the applied stress axis. These Schmid factors were compared with the maximum Schmid factors for all slip systems of the respective grains [256].

The active slip systems were determined by means of the directions of the visible slip traces at the specimen surface in combination with the rotation matrix $\underline{\underline{M}}$ (Eq.

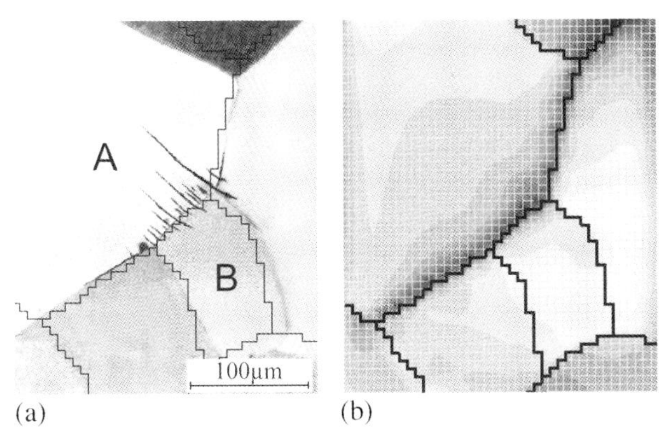

Fig. 5.11 (a) Scanning electron micrograph of transgranular slip-band cracks, initiated during cyclic deformation of the β-Ti alloy LCB at a stress amplitude of $\Delta\sigma/2 = 600$ MPa. (b) Corresponding principal stress distribution calculated by the FE method (qualitative; dark = high principal stresses).

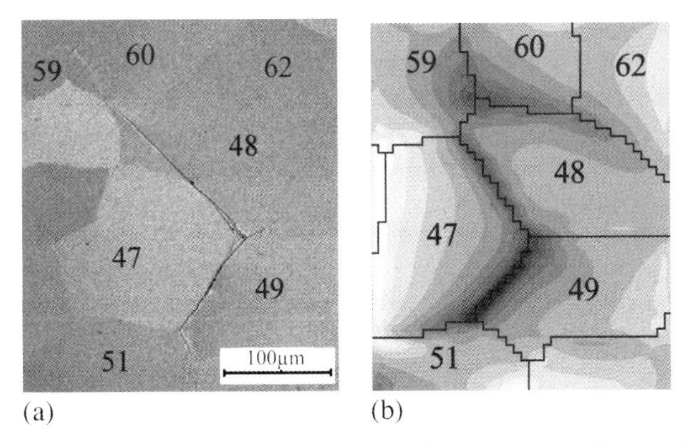

Fig. 5.12 (a) Scanning electron micrograph of an intercrystalline crack, initiated during cyclic deformation of the β-Ti alloy LCB at a stress amplitude of $\Delta\sigma/2 = 600$ MPa and (b) corresponding principal stress distribution calculated by the FE method (qualitative; dark = high principal stresses).

3.9), which can be obtained by EBSD measurements (see Section 3.3.3) and which relates the <001> crystal coordinate system with the specimen coordinate system [227]. For this purpose the ratio between the two components of the unit vector $\vec{l}_e$ of the slip trace direction

$$\lambda_{\exp} = \frac{l_y}{l_x} \tag{5.11}$$

was compared with those λ_{th} that were calculated on the basis of all 48 possible slip systems on {110}, {112} and {123} planes of the bcc lattice for the measured crystallographic orientation of the respective grain. The relationship between the normal vectors of the slip planes in the <001> crystal coordinate system $\vec{n}_{cryst}$ and the specimen coordinate system $\vec{n}_{spec} = (a_1, a_2, a_3)$ is given as

$$\vec{n}_{cryst} = \underline{\underline{M}}\vec{n}_{spec} \tag{5.12}$$

With the components of the respective slip traces $l_x^{th} = -a_2/\sqrt{a_1^2 + a_2^2}$, $l_y^{th} = -a_1/\sqrt{a_1^2 + a_2^2}$ and $l_z^{th} = 0$ the λ ratios according to Eq. (5.11) for the various slip systems can be calculated:

$$\lambda_{th} = \frac{l_y^{th}}{l_x^{th}} = -\frac{a_1}{a_2} \tag{5.13}$$

The actually activated slip systems are to be identified by the condition that both λ values are the same, i.e. $\lambda_{th} = \lambda_{exp}$. Calculating the corresponding Schmid factors M_S (see Section 4.3), related to the applied loading direction, does not result in the maximum possible ones [256], e.g., the Schmid factor for the identified {123}<111> slip system activated in the center grain of the cyclically loaded LCB specimen (Fig. 5.13) has a value of $M_S = 0.12$, while according to the maximum possible Schmid factor $M_S = 0.494$ in the respective grain the operation of the {110}<111> would be expected. Hence, the slip traces nearly parallel to the applied loading axis can be attributed to elastic anisotropy giving rise to local stress peaks, the direction of which is completely different from those of the remote applied stress.

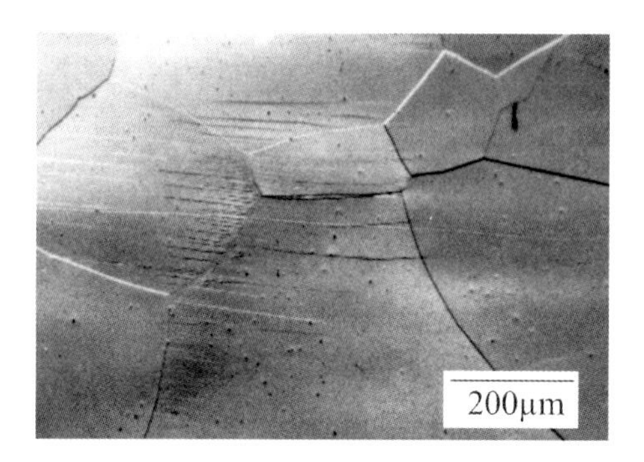

Fig. 5.13 Occurrence of slip bands being parallel to the remote stress axis (⇔) in the β-Ti alloy LCB during cyclic deformation at $\Delta\sigma/2 = 600$ MPa.

5.3.4
Analytical Calculation of Elastic Anisotropy Stresses

The significance of the elastic anisotropy during fatigue crack initiation in polycrystalline metals and alloys has been subject of a number of earlier studies. In some of them the resulting stresses were calculated using an analytical approach for simplified polycrystalline microstructures, e.g., twin boundaries in copper [250, 251] and austenitic steel [91, 257], as well as for bicrystalline copper specimens [249, 252], compare also [253].

In agreement with earlier work, e.g., Boettner et al. [258], Neumann and Tönessen [250] found predominant crack initiation in fcc materials at twin boun-

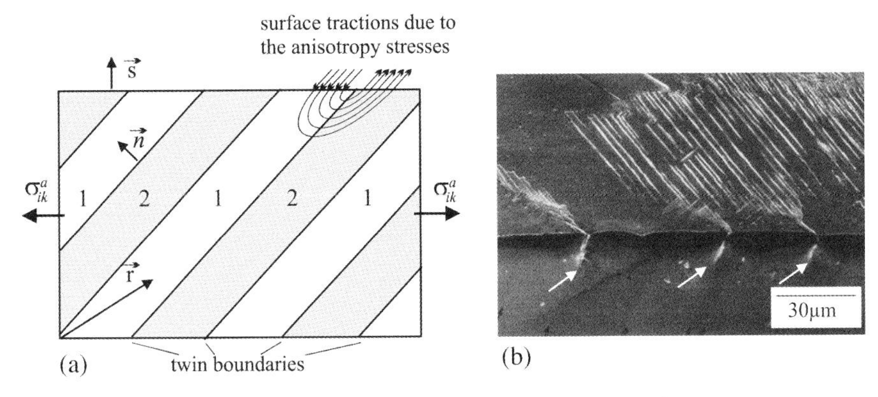

Fig. 5.14 (a) Simplified two-dimensional microstructure arrangement for derivation of an analytical solution for the mechanical stresses due to elastic anisotropy in a stack of twins. (b) Corresponding experimental observation of crack initiation at every second twin boundary (marked by arrows) in fatigued copper (after [250]).

daries. Since the structure of twin boundaries exhibits a low interface energy, one would expect a high resistance to crack initiation. However, it was shown for the case of twin-boundary stacks that each other grain boundary experienced strong slip band formation and crack initiation (see Fig. 5.14b). Again, elastic anisotropy seems to be responsible for an inhomogeneous stress distribution leading to stress peaks at the respective twin boundaries. Recently, Blochwitz and Tirschler [257] showed that, indeed, twin boundaries are the preferential crack initiation sites in fcc austenitic steel. However, once a crack is formed, further crack propagation in the bulk follows along slip bands or combinations of slip bands, according to the maximum resolved shear stress. Hence, it is only the crack initiation process that is governed by elastic anisotropy of the microstructure constituents; crack propagation, however, depends on the plastic anisotropy, i.e., the three-dimensional arrangement of the slip systems.

The local stress increase due to elastic anisotropy of a simplified microstructure consisting of a parallel arrangement of twinned grains was analytically calculated by Neumann and Tönnessen [250] in the following way. A stack of grains, separated by twin boundaries as shown in Fig. 5.14a, that is loaded by an uniaxial stress σ_{ik}^a should exhibit inhomogeneous deformation, i.e., resulting in incompatible displacements along the twin boundaries. To accommodate for this incompatibility, elastic anisotropy strains ε_i are required, explained as follows.

The local stress σ_{ik} at an arbitrary point $\vec{r}$ within the microstructure area in Fig. 5.14a can be written as

$$\sigma_{ik}(\vec{r}) = \sigma_{ik}^a + s\sigma_{ik}^b + \sigma_{ik}^c(\vec{r}) \tag{5.14}$$

The homogeneous stress σ_{ik}^a is superimposed by the anisotropy stress σ_{ik}^b, which has to be multiplied by $s = 1$ for the grains marked by 1 in the stack of twins in Fig.

5.14a, and by $s = -1$ for the grains marked by 2. The additional stress component is required to accommodate for the surface tractions, resulting from the anisotropy stresses $s\sigma^b_{ik}$ at the upper and lower boundaries of the microstructure area in Fig. 5.14a.

The elastic anisotropy strains $\varepsilon_i = n_k(\varepsilon^1_{ik} - \varepsilon^2_{ik})$ perpendicular to the twin boundaries with the normal vector $\vec{n}$ can be obtained from the material's law, taking the stiffness tensors of the grains 1 (S^1_{iklm}) and the grains 2 (S^2_{iklm}) into account. Due to the elastic anisotropy, the components of these stiffness tensors are of different values. For compatible deformation along the grain boundary planes, the anisotropy strains ε_i needs to be accommodated by the anisotropy stresses $s\sigma^b_{ik}$, and the material law can be written for the stack of twins in Fig, 5.14a as follows:

$$\varepsilon_i = \vec{n}\left(S^1_{iklm} - S^2_{iklm}\right)\left(\sigma^a_{lm} + s\sigma^b_{lm}\right) = 0 \qquad (5.15)$$

Since s changes sign at each twin boundary, fulfillment of the equilibrium conditions needs the stresses acting normal to the twin boundaries to vanish:

$$\vec{n}\sigma^b_{ik} = 0 \qquad (5.16)$$

The six equations, given by Eqs. (5.15) and (5.16), are sufficient to solve for the symmetric anisotropy stress tensor σ^b_{ik}. However, $s\sigma^b_{ik}$ yields also the occurrence of stress components (tractions) t_i, acting normal to the boundary of the microstructure segment in Fig. 5.14a (surface normal vector $\vec{s}$):

$$t_i = s\sigma^b_{lm}\vec{s} \qquad (5.17)$$

Since the surface must be free of any normal stresses, an additional stress component σ^c_{ik} according to Eq. (5.14) is required. Theoretically, the periodic change of the sign $s = +1/-1$ yields logarithmic stress singularities at these points, where the twin boundaries meet the surface. Practically, these singularities are accommodated by local plastic deformation. The height of the stress increase at the twin boundaries can be obtained by approximation for the stresses σ^c_{ik}, which are due to the change in sign only of significance up to a depth in the magnitude of the grain thickness d (according to St. Venants principle [259]). Hence, these stresses σ^c_{ik}, being pure shear stresses parallel to the twin boundaries, are given as [250]

$$\sigma^c_{ik}(\vec{r}) = w\beta(\vec{r})\left(n_i\sigma^b_{kl}s_l + n_k\sigma^b_{il}s_l\right) \qquad (5.18)$$

with n_i and n_k being the components of the grain boundary's normal vector $\vec{n}$, s_l the components of the surface normal vector $\vec{s}$ and w a parameter changing from $w = 1$ to $w = -1$ at each other grain boundary ($w = 1$, if $\vec{n}$ is directed from a grain 1 towards a grain 2, and vice versa). The function $\beta(\vec{r})$ is independent of the microstructural parameters and is a measure of the logarithmic singularity at the intersection between grain boundaries and the surface for the case of overall elastic behavior. Here, $\beta(\vec{r})$ defines the way in which the resolved shear stress τ on the slip planes in the vicinity of the twin boundaries (slip direction vector: $\vec{b}$ with the components b_i) is influenced by elastic anisotropy:

$$\tau = b_i\sigma^a_{ik}n_k + w\beta(\vec{r})b_i\sigma^b_{il}s_l \qquad (5.19)$$

Neumann und Tönessen [250] assumed a value of $\beta = 1.5$ for copper. By means of Eqs. (5.15) to (5.19) it is generally possible to calculate the stress distribution in an elastically anisotropic microstructure in an analytical way. Equation (5.19) illustrates that due to the changing sign of w the anisotropy stress σ_{ik}^{b} leads only for every second twin boundary to an increase of the resolved shear stress τ. This is in agreement with experimental results reported in [250], where local plasticity and crack initiation were observed for each other twin boundary of a stacked arrangement of twins.

It should be mentioned that the analytical solution, as introduced above, has been applied only to simple microstructures, such as periodically arranged twins or bicrystals; for a random polycrystalline microstructure the complexity of the calculation would increase exponentially, and hence the finite-element method is probably the most reasonable alternative (cf. Section 5.3.3).

5.4 Intercrystalline and Transcrystalline Crack Initiation

5.4.1 Influence Parameters for Intercrystalline Crack Initiation

As mentioned in the section above, mechanical loading of elastically anisotropic microstructures may experience high anisotropy stresses in the vicinity of grain or phase boundaries, even for low remote stress amplitudes. Besides the fundamental studies of bicrycstalline [249, 252] and polycrystalline copper [250, 251], this was shown also for polycrystalline nickel and austenitic steel by Blochwitz et al. [260] and for commercial 316L austenitic steel by Blochwitz and Tirschler [257]. Beside the anisotropy effect, Blochwitz et al. [260] found that microcracks tend to initiate at high-angle random grain boundaries (according to the CSL scheme, cf. Section 3.3.5). They identified the plastic incompatibility between the neighboring grains as a major factor determining crack initiation, and quantified this influence factor by introducing a misorientation crack factor M. This factor includes the differences in the rotation vectors $\vec{R}$ of the primary slip systems of the neighboring grains:

$$\Delta\vec{R} = \vec{R}_2 - \vec{R}_1 = \left(\vec{n} \times \vec{s}\right)_{\text{grain 2}} - \left(\vec{n} \times \vec{s}\right)_{\text{grain 1}} \tag{5.20}$$

with $\vec{n}$ being the normal vector, $\vec{s}$ the direction vector of the respective primary slip systems and β the angle between the respective grain boundary and the loading direction (see Fig. 5.15). For the latter, the relationship $\sin^2 \sigma_n/\sigma_a$ is valid, where σ_n is the component of the applied stress σ_a acting normal to the grain boundary plane. In combination with the unit vector of the grain boundary trace on the specimen surface $\vec{e}_{GB}$ the misorientation crack factor can be expressed as follows:

$$M = \sin^2 \beta \Delta\vec{R} \cdot \vec{e}_{GB} \tag{5.21}$$

Analysis of the experimental results in [260] yields an average value of $M = 0.0038$ for all grain boundaries, but $M = 0.4133$ for those grain boundaries showing cracks.

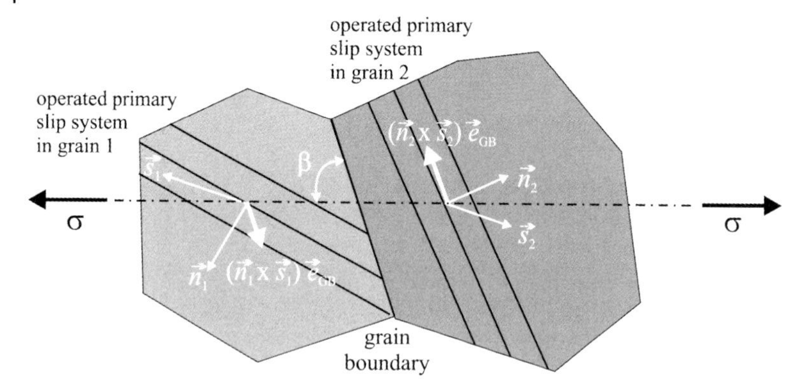

Fig. 5.15 Geometrical relationship to apply the misorientation factor scheme according to [260] (for details see text).

These results support the hypothesis that crack initiation at grain boundaries is determined by both the elastic and plastic incompatibility of adjacent grains. The influence of plastic slip incompatibility (see Kim and Laird [223, 225] and Peralta and Laird [261]) is predominant during crack initiation for high strain amplitudes, i.e., low-cycle fatigue (LCF) conditions [199] and high misorientation angles (cf. Zhang and Wang [262, 263] and Zhang et al. [264]). Elastic anisotropy becomes most significant during crack initiation at low and very low strain amplitudes [265], i.e., HCF and VHCF.

Studies of the behavior of short fatigue cracks in the β-titanium alloy LCB [227, 254] revealed that crack initiation occurs almost exclusively at the alloy's grain boundaries. Due to the elaborate vacuum metallurgy for processing the titanium, the influence of nonmetallic inclusions during fatigue damage is negligible for titanium alloys. The fraction of intercrystalline cracks x_{ic}, where crack initiation occurred by grain-boundary separation, remained approximately constant. At an applied stress amplitude of $\Delta\sigma/2 = 600$ MPa and $R = -1$ cracking started with $x_{ic} = 15\%$ at 10% of the fatigue life and increased slightly to $x_{ic} = 18\%$ at 60% of the fatigue life [254]. Evaluation of the misorientation relationship of all the grains in the shallow-notched gauge length (see Section 3.1.2) revealed that those grain boundaries that cracked in an intercrystalline manner are high-angle grain boundaries ($\Theta > 15°$) without exception. Some 65% of these grain boundaries are of a random misorientation relationship, i.e., they do not belong to the group of special CSL boundaries with Σ1 to Σ49 misorientations (cf. Section 3.3.5). However, it must be added that the fraction of random high-angle grain boundaries of all the evaluated boundaries is 75%, and hence one cannot draw the conclusion that random high-angle grain boundaries are necessarily preferential crack-initiation sites. The results of the evaluation are summarized in Table 5.3. Obviously, compared with the compatibility of slip systems of neighboring grains the structure of the grain boundaries is only of secondary significance for intercrystalline crack initiation. An exception are the Σ1 low-angle grain boundaries, but these boundaries

Table 5.3 Evaluation of intergranular initiated fatigue cracks in the β-titanium alloy LCB after 60% of the total fatigue life at $\Delta\sigma/2 = 600$ MPa and $R = -1$.

	Fraction of a total of 596 studied grain boundaries (in %)	Fraction of a total of 41 intergranular cracked grain boundaries (in %)
Small-angle grain boundaries	15	0
High-angle grain boundaries	85	100
Special CSL grain boundaries (Σ1–49)	26	35
Σ3 grain boundaries	9	20
Random high-angle grain boundaries	74	65

have generally (per definition) small angles of $\Theta < 15°$ between the adjacent slip planes.

It can be concluded that the criterion for grain boundary separation depends on the spatial position of the slip systems of the respective neighbored grains. According to Zhai et al. [266, 267] and Argon and Qiao [268], the twist misorientation between the adjacent slip systems in the grain boundary plane (see Fig. 5.16) is of major significance for the resistance of a grain boundary to intercrystalline separation. Slip transmission across the boundary can be considered as a process stepwise involving slip alternation from one slip plane to another; the larger the steps the higher the plastic incompatibility and the extrinsic resistance of the boundary. For such a situation the resistance of a grain boundary to intercrystalline crack initiation depends only on the intrinsic strength of the boundary.

Once a grain of a polycrystalline metal or alloy is locally plastified, e.g., due to elastic anisotropy, and slip transfer into the neighboring grains is hindered by plastic incompatibility, then the occurrence of intercrystalline failure depends on the stress acting on the grain boundary due to the pileup of dislocations and the intrin-

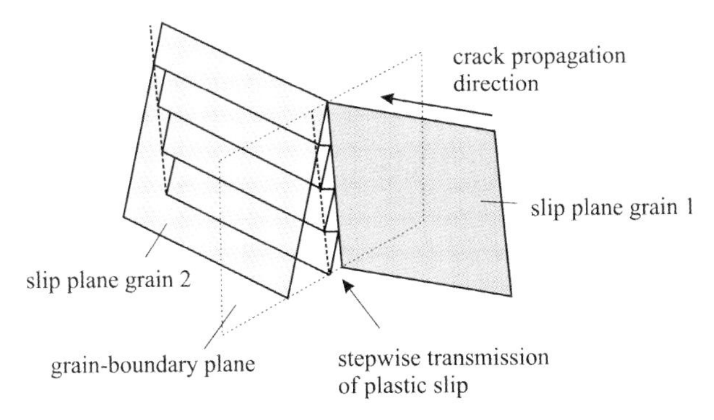

Fig. 5.16 Significance of the twist misorientation for the resistance of a grain boundary to intercrystalline separation.

sic strength (cohesion) of the respective grain boundary itself. According to Section 4.3, the stress at the boundary due to pileup $\tau_{\text{pile-up}}$ can be written as

$$\tau_{\text{pile-up}} = n\tau_a, \quad n = AL \tag{5.22}$$

Here, n is the number of dislocations along the slip band, A a constant factor depending on the type of dislocation, τ_a the resolved shear stress at the slip band and L the length of the slip band, i.e., the pileup length, that can be set equal to the alloy grain size. Hence, it is obvious that the pileup stress $\tau_{\text{pile-up}}$ must exceed a certain critical value to cause intercrystalline crack initiation. According to Gysler et al. [269], this is the case when the grain size is larger than approximately 10 μm.

However, coherent precipitates, which can be cut by dislocations, promote slip planarity and can increase the pileup stress $\tau_{\text{pile-up}}$ substantially. This is also an active mechanism in the β-titanium alloy LCB, discussed above, governed by the coherent ω phase [269, 270].

In summary, intercrystalline crack initiation at room temperature, not taking grain boundary precipitates or segregation effects into account, can be attributed to two mutual superimposing mechanisms:

1. Elastic anisotropy, locally causing high normal stresses acting on the grain boundaries, that may exceed the interfacial cohesion. Figure 5.17 shows an example for an intercrystalline crack in the β-titanium alloy LCB without pronounced plastic deformation within the adjacent grains. It should be noted here that the grain boundaries themselves, exhibiting a high dislocation density, can be considered as slip planes in a similar way, as is the case during grain boundary sliding at elevated temperatures, a mechanism that can lead to intercrystalline crack initiation without clearly visible plasticity. This is the case when the grain boundary plane itself can be considered as a slip plane. This was shown experimentally for martensitic steel by *in situ* indentation in a transmission electron microscope by Ohmura et al. [128].

2. Plastic incompatibility, giving rise to dislocation pairing along the active slip band and/or dislocation pileup at the grain boundaries under planar-slip conditions, eventually leading to the formation of slip steps. This form of intercrystalline crack initiation can be seen in Fig. 5.18, showing the polished surface (Fig. 5.18a) and the fracture surface (Fig. 5.18b) of a fatigued specimen of the β-titanium alloy LCB.

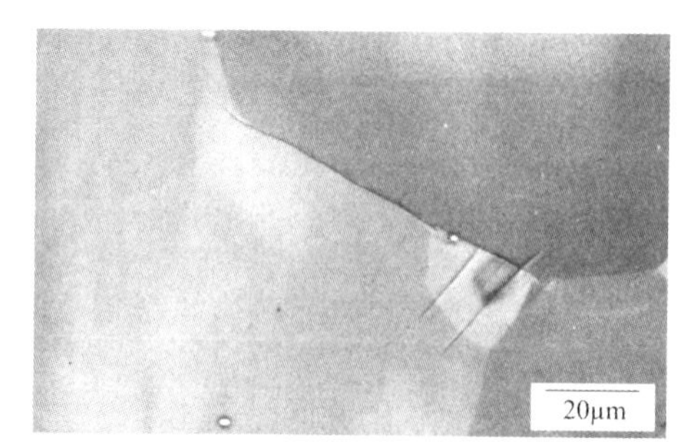

Fig. 5.17 Intercrystalline crack initiation without pronounced plastic deformation in the β-titanium alloy LCB during fatigue at a stress amplitude of $\Delta\sigma/2 = 600$ MPa ($R = -1$, loading direction: ⇔).

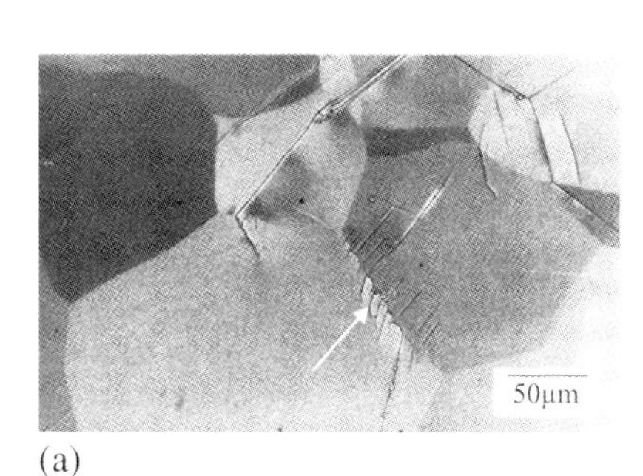

(a)

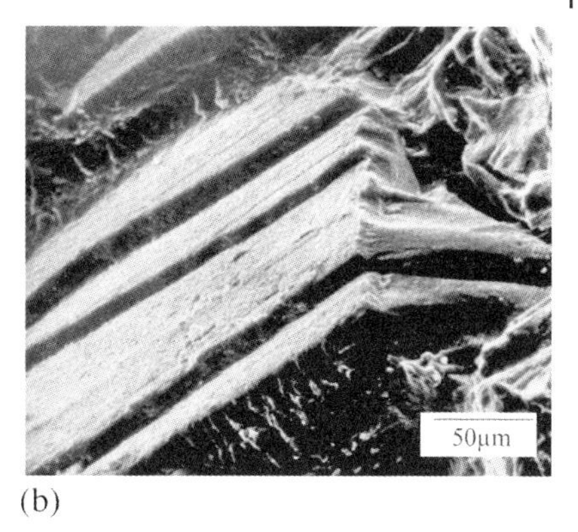

(b)

Fig. 5.18 (a) Intercrystalline crack initiation governed by slip-step formation (marked by arrow) in the β-titanium alloy LCB during fatigue at a stress amplitude of $\Delta\sigma/2 = 600$ MPa ($R = -1$, loading direction: ⇔). (b) Slip steps in the fracture surface of a separated grain boundary (LCB, $\Delta\sigma/2 = 600$ MPa).

5.4.2
Crack Initiation at Elevated Temperature and Environmental Effects

While at room temperature crack initiation is mainly governed by the generation and motion of dislocations leading to microstructural changes, at elevated temperatures diffusion and corrosion processes are often the key factors. As an example, the fatigue fracture mode of polycrystalline nickel-based superalloys changes from ductile trangranular to quasi-brittle intergranular at temperatures above approximately 550 °C. While this form of environmentally assisted intercrystalline crack propagation is the subject of Section 6.5, the present section focuses on the mechanisms leading to intercrystalline crack initiation at elevated temperatures.

Generally, one must distinguish between creep damage and chemical attack, both processes being thermally activated, and hence promoted by higher temperatures. At very high temperatures (above approximately 800 °C) high-temperature corrosion products frequently act as crack nuclei. For instance, internal oxidation, carburization or nitridation start preferentially along the alloy's grain boundaries due to the eased nucleation and/or fast interfacial transport of the reactants oxygen, carbon or nitrogen, respectively. This causes a substantial deterioration of the mechanical properties of the alloy's surface layer by weakening the grain boundaries by brittle corrosion products (see e.g. [271]). Figure 5.19 shows, as an example, an intercrystalline crack path through an internally nitrided (TiN and AlN precipitates) surface layer in the nickel-based superalloy alloy 80A [272]. Also during formation of protective oxide scales, Cr_2O_3 and Al_2O_3, the grain boundaries play a significant role in the transport processes of the corroding species and nucleation processes (cf. [273]), often resulting in local enrichment of corrosion products, that may act as crack initiation sites when the respective component/specimen is mechanically loaded.

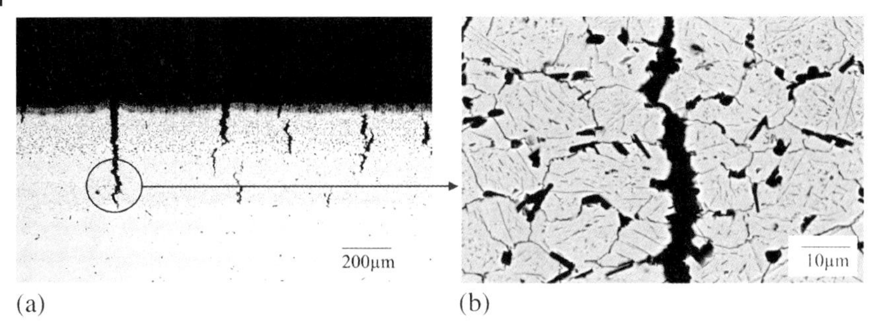

Fig. 5.19 Occurrence of intercrystalline cracks along grain boundaries decorated by internal TiN (small needle-like) and AlN precipitates (large blocky) in the nickel-based superalloy alloy 80A, nitrided in a He–N_2–H_2 atmosphere at 100 °C after tensile testing. Micrographs were taken from the gauge length (after [272]).

Also in the case of single-crystalline superalloys that are used for precision-cast turbine blades of the first stages of aeroengines or land-based gas turbines, a thermally grown oxide (TGO) scale (for very high-temperature applications Al_2O_3-forming alloys are required) or an applied protective coating system, like a thermal barrier coating (TBC) in combination with a Ni(Co)CrAlY bond coat, may cause crack initiation by transferring brittle cracking of the scale into the substrate. This was shown, for example, by studies of Bressers et al. [109] for the single-crystalline nickel-based superalloy SRR99. In general, the fatigue life of components with a brittle scale on top is shorter than that of the uncoated material. However, coatings on high-temperature materials increase the possible service temperature and the service life that is determined by degradation by high-temperature corrosion attack rather than intrinsic fatigue damage.

Besides high-temperature corrosion effects, creep is the second key factor determining the service life of high-temperature materials. During creep, intercrystalline crack initiation is determined by the combination of grain boundary sliding and vacancy diffusion towards areas of high hydrostatic tension [80, 175]. This leads to the formation of pores along the grain boundaries and grain boundary triple lines exhibiting local stress maxima, the latter being caused by elastic and plastic anisotropy and grain-boundary sliding (see [228]).

A pore of radius r is able to grow when the total increase in energy ΔE

$$\Delta E = A_P \gamma_P - \sigma_T V_P - A_{GB} \gamma_{GB} \tag{5.23}$$

composed of the work put into the pore volume V_P by the tensile stress σ_T, the surface energy γ_{GB} being released during vanishing of the grain boundary section A_{GB}, and the energy γ_P of the new formed pore surface A_P, exceeds its maximum ($\partial \Delta E / \partial r = 0$). Coalescence of several pores leads eventually to crack initiation.

As already mentioned, grain-boundary sliding is, in addition to pore formation, an important creep damage mechanism. Grain-boundary sliding depends strongly on the grain boundary geometry with respect to the direction of the local stress σ, which can considerably deviate from the remote stress due to elastic anisotropy. According to [175], the strain rate due to grain-boundary sliding can be written as

$$\dot{\varepsilon}_{GB} = \text{const.}\,\frac{\delta\sigma^{n}}{\eta_{GB}d} \tag{5.24}$$

with δ being the grain boundary width, in the literature often estimated to be of the order of 0.5 nm (cf. [274]), n the stress exponent, d the mean grain size and η_{GB} the grain boundary viscosity, the latter being a measure for the resistance of a grain boundary to sliding. The grain boundary viscosity depends on the degree of segregation by impurities, the presence of nonmetallic inclusions, and especially on the diffusion coefficient of grain-boundary short-circuit diffusion D_{GB}. The latter is determined by the structure of the grain boundary [275], i.e., low-angle and special CSL grain boundaries of $\Sigma \leq 29$ having a relatively low defect concentration exhibit a particular low diffusivity D_{GB} (cf. Section 6.5.4). A low grain-boundary diffusion coefficient D_{GB} increases the grain boundary viscosity, and hence leads to a lower grain-boundary-sliding rate. For cyclic deformation, quantification of grain-boundary sliding as a mechanism contributing to fatigue crack initiation is difficult, since grain-boundary sliding can be partly reversible (cf. [276]).

Figure 5.20 summarizes schematically the main factors influencing crack initiation at elevated temperatures.

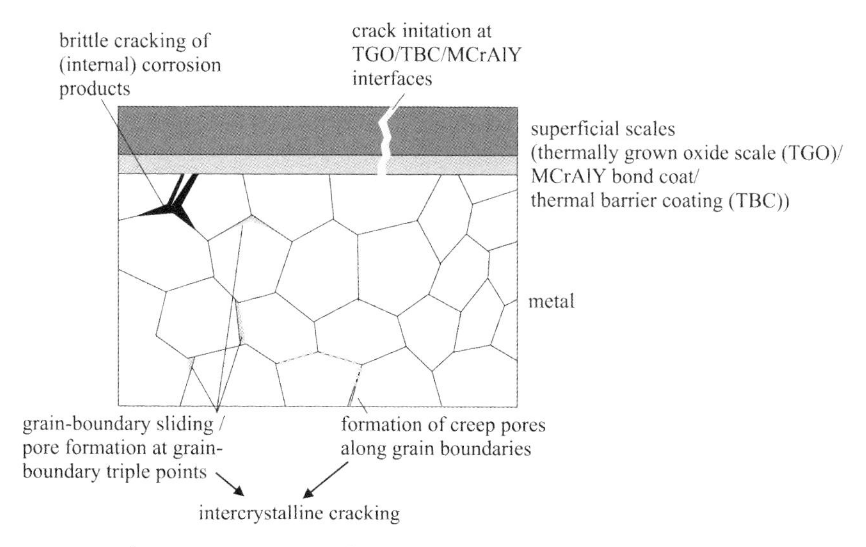

Fig. 5.20 Schematic representation of the factors influencing crack initiation at elevated temperatures.

5.4.3
Transgranular Crack Initiation

As discussed in the preceding sections, crack initiation along slip bands depends on the barrier strength, the crystallographic misorientation and the structure of grain boundaries. Once the local resolved shear stress, affected by the applied stress amplitude and elastic anisotropy, exceeds the critical shear stress (cf. Section 4.2), dislocation motion and slip band formation set in. Certainly, this process depends on the local geometrical arrangement of the slip systems. Summarizing this in a simplified statement, all slip planes that are inclined by approximately 45° with respect to the loading axis are particularly susceptible to plasticity and slip-band formation (see [277, 278]). If the resistance of the grain boundary to (1) slip transmission (plastic incompatibility) and (2) interfacial separation is sufficiently high, then the grain boundary will increase the back stress encouraging dislocation pairing and/or dislocations pileup, eventually causing slip-band separation, i.e., transgranular crack initiation. Figures 5.11 and 5.21 show examples of transgranular crack initiation at the surface of fatigued specimens of the β-titanium alloy LCB (Figs. 5.9 and 5.21a) and an austenitic–ferritic duplex steel (Fig. 5.20b).

Of particular significance for transgranular crack initiation are triple points (at the alloy's surface). Often, triple points are sites of highest stresses. These stresses may be further increased by elastic anisotropy and intercrystalline damage. Each separated grain boundary terminates at a triple line causing a substantial increase in the stress intensity.

Besides the grain boundaries, intersection points of operated slip bands exhibiting local maxima of plastic deformation may act as crack nucleation sites [279]. This was observed by several authors studying the deformation-induced martensite formation in metastable austenitic steels [280, 281]. They found the formation first of nuclei of ε martensite, the hexagonal transition phase in the fcc γ austenite

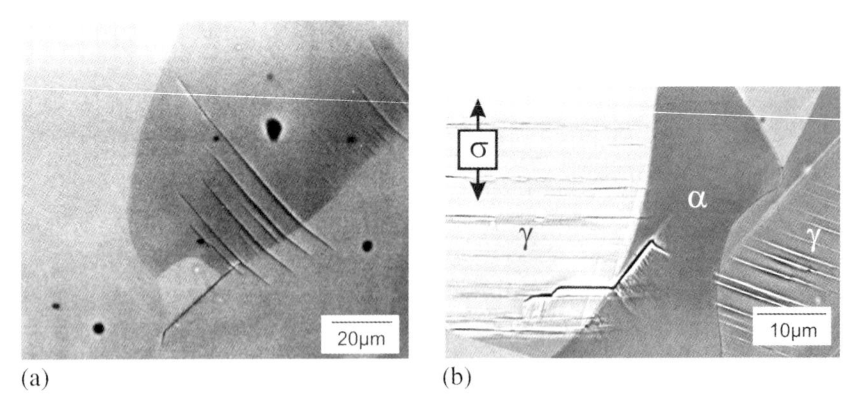

Fig. 5.21 Transgranular slip band cracks initiated at a grain boundary (a) in the β-titanium alloy LCB ($\Delta\sigma/2 = 600$ MPa, $R = -1$, loading direction: ⇔) and (b) in austenitic ferritic AISI F51 duplex steel ($\Delta\sigma/2 = 400$ MPa, $R = -1$).

to bct α′ martensite, preferentially taking place at slip-band intersection points. A very similar behavior was also observed by Long et al. [282] for crack initiation in β-titanium alloys in the single-phase bcc condition, contrary to the results obtained for the β-titanium alloy LCB, where cracks were found to initiate exclusively at grain boundaries.

5.5 Microstructurally Short Cracks and the Fatigue Limit

The mechanisms of fatigue-crack initiation are of particular significance for the discussion of the fatigue limit of metals and alloys. Accordingly, the fatigue limit can be defined as the load level below which cracks are not able to grow. From a technical point of view, this means that a component or a specimen must survive at least $N = 2\times10^6$ cycles without failure. Strictly speaking, the assumption of this value is limited for the fatigue limit definition of bcc carbon steel. Most fcc materials do not exhibit such a kind of fatigue limit, plasticity cannot be completely suppressed, so that even for very high numbers of cycles slip accumulation causes crack initiation. Therefore, a technical fatigue limit for Al alloys and austenitic steels is used for a fatigue life of $N = 10^7$ to 10^9 cycles, during which no failure occurs. Probably, crack initiation is never fully suppressed, but all the cracks are stopped at a microstructural barrier [283], e.g., a grain or a phase boundary. The variation in the barrier strength can be considered as one reason for the scatter as well as for the overload sensitivity of the fatigue limit. According to [284], cracks can grow to a length of 10 to 15 grain diameters within the fatigue limit.

Many technically important components, which experience high-frequency loading or which require a very long service life, need to withstand $N = 10^9$ and more cycles. Relatively new research in the field of UHCF or VHCF during the last 15 years (see [212]) has shown that under these conditions the assumption of a fatigue limit can be misleading, i.e., the respective total-life design can be nonconservative. For very high numbers of cycles a continuous decrease in the capable stress amplitude was observed until at $N > 10^{10}$ cycles an absolute value of the fatigue limit is reached (irreversibility limit, cf. [285, 286], see Fig. 5.22), which depends on the size of the nonmetallic inclusion and the conditions for plastic-slip reversibility.

The conditions where the conventional assumption of the fatigue limit is not applicable were summarized and critically discussed by Miller and Donnell [287, 288]. At extremely high numbers of cycles, the crack-initiation mechanisms change. Due to the very small extent of plasticity the significance of plane stress condition at the surface becomes negligible and small material inhomogeneities, such as large grains, pores and nonmetallic inclusions, in the bulk increasingly govern the fatigue-crack-initiation process [289]. Also the barrier effect of grain and phase boundaries is only of limited significance under such conditions. High-frequency vibrations, time-dependent plastic deformation (creep) or time-dependent corrosion attack may form the bridge from a nonpropagating crack (local fatigue limit) to resuming crack advance [290]. Such a behavior is represented schematical-

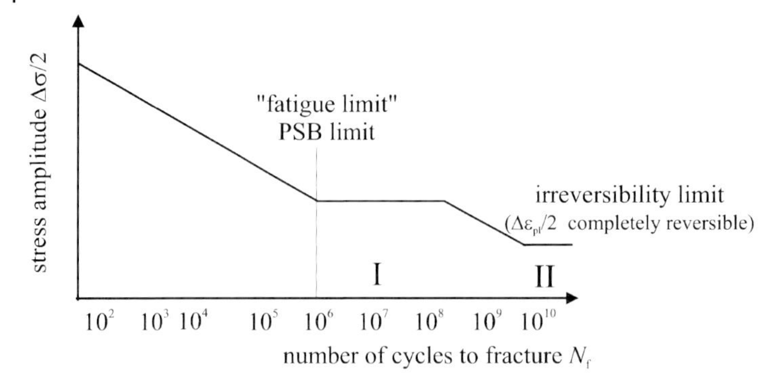

Fig. 5.22 Two-stage S/N diagram (Wöhler diagram) covering also the VHCF regime.

ly in Fig. 5.23, showing a crack that is blocked by a microstructural barrier, i.e., the crack-propagation rate approaches $da/dN = 0$ at a crack length of a_{FL}. Under normal conditions the material would have reached the fatigue limit. By the time-dependent mechanisms mentioned above, however, the crack can grow to a critical crack length of a_{crit}, which is required to overcome the barrier. Then, fatigue-crack propagation may resume, and, consequently, the fatigue limit disappears. In the Wöhler diagram (S/N curve) this would correspond to a continuously decreasing stress value.

This modified analysis of the fatigue limit manifests the technical importance of a detailed understanding of the fundamental mechanisms leading to technical crack initiation. Together with the discussion on crack propagation mechanisms in Chapter 6, the main microstructural parameters are reviewed and at least qualitatively evaluated in order to provide a sound basis for mechanism-based modeling approaches, which are introduced in Chapter 7 and which are going to serve as important tools for future development and evaluation of fatigue-resistant materials and new methods of service-life prediction.

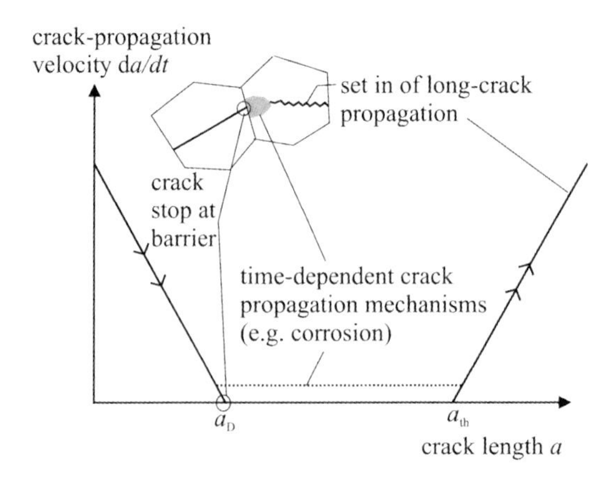

Fig. 5.23 Elimination of the fatigue limit by superimposition of time-dependent damage processes: schematic representation of the crack propagation rate da/dN vs. the crack length a (according to [287, 288]).

5.6 Crack Initiation in Inhomogeneous Materials: Cellular Metals

The previous remarks on crack initiation in this chapter were restricted to metals and alloys that can be considered as homogeneous, except for variation of crystallographic orientation and the presence of secondary phases. The latest technical developments require more and more materials that fulfill special functionalities or low weight restrictions. Hence, materials research moves considerably to "modern" material concepts, e.g., composite materials, smart materials or cellular materials. All these structures show special kinds of damage mechanisms, which often have to be discussed on the basis of different length scales. In consideration of a class of inhomogeneous materials, in the following section some special features of the fatigue-damage behavior of cellular metals are briefly discussed. These materials exhibit the lowest density of all metallic materials, and have a strong potential for lightweight or energy-absorbing applications, but also for functional structures, e.g., heat exchangers or catalytic converters. A good overview is given in [291]. Generally, one distinguishes between open-cell metal sponges and closed-cell metal foams. Open-cell metal sponges can be produced, for example, by precision casting using a polyurethane precursor ([292], Fig. 5.24a), or by vapor deposition [293], while the production of closed-cell foams is either based on (1) a powder metallurgical route using a foaming agent, e.g., titanium hydride (about 0.5% TiH_2), that creates, when dissociating, gas bubbles in the substrate material (mostly Al) that has to be heated up above the solidus temperature ([293], Fig. 5.24b), or on (2) a melting route, where gas is blown into the melt, forming bubbles that are stabilized by increasing the viscosity by means of SiC particles ([294], Fig. 5.24c).

When talking about cracks in such a cellular structure, there are generally two different length scales to be considered. Microcracks are generated within individual cell struts or cell walls that are subject to particularly high mechanical stress. On the other hand, crack initiation can be understood as gradually breaking cell struts as a whole. This separation is represented schematically in Fig. 5.25. From the point of view of continuum mechanics, modeling is difficult and there is no general idea as to how to handle structural integrity for large cellular structures. FE models quickly reach the limit of processing power when applied to real cellular structures [111]. Simple structure models cannot account for the cell strut geometry and microstructure in a straightforward way. As a promising alternative approach to model cellular structures, homogenization methods are going to be developed where a small representative volume element (RVE) is defined and analyzed with respect to its stress–strain behavior. The RVE is considered as the smallest repeat entity and set equal to a virtual homogeneous medium that exhibits the same strain energy and the same deformations [295]. Eventually, complex structures made of the cellular material can be handled by applying conventional FE techniques to the RVE.

It was shown that elastic deformation is considerably nonlinear and anisotropic. The derivation of elastic moduli is difficult, since even for small strains of the overall structure some cell struts may be deformed plastically. Plastic deformation of

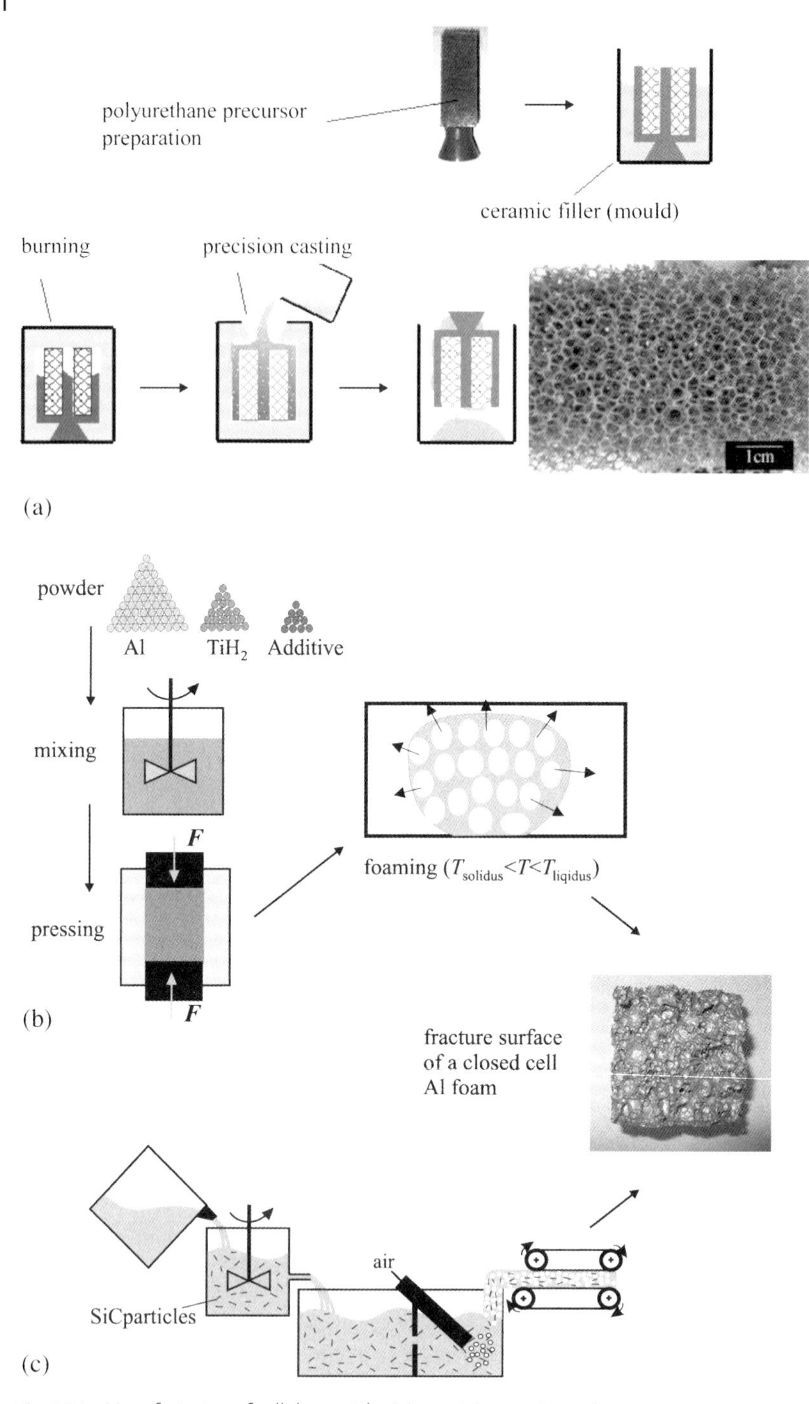

Fig. 5.24 Manufacturing of cellular metals: (a) precision casting using polyurethane precursor, (b) powder metallurgical route, (c) melting route.

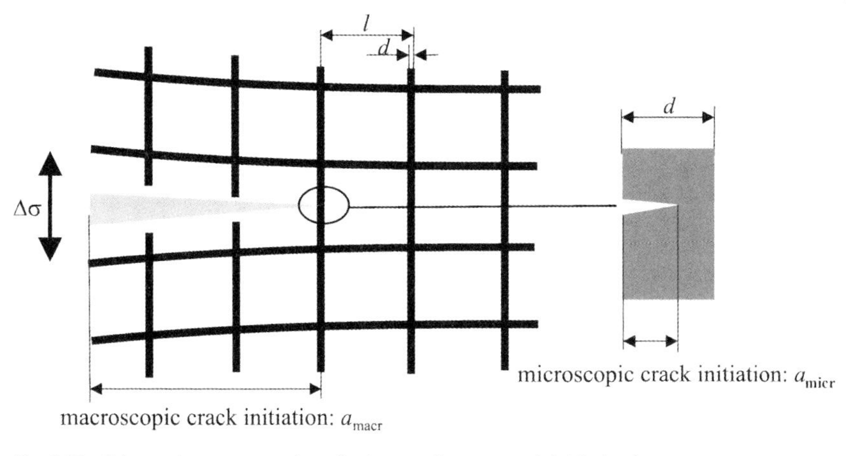

Fig. 5.25 Schematic representation of micro- and macrocrack initiation in cellular metals (*l*, cell size; *d*, cell-strut diameter).

cellular metals (in compression) sets in locally at deformation bands that collapse after exceeding a critical local strain. Plastic deformation resumes by activating further deformation bands. The behavior in tension is completely different. It is determined by alignment of the cell struts towards the applied load axis and crack propagation. Figure 5.26 shows schematically the stress–strain behavior of cellular metals under monotonic compression and tension.

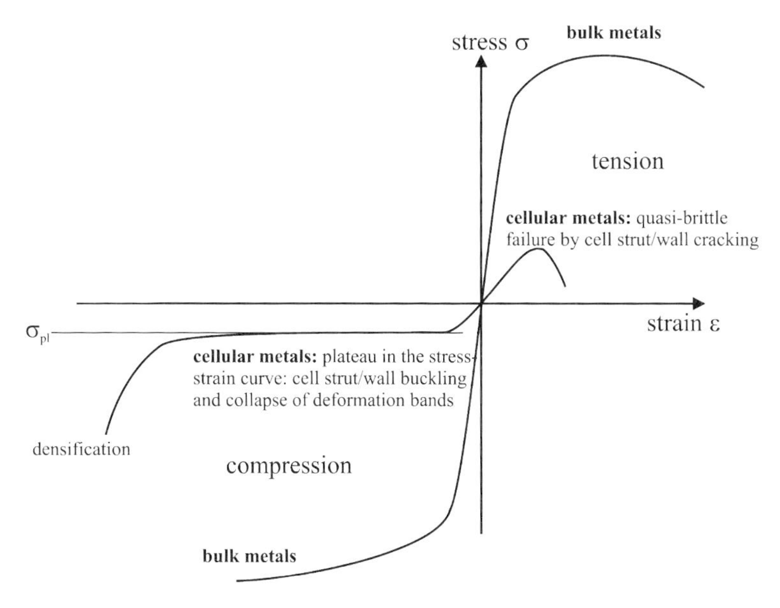

Fig. 5.26 Schematic representation of the monotonic tensile and compression behavior of cellular metals.

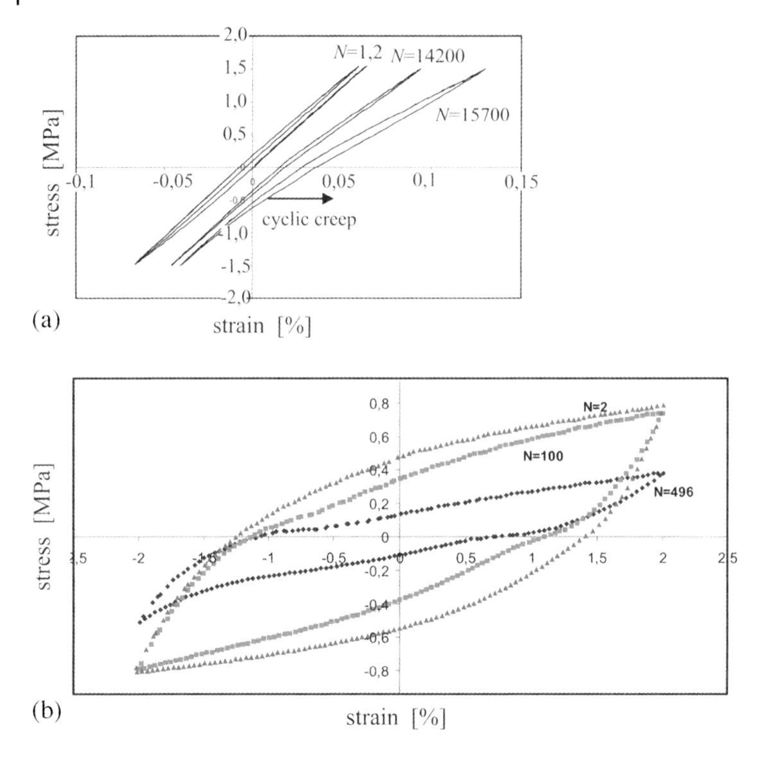

Fig. 5.27 Stress–strain hysteresis loops of (a) a stress-controlled fatigue test at room temperature for closed-cell Al foam [296] and (b) a strain-controlled fatigue test at $T = 200\,°C$ for open-cell brass sponge [297].

Evolution of fatigue damage is preferentially determined by the tensile parts of the load spectrum. Gradually cracking of cell struts leads to a decrease in the structure stiffness that can be observed as cyclic creep and decrease in the slope of the tensile branch of the stress–strain hysteresis loops, as shown for the example of cyclic loading of a closed-cell AlSi7Mg+15Vol.%SiC foam HAL 175/4/1 (Hydro) and an open-cell Cu-11%Zn sponge in Fig. 5.27 [296]. The kink in the slope of the hysteresis loop in the compression regime (see circle) can be attributed to a delayed contact of the broken cell struts due to strong plastic deformation [297, 298].

At fatigue crack initiation sites, grain boundaries and large precipitates could be identified. Figure 5.28a shows for the example of an open-cell Al sponge, grain sizes of the same order of magnitude as the cell strut size, hence making the grain boundary the weakest link prone to fracture [298]. In a similar way, the SiC precipitates in the closed-cell Hydro foam accumulated at the surfaces of the cell walls can be considered as crack initiation sites (Fig. 5.28b). Nevertheless, fatigue crack propagation is not only a sequence of brittle cell-strut-cracking events; the fatigue striations shown in Fig. 5.29 prove that similar crack propagation mechanisms are active as in the case of bulk materials [297].

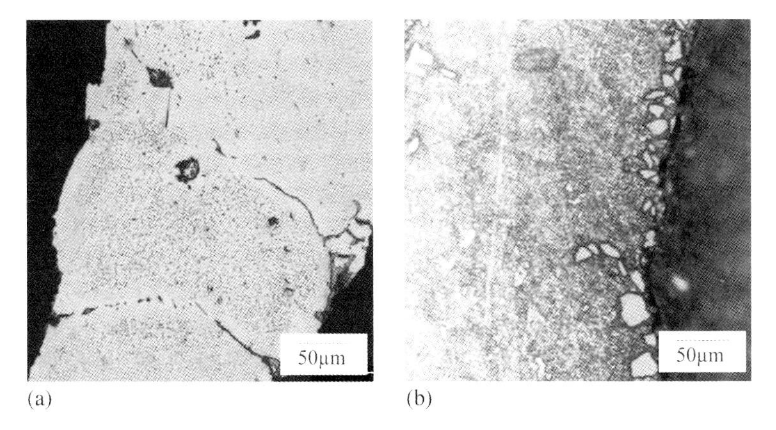

Fig. 5.28 Microstructure of (a) an open-cell Al sponge with Θ′ precipitates and grain diameter of the order of the cell strut diameter and (b) a closed-cell Al foam HAL 175/4/1 with SiC particles enriched at the cell walls.

Fig. 5.29 Fatigue striations in the fracture surface of a single cell strut after fatigue testing of open-cell Al sponge.

The crack initiation process during fatigue loading of cellular metals is obviously governed by (1) microstructural factors, e.g., very large grains, (2) a pronounced structure-dependent stress triaxiality, and (3) the macrostructure of the cellular metal. The latter plays a key role in very inhomogeneous materials. It was shown by fracture surface analysis that the final fatigue crack always propagates along cross sections that exhibit the largest pores [299].

6 Crack Propagation: Microstructural Aspects

6.1 Special Features of the Propagation of Microstructurally Short Fatigue Cracks

Again, one could start this section with the question about the length scale, where we distinguish between microstructural defects, crack initiation and crack propagation. As a logical continuation of the preceding chapter, we consider microscopic crack initiation, due to local stress concentration, plastic deformation and early growth up to the first microstructural obstacle, as to be given and belonging to the crack initiation phase. Propagation of microstructurally short fatigue cracks (microcracks) is governed by interactions between local microstructural features and the resolved stress state acting as crack driving force at the crack tip. Normally, the significance of these interactions extends only for several grain diameters, and hence microstructurally short cracks cannot be detected by conventional methods of nondestructive materials testing, like ultrasonic testing or X-ray analysis, which are typically of a resolution of the order of 0.5 mm. The technical relevance of such kinds of short cracks increases with increasing strength, higher surface finishing and for high-cycle fatigue (HCF) and very-high-cycle fatigue (VHCF) loading conditions. For soft materials, high stresses or for variable amplitude loading with overloads in the early stage of service, crack propagation is the predominant mechanism during fatigue life. Otherwise, according to [300], the regime of microcrack propagation prevails during 65–90% of the fatigue life, while only the last 5–10% of the fatigue life can be considered as to be determined by long cracks and treated by linear elastic fracture mechanics (LEFM). The remaining 5–25% are determined by the mechanisms of crack initiation, i.e., stress concentration and irreversible cyclic slip accumulation (cf. Fig. 5.1).

Taking this into account, it is obvious that a fundamental improvement of methods for service life assessment does not only require a sound understanding of the mechanisms of microcrack propagation, but also nondestructive crack or damage detection methods with a sensitivity at the micrometer length scale.

Fatigue Crack Propagation in Metals and Alloys: Microstructural Aspects and Modelling Concepts. Ulrich Krupp

ISBN: 978-3-527-31537-6

6.1.1 Definition of Short and Long Cracks

A clear separation between microstructurally short cracks, physically short cracks and mechanically short cracks is sometimes an object of confusion. Hence, at the beginning of the present section a definition of the crack length accounting for the respective propagation mechanisms is provided. According to the definition of Suresh and Ritchie [36, 301], the propagation rate da/dN, immediately after crack initiation, is determined by interactions with local microcstructural features, being characteristic for the material, e.g., grain and phase boundaries, precipitates and pores. Such cracks are termed *microstructurally short cracks*. Once the crack length exceeds several grain diameters (for numbers see Section 6.4), the strong influence of the microstructure vanishes, and crack propagation is driven by the plastic zone ahead of the crack tip, i.e., crack-tip-opening displacement (CTOD). These cracks are termed *mechanically short cracks*. Transition to *physically short cracks* is associated with the condition that the size of the plastic zone at the crack tip is negligibly small, as compared to the crack length, and hence the concepts of LEFM are applicable. A fatigue crack is considered to be a *long crack* when, besides the intrinsic crack-driving force at the crack tip, the extrinsic influence factors are fully developed. Extrinsic factors are mainly crack-closure effects (see Section 6.3) in the wake of the crack tip altering the crack-driving force. The propagation rate of long fatigue cracks can be predicted by Paris' law (Eq. 2.5). Suresh [36] also uses the term *chemically short cracks*. Even though these cracks belong to the group of physically short or long fatigue cracks, they may exhibit an abnormal propagation behavior due to chemical reactions with an aggressive environment.

The classification of short and long fatigue cracks together with their characteristic features is summarized in Table 6.1.

A common feature of short cracks is their abnormal propagation behavior as compared with the "normal" Paris-type behavior, where da/dN is an exponential function of the range of the stress intensity factor ΔK that applies almost generally to long fatigue cracks in metals and alloys. Contrary to the prediction of the Paris law, for microstructurally short cracks, e.g., in steels, aluminum alloys and nickel-based alloys [302], an initially decreasing crack-propagation rate has been observed. The abnormal behavior of microstructurally short fatigue cracks and its significance for the overall fatigue damage process has been subject of many studies, e.g., Lindley and Nix [200], Tokaji and Ogawa [303], Taylor and Knott [304] or Rodopolous and de los Rios [305].

However, a consistent and quantitative definition of short and microstructurally short cracks does not exist. According to ASTM a crack is considered as a short crack as long as the radius of the plastic zone is larger than one-fiftieth of the crack length ($r > 1/50a$). A high material strength and large grain sizes seem to promote short-crack behavior [305]. Generally, below a critical crack length, which lies within the length scale of the microstructure constituents, a material cannot be considered as a continuum [304]. Even though this relationship is plausible and quite trivial, it is not accounted for by many modeling approaches applying the concept of LEFM to short-crack problems (cf. Chapter 7).

Table 6.1 Definition and characteristics of fatigue cracks of different lengths.

Abnormal crack-propagation behavior		K-determined crack-propagation behavior	
Microstructurally short cracks	**Mechanically short cracks**	**Physically short cracks**	**Long cracks**
• Strong influence of the microstructure • Mainly mode II crack propagation (stage Ia) and transition to mode I (stage Ib, see text) • Roughness-induced crack closure	• No pronounced crack-closure effects • Little influence of the microstructure • Mainly mode I crack propagation • Large plastic zone ahead of the crack tip (relative to the crack length)	• Little influence of the microstructure • Plasticity-induced crack closure not completely developed • Mainly mode I crack propagation • Negligibly small plastic zone ahead of the crack tip (relative to the crack length)	• Completely developed plasticity-induced crack closure (plastic wake) • Macroscopically long cracks (crack length of the order >0.5 mm)

In the following section the most prominent features and influence factors of the propagation behavior of short and microstructurally short cracks are summarized:

- Often, microstructurally short cracks grow much faster than one would expect on the basis of long-crack data. This was clearly pointed out by Pearson in 1975 [31] for the first time.
- Microstructurally short cracks can grow far below the threshold value for the range of stress-intensity factor ΔK_{th} for long fatigue cracks. Such effects were also observed at high temperatures for a single-crystalline nickel-based superalloy CMSX-2 [306]. As shown in Section 6.3, this effect can be partially attributed to differences in the crack-closure behavior.
- Immediately after crack initiation, the crack-propagation rate drops continuously. This is due to microstructural barriers, grain or phase boundaries, e.g., pearlite lamellae [283] and cementite plates [302] in steels. If such a barrier cannot be overcome, the crack is stopped. If this is the case for all cracks, the fatigue limit is reached. Further reasons for a decrease in the propagation rate of microstructurally short fatigue cracks are crack-tip shielding, crack branching and crack coalescence, addressed in Section 6.4.
- Once a short fatigue crack is not ultimately stopped by a microstructural barrier, the crack-propagation rate will – after passing through a minimum – increase again. This is often the case when the crack length has reached the dimension of the grain size (or other characteristic microstructural features). The minimum of the crack-propagation rate is determined by the degree of plasticity in a grain, induced by crack propagation in the adjacent grain containing the crack. The degree of plasticity transmitted across the grain boundary depends considerably on the crystallographic misorientation between the slip planes of neighboring grains. For small misorientation angles, e.g., low-angle grain boundaries, crack propagation is not impeded [201, 202].

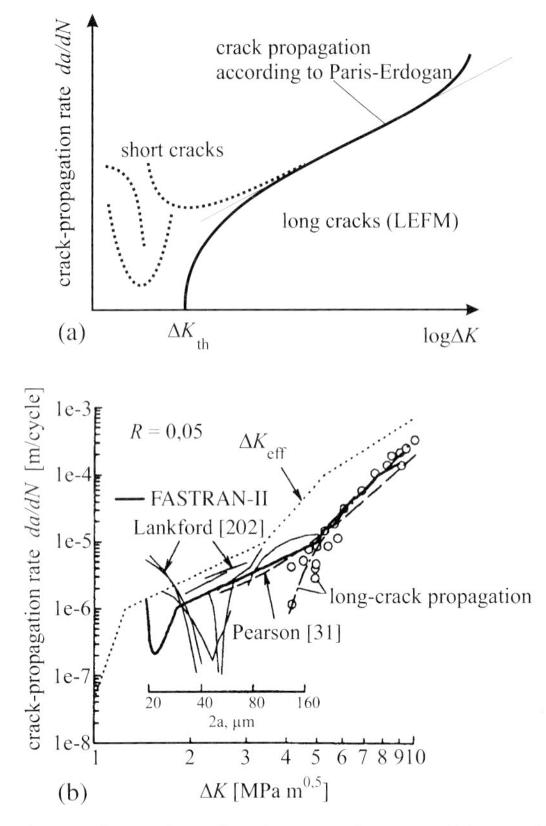

Fig. 6.1 Comparison of crack propagation rates of short and long fatigue cracks (a) schematically, and (b) measured and predicted data (FASTRAN II [307]) for cracks in a 7075 T6 aluminum alloy (after [201]).

It can be concluded that the abnormal crack-propagation behavior of microstructurally short cracks is determined by (1) metallurgical effects, such as the size and orientation distribution of the grains and precipitates and (2) by a discontinuously changing contribution of crack-tip shielding due to crack-closure effects. Since for the length scale of microstructurally short cracks the material cannot be considered to be a continuum, the concept of transferability of LEFM ($K/\Delta K$ concept), where the crack-propagation rate depends on the combination of crack length and the remote mechanical stress, is not applicable [201].

Figure 6.1 shows schematically (Fig. 6.1a) and for the example of the heat-treatable Al alloy Al7075-T6 (Fig. 6.1b) the correlation between long and short fatigue crack-propagation rate versus the range of the stress intensity factor ΔK.

Among the numerous publications that deal with the short-crack problem, a few conference proceedings should be pointed out, which are devoted solely to various aspects of short-fatigue-crack behavior [308–311]. Furthermore, the comprehensive review articles by Hudak [312], Suresh and Ritchie [301], Ritchie and Peters [313], Miller [314], McEvily [315] and Lukas [199] are worth mentioning.

6.2 Transgranular Crack Propagation

6.2.1 Crystallographic Crack Propagation: Interactions with Grain Boundaries

For most metals and alloys, crystallographic crack propagation is characteristic of the early stage (stage I) of fatigue damage near the threshold of the stress-intensity range ΔK_{th} for long fatigue crack initiation [316]. These cracks grow along crystallographic slip planes, mostly inclined by an angle close to 45° with respect to the applied stress axis, driven by the resolved shear stress (loading mode II). With increasing crack length or increasing ΔK, the normal-stress contribution becomes more significant, forcing the crack path to a direction perpendicular to the applied load axis (loading mode I) by operating various slip systems. An example for such a situation is given in Fig. 6.2, showing a microstructurally short fatigue crack in an austenitic–ferritic duplex steel AISI F51 (X2CrNiMoN 22 5 3), initially grown along a slip band in a ferrite grain and having changed to multiple slip propagation in the adjacent austenite grain. Details about heat treatment, application, etc., of this kind of high-strength stainless steel can be found elsewhere [317].

This kind of crack propagation, driven by double slip, may appear very similar to the stage II crack-propagation mechanism, as discussed in Section 6.2.2. It must be emphasized, however, that it clearly belongs to the short-crack regime. Hence, it makes sense to term it *stage Ib* crack propagation [318]. While each of the fatigue striations formed during stage II crack propagation correspond to only one single cycle, slip accumulation during many cycles on various slip planes is characteristic for stage Ib crack propagation.

Similar to the crack-initiation process, crystallographic crack propagation (microstructurally short cracks) is determined by the spatial arrangement of the slip planes with respect to the local stress direction, the latter influenced by the material's elastic anisotropy (Section 5.3). However, for this situation calculation of the mechanical stresses by the finite-element method becomes much more difficult, since crystal plasticity needs to be taken into account [248, 255]. An alternative method to quantify the local driving force of short cracks was suggested by Ravichandran [319], who considered locally different stress conditions by means of

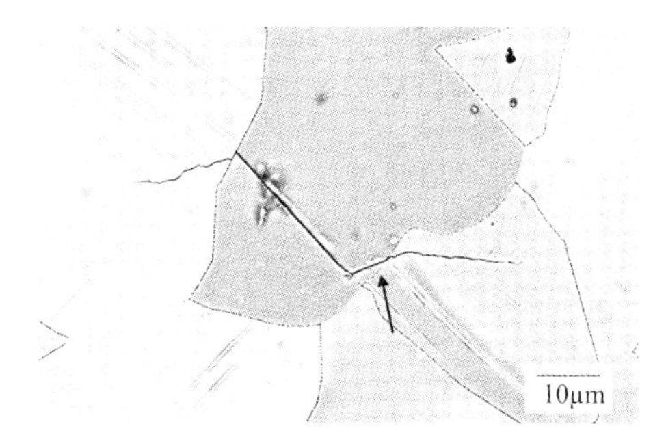

Fig. 6.2 Microstructural fatigue crack propagation in AISI F51 duplex steel: transition from single-slip to double-slip crack propagation (marked by arrow).

a variable stress-intensity factor K along the crack front (cf. also Newman and Raju [320]). However, such an approach is only of limited applicability with respect to microstructural cracks, since the conditions for the application of linear-elastic fracture concepts do not apply.

Once the critical resolved shear stress on a slip plane in a favorably oriented grain is exceeded (cf. Section 4.2), plastic deformation by dislocation motion sets in. However, this deformation is constrained by the grain boundaries of the respective grain. When transmission of plasticity is hindered, e.g., by a large misorientation between the neighboring slip systems, the dislocations can be piled up along the slip band. Such a situation is shown in Fig. 6.3 for cyclic deformation of AISI F51 duplex steel (Fig. 6.3a) and by the simplified schematic representation in Fig. 6.3b. It should be mentioned here that one should always be careful with the interpretation of dislocation arrangements in a thin foil, since handling and stress relief in the polycrystalline microstructure can cause the generation of artifacts.

Pileup or pairing of dislocations along a slip band leads to a considerable stress increase at dislocation sources in the adjacent grains and, eventually, when a critical stress is exceeded, to slip transmission or slip continuation in the neighboring grain. The influence of the crystallographic orientation, i.e., the spatial arrangement of the slip planes in the adjacent grains, is obvious. Assuming planar slip prevailing during fatigue, as it was shown using transmission electron microscopy

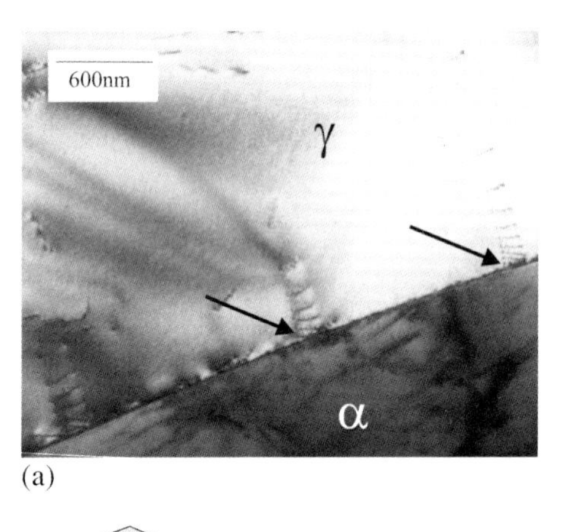

(a)

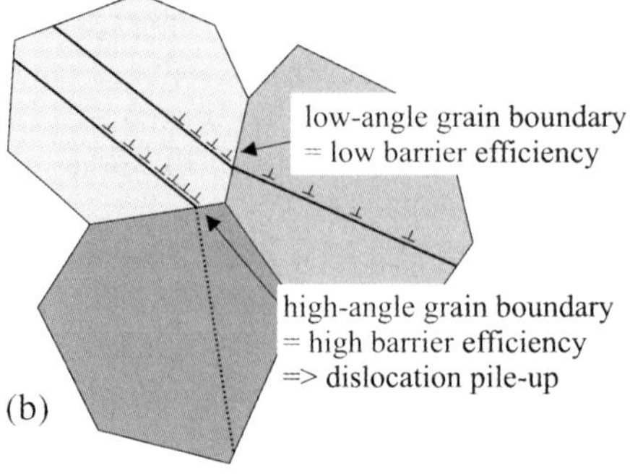

(b)

Fig. 6.3 (a) Transmission electron micrograph of dislocation pileup at a $\gamma\alpha$ phase boundary in AISI F51 duplex steel after cyclic deformation. (b) Greatly simplified representation of the relationship between the barrier strength of a grain boundary and the misorientation between neighboring slip planes.

(TEM), e.g., for AISI F51 duplex steel [317], it is evident that with increasing grain size a higher number of dislocations can be either piled up or paired along the slip band, leading to a stress increase until either the grain boundary becomes ineffective as a barrier for the succeeding dislocations or dislocation sources in the adjacent grains are activated. This relationship is accounted for by the well-known Hall–Petch relationship [180, 181] (cf. Section 4.3):

$$\sigma_Y = \sigma_0 + \frac{k}{\sqrt{d}} \tag{6.1}$$

where the yield strength σ_Y correlates with the material's grain size d by means of the Hall–Petch constant k, plus the critical stress σ_0, representing the sum of all obstacles impeding the free movement of dislocations in a single crystal, like coherent small precipitates, forest dislocations and the Peierls–Nabarro stress (cf. [321] and Section 4.2). An application of the modified Hall–Petch relationship to derive fatigue-crack-propagation modeling data for two-phase materials is introduced in Section 7.3.

The dislocation pileup concept can be adopted to explain intercrystalline separation of grain boundaries in competition with transgranular crack propagation across grain boundaries with oscillating crack-propagation rates. If on the one hand the stress increase due to dislocation pileup at the grain boundary cannot be released by slip transmission into the neighboring grain, because of unfavorable positions of the slip systems, but on the other hand it is large enough to overcome the energy γ_0 to form new surfaces, then intercrystalline crack propagation occurs. Liang and Laird [322] found by experiments on coarse-grained copper that also multiple slip can promote intercrystalline cracking by maximizing cooperative slip between the adjacent grains.

The dislocation pileup ahead of an existing transgranular slip band crack does not only cause stress increase in the neighboring grain; cyclic plasticity also drives crack propagation. However, with decreasing distance between crack tip and adjacent grain boundary, the dislocation density along the slip band increases and the dislocations become increasingly immobile. This causes a higher resistance to crack advance; i.e., the crack propagation rate decreases. Only when the neighboring grain becomes plastified, is cyclic slip displacement on the respective slip planes promoted, and the crack-propagation rate increases. The consequence is an oscillation in the crack-propagation rate, which has been observed for microstructurally short fatigue cracks in several studies, e.g.. [323], and that is also the basis of the short-crack model of Navarro and de los Rio [324] discussed in detail in Section 7.2. Figure 6.4 shows an example for interactions between short-fatigue-crack propagation and grain boundaries during HCF loading of an Al2024 T3 aluminum alloy.

If the spatial angle between the slip systems of neighboring grains is large, then the respective grain boundary forms an effective barrier against fatigue-crack propagation, as long as the crack is short and the stress amplitude small. If adjacent slip systems are lying virtually in the same plane, as it is the case for low-angle grain boundaries, direct slip transmission is possible [127, 325], and the crack-propaga-

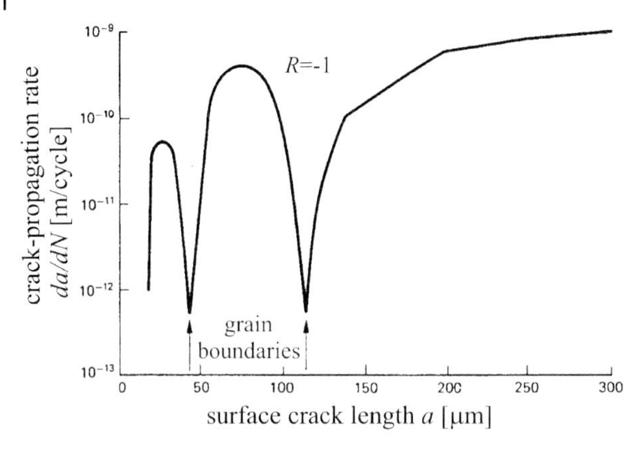

Fig. 6.4 Oscillation in the crack propagation rate (2024 T3 aluminum alloy); minima are due to the barrier effect of grain boundaries (after [105, 323]).

tion rate remains almost constant. An example for the relationship between crystallographic misorientation and early fatigue-crack propagation is shown in Fig. 6.5 for the example of the β-titanium alloy LCB [227]. A slip-band crack initiated at a grain boundary grows across a low-angle grain boundary (position x in Fig. 6.5a) neither changing the crack path nor the crack-propagation rate (Fig. 6.5b). A similar behavior was also reported by Long et al. [282] for low-angle grain boundaries in similar β-titanium alloys. At the subsequent high-angle grain boundary (position y

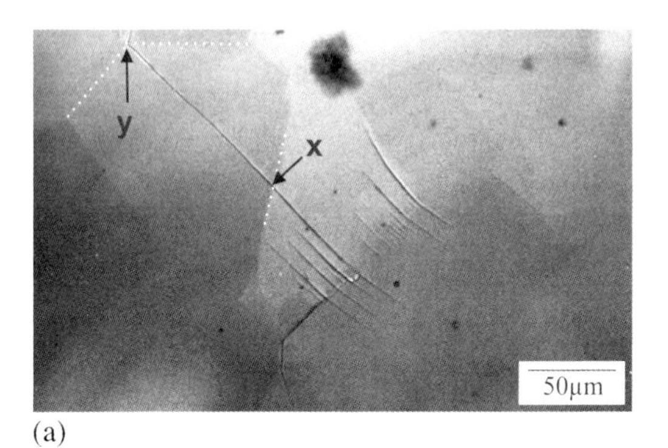

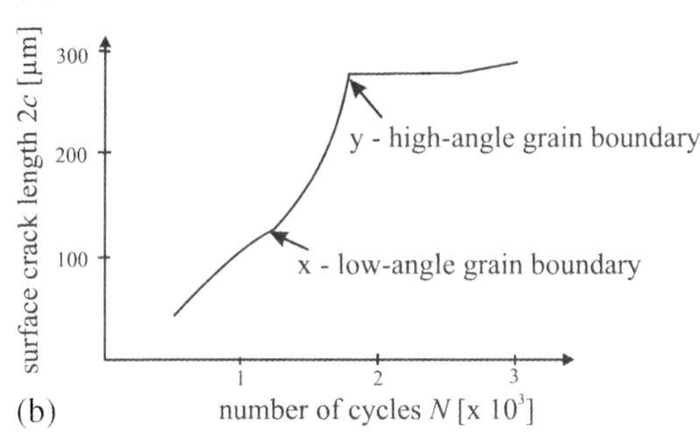

Fig. 6.5 (a) Slip-band cracks in the β-titanium alloy LCB ($\Delta\sigma/2$ = 600 MPa, loading direction: ⇔); (b) crack length vs. number of cycles for the crack marked by arrows.

in Fig. 6.5a), however, the crack is stopped for more than 800 cycles, until it resumes growing in a completely different direction and at a lower rate.

Preferential propagation of microstructurally short cracks along low-index slip planes during the early stage of fatigue damage was reported also for various other alloys, e.g., by Ruppen et al. [326] for the titanium alloy Ti-6Al-4V, by Blochwitz et al. [327] for 316L austenitic steel, by Turnbull and de los Rios [328] for an Al-Mg alloy or by Tokaji and Ogawa [303] for several steels, titanium and the aluminum alloy 7075-T6. Tokaji and Ogawa found that crystallographic crack propagation along slip bands is particularly promoted by the compression part of the fatigue loading spectrum, due to the absence of the mode I crack-opening contribution during remote tension. This relationship becomes evident by a high fraction of slip-band facets in the fracture surface, in a similar way as is shown in Fig. 6.6 for a fatigued sample of the β-titanium alloy LCB. Since slip-band cracks are driven by cyclic plasticity along different parallel slip planes, these facets are never perfectly flat. They rather exhibit small slip steps or traces of intersections with secondary slip systems.

The influence of the grain size on the early crystallographic crack propagation (microstructurally short cracks) is discussed, e.g., by Miller [314] or Turnbull and de los Rios [328]. According to their results, the transition of the plastic zone from the crack-containing grain to the neighboring uncracked grain is eased by the larger slip length in coarse-grained materials. Even for high crystallographic misorientations, the barrier effect of the grain boundaries decreases. The consequence is a lower fatigue limit, which can be correlated directly with the strength of microstructural barriers (cf. Section 5.5), and a trend towards pronounced crack deflection. In contrast, in the case of fine-grained materials the scatter in short crack-propagation rates is higher, due to the greater efficiency of the grain boundaries as microstructural obstacles (see also [303]). When discussing grain size effects, one should take a possible superimposition by texture effects into consideration.

It is a common assumption that the barrier strength of a grain boundary is determined by the respective crystallographic misorientation [204, 329]. However, a closer inspection reveals that the resistance of a grain boundary to slip transmis-

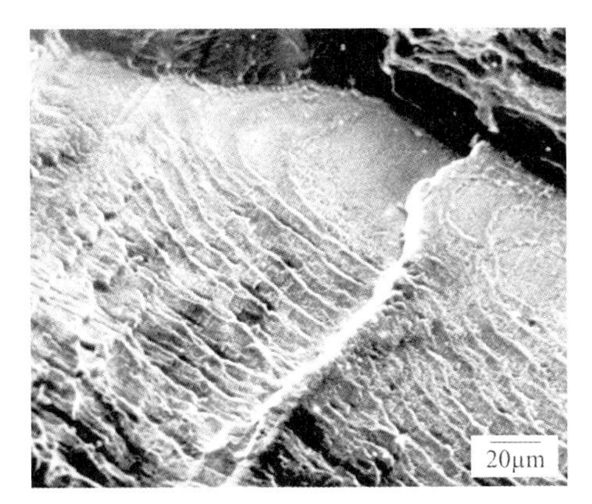

Fig. 6.6 Slip-band facets (microcleavage) in the fracture surface of a fatigued specimen of the β-titanium alloy LCB ($\Delta\sigma/2 = 600$ MPa).

sion is determined by multiple parameters. As mentioned in Chapter 3, the misorientation angle Θ and direction <*uvw*> scheme does not account for the three-dimensional position of the grain-boundary plane defined by its normal vector (*hkl*). Most important is the spatial misorientation of the neighboring slip planes (Fig. 6.7). From surface observations we can obtain only the traces of operated slip systems, which provides the tilt misorientation angle Φ. However, the intrinsic resistance of a grain boundary depends also on the twist misorientation angle ξ within the grain boundary plane (cf. Zhai et al. [266, 267]) and the structure of the grain boundary itself (see, e.g., [129]). The tilt/twist misorientation relationship between two neighboring grains containing a slip band crack and two activated slip planes is represented schematically in Fig. 6.7.

The lowest barrier effect for propagation of microstructurally short cracks can be expected for a misorientation relationship where the twist angle ξ is close to zero and the tilt angle Φ is very small. From the simplified schematic in Fig. 6.7 it is obvious that not only an increase in tilt misorientation Φ, but also an increase in the twist misorientation ξ restricts the slip compatibility across a grain boundary. The latter affects also the process of actual crack propagation across the boundary, since the areas in the grain boundary plane, which do not match to the crack path, need to be overcome. According to Argon and Qiao [268], this occurs by staircase-type cracking of these areas along the intersection line between slip-band crack and grain-boundary plane. The results of Zhai et al. [266, 267] for an Al–Li alloy allow the conclusion that high values of the twist misorientation angles ξ may cause blocking of crack propagation at the respective grain boundary.

While the preceding sections are focused on the factors influencing the initiation and propagation behavior of microstructurally short fatigue cracks, in the following the more fundamental mechanisms of crystallographic fatigue-crack propagation are discussed (a more quantitative analysis is provided in Chapter 7). Contrary to mode I crack propagation by plastic blunting (see Section 6.2.2), crack propagation along slip bands is driven by the resolved shear stress (mode II load-

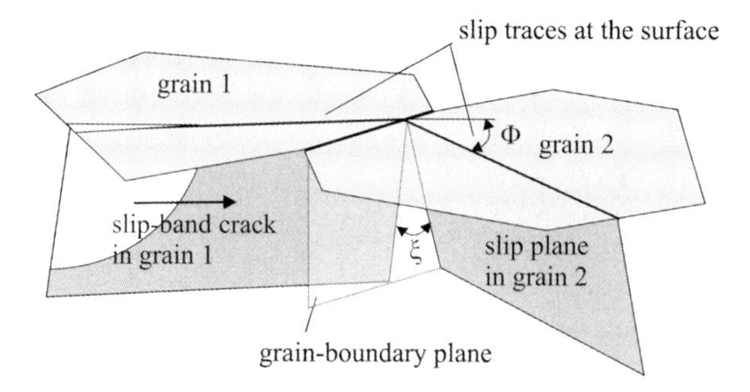

Fig. 6.7 Crack propagation along slip bands and across a grain boundary: (a) the barrier effect is a consequence of the tilt (Φ) and twist (ξ) misorientation; (b) the intrinsic misorientation is determined rather by the twist misorientation angle (ξ) in the grain boundary plane and the grain boundary structure.

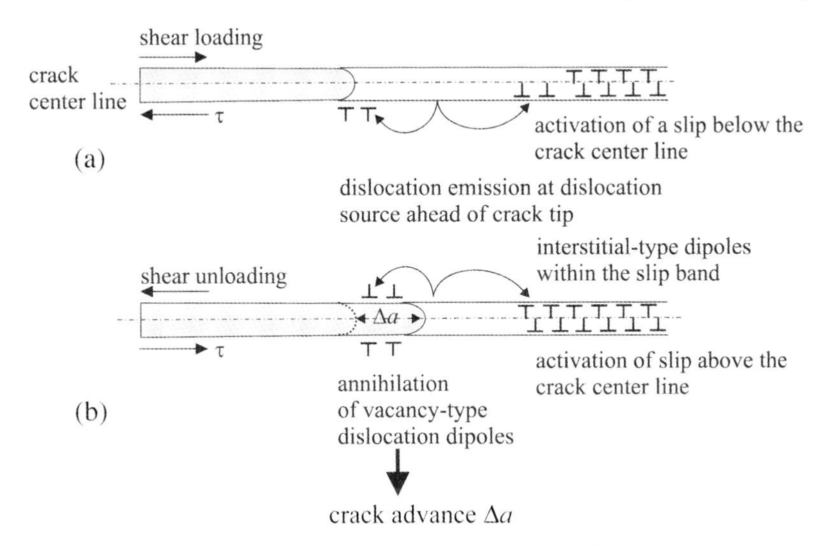

Fig. 6.8 Crack propagation due to cyclic slip irreversibility (according to [330]). The mode I contribution in the stress field at the crack tip leads to a preferential activation of slip planes (a) above the crack center line during loading and (b) below the crack center line during unloading.

ing condition) that can be approximated by Schmid's law (Eq. 4.22). However, almost generally, the slip band experiences some kind of normal stress contribution (mode I loading condition) that leads to a certain degree of asymmetry of the shear stresses acting on the slip band. According to Wilkinson et al. [330], this asymmetry causes generation and motion of dislocations along individual slip planes above and below the center line of the crack/slip band during loading and unloading (or reverse loading), respectively. As a consequence, dislocation dipoles of the interstitial type are emitted from the crack tip into the slip band, while dislocation dipoles of the vacancy type are absorbed by the crack tip leading to infinitesimal crack advance. This mechanism, which is closely related to the fundamental planar crack-initiation mechanism by Fujita [277], is schematically represented in Fig. 6.8.

Also, Weertmann [331] and Caracostas et al. [332] attribute crystallographic crack propagation to the activity of parallel operating slip planes. However, they assume a dislocation distribution function following the BCS-crack approach (after Bilby, Cottrell and Swinden [71], cf. Section 7.2).

6.2.2
Mode I Crack Propagation Governed by Cyclic Crack-Tip Blunting

With increasing crack length, superimposition by mode I loading during shear-controlled crack propagation becomes increasingly significant. When the crack tip reaches a grain with slip planes unfavorably oriented with respect to the applied stress state, the crack tends to continue its propagation perpendicular to the re-

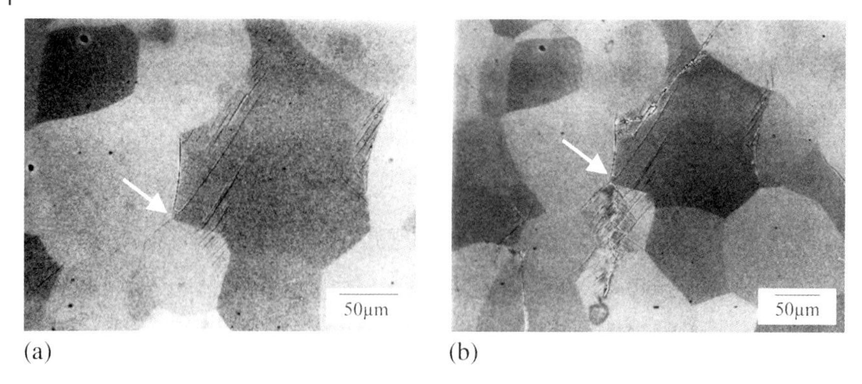

Fig. 6.9 (a) Intercrystalline microcrack (arrow) after 10% of fatigue life in the β-titanium alloy LCB; (b) transgranular crack propagation starting from the marked triple point and governed by alternate operation of two slip systems ($\Delta\sigma/2 = 600$ MPa, loading axis: ⇔).

mote loading direction (mode I) under operation of alternate slip systems. Such a situation is shown in Fig. 6.2 (AISI F51 duplex steel) and in Fig. 6.9 (β-titanium alloy LCB). In the latter case, the mode I loading contribution at the triple line coinciding with the intercrystalline crack marked by an arrow (Fig. 6.9a) is obviously high enough to force the path of subsequent crack propagation perpendicular to the loading axis, operated by two alternate slip systems (Fig. 6.9b).

According to Richter et al. [123] and Laird [333] multiple-slip crack propagation can be understood simply as an additive combination of Burgers vectors of dislocations moving on the activated slip systems, yielding an effective Burgers vector $\vec{b}_{\mathrm{eff}}$. This idea is represented schematically in Fig. 6.10a for slip occurring on four {111}<110>-type slip systems in the face-centered cubic (fcc) lattice structure leading to $\vec{b}_{\mathrm{eff}}$ [334]. As an example, Fig. 6.10b shows such a situation for short-crack propagation governed by multiple slip on alternating (111) and $(11\bar{1})$ planes. The

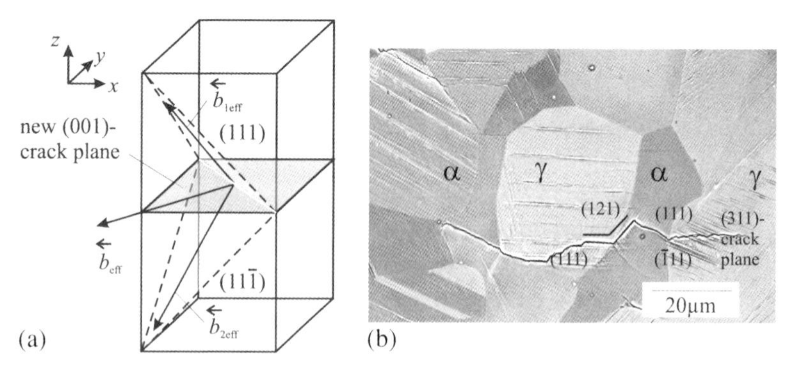

Fig. 6.10 (a) Net crack propagation due to additive superimposition of Burgers vectors of two slip systems; (b) microcrack path in AISI F51 duplex steel ($\Delta\sigma/2 = 400$MPa, $N = 170\,000$ cycles, loading axis: ⇕; after [334]).

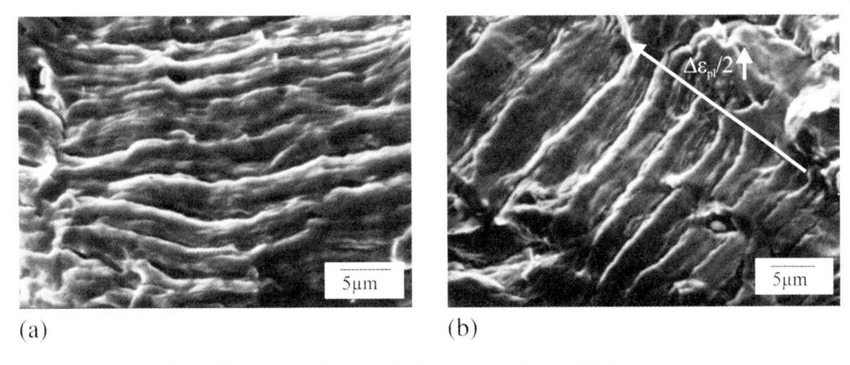

Fig. 6.11 Examples of fatigue striations in the fracture surfaces of fatigued copper: (a) for constant $\Delta\varepsilon_{pl}/2$ and (b) for incremental step test.

observed inclination of the crack path at the surface of −1° with respect to the load axis vertical, corresponding to a (311) plane, can be obtained by the respective Burgers vector $\vec{b}_{eff} = 2\vec{b}_{1,eff} + \vec{b}_{2,eff}$ according to Fig. 6.10a [334].

Crack propagation by alternate activation of slip systems leads to the formation of so-called striations in the fracture surface. Originally, they were reported in the early paper by Forsyth and Ryder [22] as to occur in a cycle-by-cycle manner; i.e., each striation in the fracture surface corresponds to one complete fatigue cycle. Forsyth [335] distinguishes between stage I fatigue crack propagation favoring single slip and stage II crack propagation leading to striations by activation of complex triaxial slip. The formation of fatigue striations was confirmed later by many other studies, e.g., [336, 337]. Figure 6.11 shows an example of such kind of fatigue striations in polycrystalline copper for constant amplitude loading (Fig. 6.11a) and for an incremental step test with gradually increasing load amplitude (Fig. 6.11b).

However, the fatigue striations observed during microcrack propagation (cf. Figs. 6.2, 6.9 and 6.10b) do obviously not correspond to individual cycles. It is assumed that net crack advance is due to a very similar mechanism to that introduced in the previous section for slip-band crack propagation. The main difference is that cyclic slip, pileup and work hardening occur in an alternate manner on multiple slip systems, virtually according to Neumann's model (see Fig. 6.14 [24]). Hence, this stage of the fatigue damage process should be termed stage Ib crack propagation. This implies that even those short cracks that grow by multiple slip (mode I) at the crack tip are considered as to be microstructurally short cracks, and, in fact, it has been observed that depending on the position of the slip systems the crack-propagation mode may switch back from mode I (stage Ib) to mode II (stage Ia) crack propagation. Figure 6.12 shows cracks propagated in stage Ia and in stage Ib in the fracture surface (Fig. 6.12a) and on the electropolished surface (Fig. 6.12b) for the example of AISI F51 duplex steel.

Only when the plastic zone size ahead of the crack tip is large, i.e., many independent slip systems are activated in such a way that plastic deformation can be considered to be "smeared" and no longer depending on the local crystallographic orientation, does the mechanism of stage II crack propagation prevail. The respec-

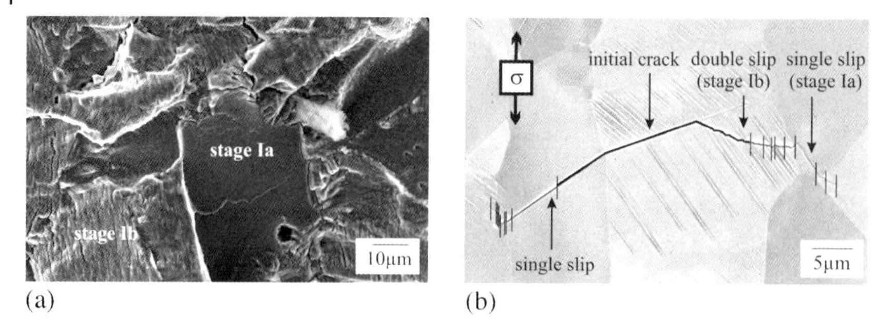

Fig. 6.12 Transitions from stage Ia to stage Ib in (a) the fracture surface and (b) the electropolished surface of AISI F51 duplex steel.

tive mechanism is described in a very clear way by the model of Laird [23, 338]. Activation of two slip systems inclined 60° with respect to the load axis (maximum shear stress) leads to blunting of the crack tip during the tensile half cycle (Fig. 6.13a–c). Due to the irreversibility of plastic deformation, blunting cannot be cancelled completely during unloading, and hence re-sharpening of the crack produces crack advance by Δa and the formation of one striation (Laird termed the striations originally "ripples" [338]) corresponding to one fatigue cycle (Fig. 6.13d–f).

Fatigue striations provide worthwhile information for the analysis of failure cases and fatigue experiments, since they reflect the load history. For instance, one can correlate the striation spacing with the crack-propagation increment per cycle: Measuring striation spacing at different locations of the crack path yields the crack propagation vs. crack length/stress-intensity factor curve.

Fatigue crack propagation and the origin of striations in single-crystalline copper specimens is described in detail by Neumann [24, 339] owing to alternate operation of slip systems according to the transition from stage I to stage II (stage Ib). Figure 6.14 represents the general idea of Neumann's model: the alternating change of the operating slip band due to work hardening along the slip bands.

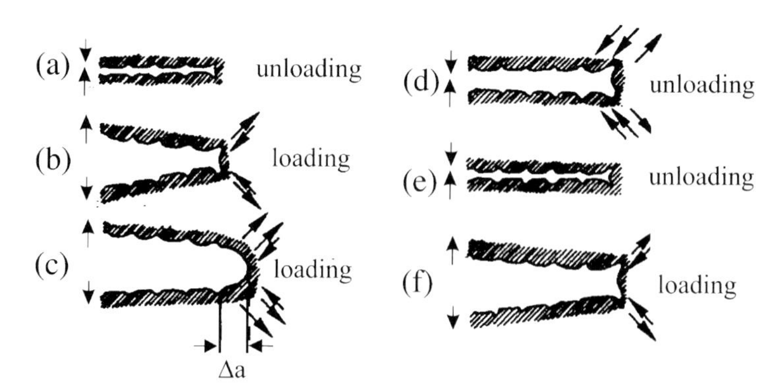

Fig. 6.13 Schematic representation of the model of Laird (after [23]) for one complete unloading (a, d, e) and loading (b, c, f) cycle (for details see text).

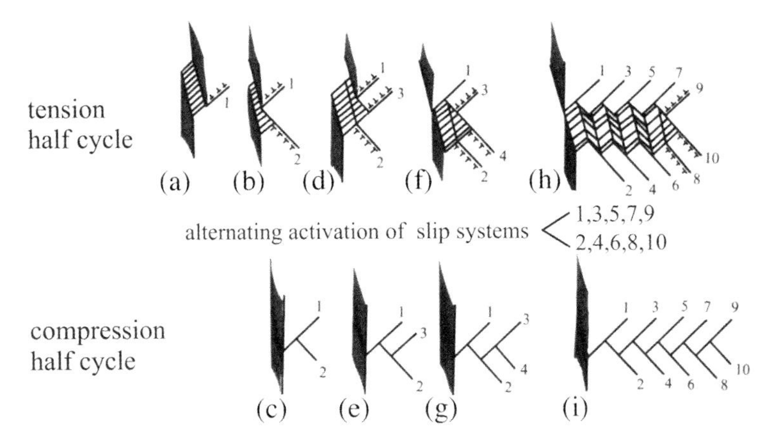

Fig. 6.14 Schematic representation of the model of Neumann (after [24]); for details see text.

Slip displacement during tension leads to work hardening and, correspondingly, to a stress increase at the slip band. This causes activation of a second slip band on the opposite site, which is stressed in a similar way but exhibiting a lower resistance. While slip resumes, also the second slip band experiences work hardening in combination with a stress increase. Hence, activation of the slip systems does alternate. During load reversal subsequent plastic deformation will avoid the work-hardened direction leading to the activation of alternate slip bands, eventually resulting in the situation shown in Fig. 6.14. Characteristic for this kind of stage Ib/II fatigue-crack propagation (physically short cracks) is, particularly in the case of ductile fcc materials, a macroscopically very flat fracture surface with striations. The roughness of the striations depends, in accordance with Neumann's model [24] and with Pippan [340], on the degree of cold work applied to the material prior deformation. It should be noted here that the Neumann model refers to both stage Ib crack propagation idealized for the generation of one striation per complete cycle or stage II crack propagation, where the slip band is formed by "smeared" dislocation motion with plenty of different Burgers vectors.

6.2.3
Influence of Grain Size, Second Phases and Precipitates on the Propagation Behavior of Microstructurally Short Fatigue Cracks

At least at low temperatures, where creep effects are negligible, the crack-propagation rate increases with increasing grain size. This is due to the free slip length that depends on the alloy grain size and geometry. For coarse-grained materials or shear deformation parallel to the rolling direction, i.e., parallel to grain elongation, the free slip length is large and can accommodate a large number of dislocations. Such a situation causes large mechanical stresses acting on dislocation sources in the adjacent grains, and hence transmission of plasticity and cracks across the grain boundaries is promoted. This mechanism was observed in several earlier studies, e.g., by Gysler et al. [269] for precipitation-strengthened α- and β-titanium alloys. They estimated the fracture stress as inversely proportional to the grain size. Similar effects were found by Briggs et al. [341] for Ti 15 3 or Nakajima et al. [121] for various α+β-titanium alloys. In the latter study the maximum fatigue-crack length, for which a strong grain size effect can be observed, is given by $a < 200$ μm. A general statement, however, is difficult, since the propagation behavior of microstructurally short cracks depends not only on the alloy grain size and geometry or macro- and microtexture effects, but also on the material's yield strength that might be different for two- or multiphase alloys. As shown in Section 6.3, there is an additional indirect effect of the grain size on the microcrack propagation rate by means of roughness-induced crack closure.

Many technical materials are composed of various phases or contain intermetallic or nonmetallic precipitates. Generally, the individual phases exhibit various elastic and plastic properties, leading to a scatter in the resistance of the phase and grain boundaries to microcrack propagation. In particular for materials that contain major phase constituents of different lattice structures, like α+β-titanium alloys (hexagonal close-packed (hcp) and body-centered cubic (bcc)), TRIP steels (bcc ferrite, retained fcc austenite, bainite and bct martensite), duplex steels (fcc and bcc) or γ′-strengthened nickel-based alloys (ordered fcc and fcc), the transmission of slip is much more complex than in the case of single-phase materials. Again, the transmission depends primarily on the spatial arrangement of the slip systems of neighboring grains and the resolved shear stress acting on them. But in addition, the crystallographic relationship between neighboring phase patches of different crystal structure has to be taken into account, e.g., the Kurdjumov–Sachs relationship between fcc and bcc lattices, being relevant for fcc/bcc duplex steel (<111> directions in the fcc γ austenite are parallel to <011> directions in the bcc α ferrite). Such crystallographic relationships are of particular significance for (1) materials that experience deformation-induced phase transitions, e.g., TRIP steels or metastable austenitic steels that form ε and subsequently α′ martensite as a consequence of large strains, and (2) direct transmission of dislocations.

Since the local microcrack-propagation rate is considerably affected by the induced plasticity of the neighboring grain, its value may increase even before reaching the boundary from a high-yield-strength grain to a low-yield-strength grain, and vice versa. This was shown by Pippan et al. [342, 343] for bimaterials, consisting of weak ARMCO iron and 340 ferritic steel, the latter having a yield strength three times higher but the same elastic properties. Additionally, Pippan et al. [342, 343] found a tendency to crack branching when cracks approach boundaries of the plastically weaker material.

Differences in the barrier strength of phase and grain boundaries coupled with different values of the yield strength of the respective phases have also been observed during fatigue of the two-phase austenitic–ferritic AISI F51 duplex steel. An example is shown in Fig. 6.15 [344]. A microcrack that was grown from the harder α phase into the softer γ phase across an αγ phase boundary did not experience a significant barrier effect (Fig. 6.15a, position A) while it was effectively blocked later by an αγ phase boundary towards the α phase (Fig. 6.15b, position B). Due to the high resistance to slip transmission of the phase boundary, a secondary upper branch popped out (Fig. 6.15b, arrow) and propagated into the neighboring γ grain without being considerably blocked by the $\gamma\gamma$ grain boundary. The corresponding crack length for the two crack branches vs. the number of cycles is represented in Fig. 6.15d. As is also shown in Section 7.3.3, the intrinsic barrier effect of the $\alpha\gamma$ phase boundaries is substantially greater than that of the $\alpha\alpha$ and $\gamma\gamma$ grain boundaries [344], while the overall barrier effect is furthermore influenced by the respective critical cyclic yield strength of the phase of the adjacent grain.

According to Stolarz et al. [345, 346] and Johanssson and Oden [347], who also investigated the fatigue and microcrack-propagation behavior in duplex steels, the higher strength of the α-ferrite phase concentrates plastic deformation in the austenitic γ phase. Depending on the local phase arrangement and the respective load sharing [348], crack propagation can be stopped when reaching the first micro-

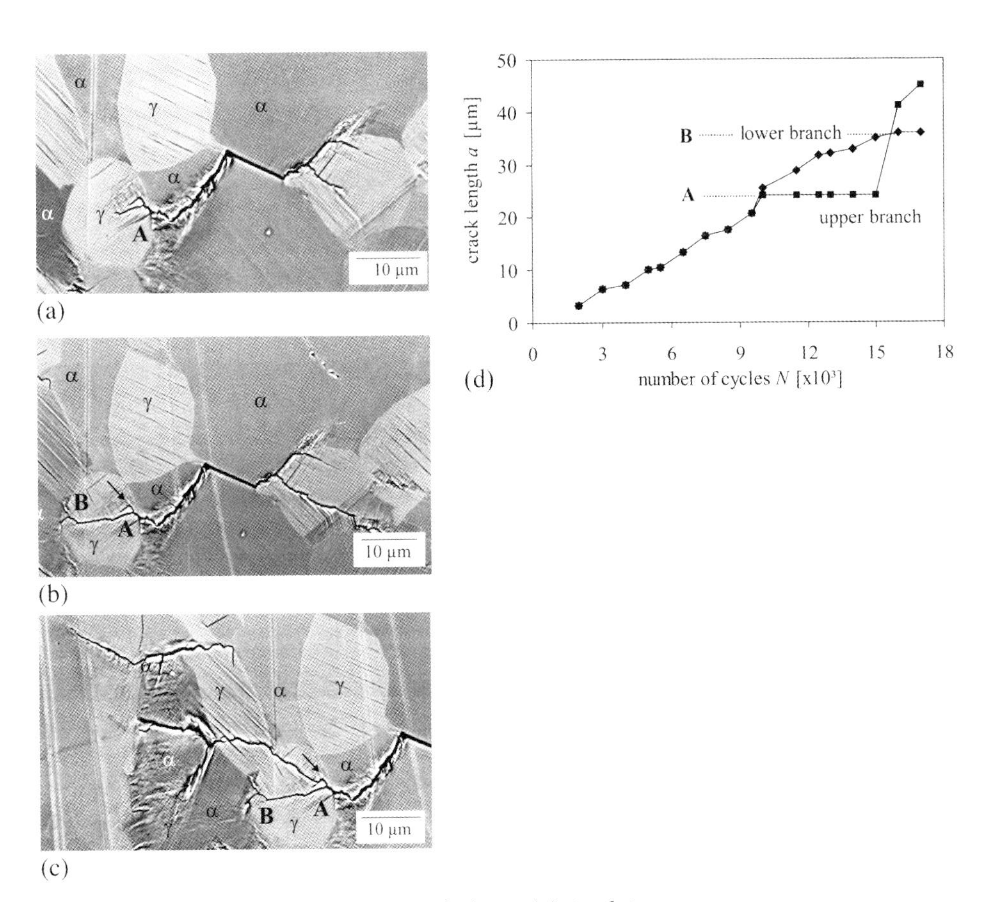

Fig. 6.15 Microcrack propagation in AISI F51 duplex steel during fatigue at $\Delta\sigma/2 = 550$ MPa (loading axis: ⇕) after (a) 10 500 cycles, (b) 15 000 cycles, (c) 17 000 cycles. (d) Summary of the corresponding crack lengths vs. numbers of cycles data (for details see text).

structural barrier. The strength in both phases can vary within a large scatter band: in the austenitic γ phase mainly due to work hardening, in the ferritic α phase due to precipitation of the brittle α' phase (low-temperature aging [349]). This may cause completely different scenarios of the fatigue-damage process [350, 351]. As shown schematically in Fig. 6.16, embrittlement of the α ferrite phase gives rise to microcleavage, i.e., microcracks that are grown crystallographically in a brittle manner and that exhibit a fracture energy that has to be accommodated by plastic deformation in the ductile austenite phase.

Further significant effects during microcrack initiation and propagation are caused by nonmetallic or intermetallic precipitates. These depend on the degree of coherence and the elastic–plastic properties of the respective particles with respect to the surrounding substrate material. Without going into detail, it should be noted here that coherent small particles cut by dislocations promote slip planarity of a material, while large and noncoherent precipitates favor the Orowan mechanism forcing dislocations to cross-slip or at elevated temperatures to climb. This behavior was shown by Hornbogen and Zum Gahr [352] for an austenitic Fe–Ni–Al steel with different kinds of microstructures. The relationship between precipitates and slip planarity is supported by studies on aluminum alloys in various heat-treatment conditions. It was observed that in the under-aged condition with finely dispersed coherent particles, planar slip prevails and microcrack propagation is mainly along crystallographic slip planes. In the over-aged condition with large particles, the fatigue damage process is governed by multiple slip, leading to a flat smooth-appearing stage II crack-propagation plane.

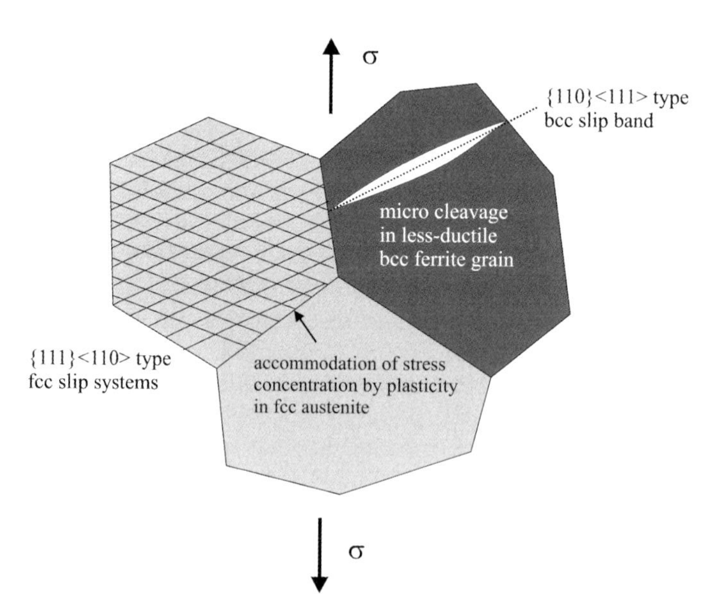

Fig. 6.16 Schematic representation of the influence of load sharing between austenite and ferrite grains in AISI F51 duplex steel.

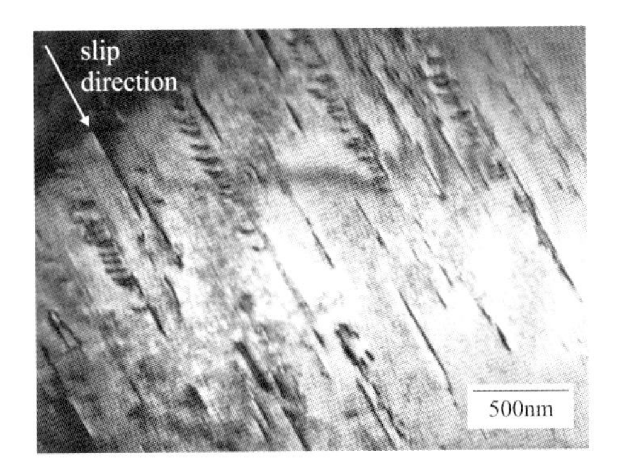

Fig. 6.17 Planar dislocation arrangement after cyclic deformation of the β-titanium alloy LCB ($\Delta\sigma/2 = 600$ MPa, $\vec{g} = [001]$; after [90]).

TEM examinations of fatigued specimens of the β-titanium alloy LCB in the solution heat-treated condition [90] and the AISI F51 duplex steel [317] supports the assumption that slip behavior is predominantly planar (Fig. 6.17). In the case of the LCB titanium alloy this behavior should be promoted by tiny precipitates of the ω phase (cf. [269, 270]). Technical β-titanium alloys with incoherent α precipitates seem also to exhibit planar slip behavior [341].

6.3 Significance of Crack-Closure Effects and Overloads

6.3.1 General Idea of Crack Closure During Fatigue-Crack Propagation

Normally, the term crack closure refers to premature contact of the crack faces during unloading from tension. Consequently, the crack-driving force at the crack tip is not active during the complete fatigue cycle. Strictly speaking, only the part of the fatigue cycles during which the crack is open contributes to crack propagation. It was Elber in the 1960s who directed the research in metal fatigue to crack-closure effects [29, 30, 353]. Later, even complete conference sessions were devoted to this phenomenon (see e.g. [354]).

Ideally, if a crack is perfectly sharp and behaves completely elastically, then a crack should close only when unloading from tension is complete, i.e., when reaching $\sigma = 0$. However, such a situation requires a perfectly sharp and completely elastic crack. Generally, fatigue-crack propagation in metals and alloys proceeds with pronounced plastic deformation ahead of the crack tip (see Fig. 6.18a). Due to the increase in the range of the stress-intensity factor ΔK with increasing crack length and a constant remote stress amplitude, $\Delta\sigma/2$ = constant, the size of the plastic zone ahead of the crack tip increases (see Fig. 6.18b and c). During crack advance the plastic zone remains in the wake of the actual crack tip. Crack advance occurs during loading up to peak tension within the plastically deformed zone leading mainly to a relief in elastic energy. When unloading, the stretched crack

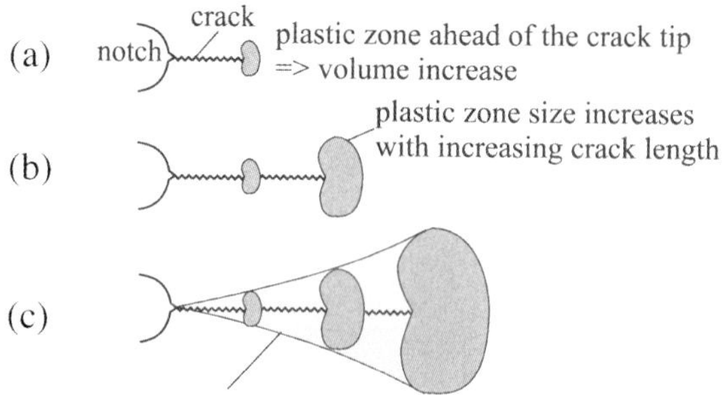

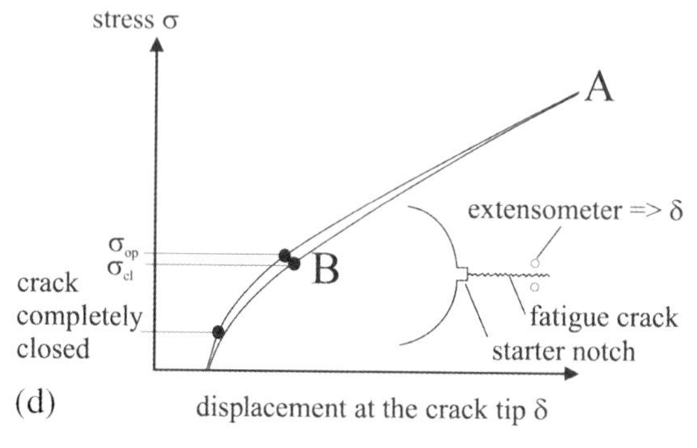

Fig. 6.18 Schematic representation of the mechanism of fatigue-crack closure according to Elber [30]: (a–c) development of a plastic wake, (d) corresponding stress vs. crack-opening diagram.

faces comes into contact before complete unloading from remote tension. The respective stress is defined as the crack-closure stress σ_{cl} (point B in Fig. 6.18d). Further unloading leads to gradual closure of the complete crack, and hence the slope of the load–displacement curve shown in Fig. 6.18 approaches the behavior of the material without a fatigue crack. The behavior described above corresponds to the original experiment of Elber [29, 30], who attached an extensometer close to the tip of a propagating crack, as shown schematically in Fig. 6.18d. This kind of crack-closure phenomenon has been termed plasticity-induced crack closure.

Today, plasticity-induced crack closure is the crack-closure mechanism that is most widely integrated in various methods of fatigue-life assessment (see e.g. [355] for short- and long-crack propagation in steels). However, it is not necessarily the closure of the highest physical and technical significance. Especially in the case of short fatigue cracks, it seems obvious that the plastic wake is short per definition, and hence its influence is quite limited [356].

Beside plasticity-induced crack closure, there are several further mechanisms that cause premature contact of the crack faces, i.e., crack closure:

- volume increase in the vicinity of the crack tip due to deformation-induced phase transformation (transformation-induced crack closure),
- oxidation of the new surfaces generated by crack advance (oxidation-induced crack closure),
- penetration of a highly viscous fluid into the crack (fluid-induced crack closure),
- geometric incompatibilities of the generated crack surfaces (roughness- or geometry-induced crack closure).

The basic ideas of theses mechanisms are represented schematically in Fig. 6.19.

The technical significance of crack-closure effects becomes obvious when following the common understanding that fatigue cracks can grow only when they are completely open. This is shown schematically in Fig. 6.20. Thus, the range of the stress-intensity factor ΔK needs to be reduced by the part of the stress-intensity range where the crack is closed, ΔK_{cl}, resulting in the effective range of the stress intensity factor ΔK_{eff} actually driving stable propagation of long fatigue cracks:

$$\Delta K_{eff} = K_{max} - K_{cl/op} = \left(\sigma_{max} - \sigma_{cl/op}\right)\sqrt{\pi a}Y \tag{6.2}$$

Certainly, this is only a very simplified representation of a phenomenon that is very complex under real circumstances, where, for example, the superposition of overloads or the influence of the environment affect the crack-closure-induced retardation of fatigue crack growth.

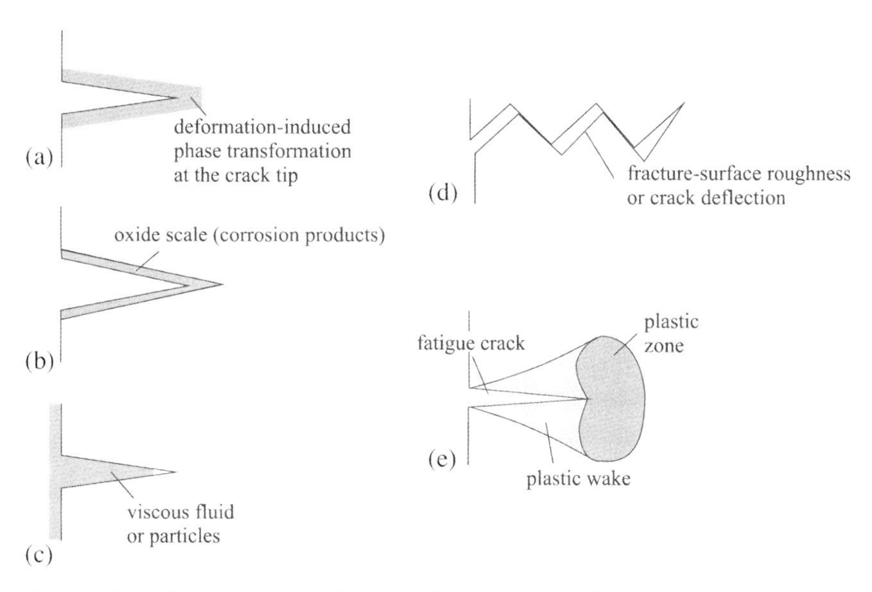

Fig. 6.19 General crack closure mechanisms (after [36]): (a) transformation-induced crack closure, (b) oxidation-induced crack closure, (c) fluid-induced crack closure, (d) roughness- or geometry-induced crack closure and (e) plasticity-induced crack closure.

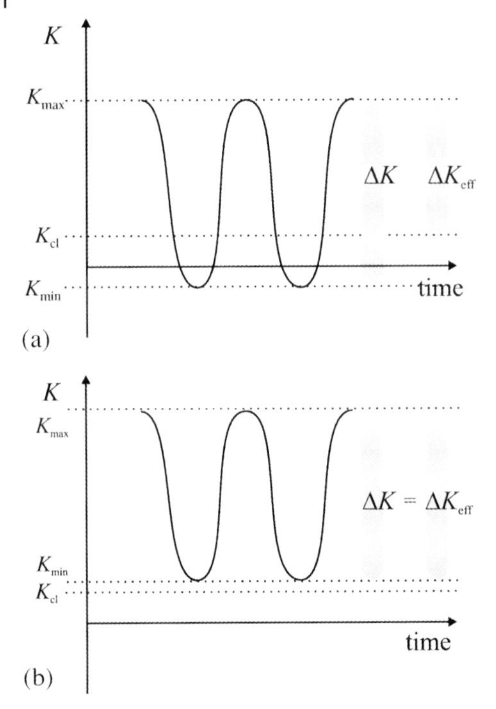

Fig. 6.20 Significance of the effective stress intensity factor ΔK_{eff} for (a) small and negative stress ratios R and (b) for higher stress ratios R. Only a portion of ΔK contributes to fatigue damage where the crack is open.

Such effects are addressed by various review articles, e.g., by Schijve [357], Liaw [358] (with the focus on crack-closure effects close to the threshold ΔK_{th} for stable fatigue-crack propagation) and Fleck [359].

The main technical significance of crack-closure effects is the retardation of fatigue-crack propagation that, however, depends strongly on the loading conditions (cf. Figs. 2.3 and 2.15). Hence, a quantitative knowledge of crack-closure effects is important to avoid nonconservative fatigue-life prediction for loading conditions where crack-closure effects are less developed than for reference laboratory tests.

Beside the inclusion of crack-closure effects into the concepts of structural integrity, they can be used also for the repair of fatigue cracks by metal infiltration [360].

In spite of numerous scientific and technical publications, the phenomenon of crack closure belongs to that group of basic fatigue mechanisms that are not fully understood yet, particularly with respect to implications in the short-fatigue-crack regime. Consequently, the significance of fatigue-crack closure is the subject of ongoing controversial discussion. The main aspects of fatigue-crack closure, and the various discussions are summarized in the following sections.

6.3.2 Plasticity-Induced Crack Closure

Even though Elber's idea of premature contact of crack faces due to the plastic wake seems to be obvious, the question arises as to where the additional material in the wake comes from. Under plane-stress conditions, prevailing in the case of thin-walled components, stretching in the plastic zone at the crack tip leads to local necking of the component (Fig. 6.21a). During crack propagation this process results in the formation of a wedge, which is pushed between the crack flanks. The consequence is premature contact of the crack faces.

In the case of compact components with small cracks, plane-strain conditions are predominant; i.e., the strain within the crack-propagation plane must be zero by definition. Hence, wedge generation by necking effects is not possible, and therefore several authors doubt a significant contribution of plasticity-induced crack closure (e.g., [361; cf. Section 6.3.6). Pippan et al. [362–364] demonstrated by means of modeling discrete dislocation motion along inclined slip planes in the vicinity of the crack tip (cf. [365], see Fig. 6.21b) that material transport from the surface into the crack is possible. As represented schematically in Fig. 6.21b, dislocation emission into two symmetrically arranged slip planes with respect to the crack tip leads to an increasing inclination of the lattice planes towards the crack-

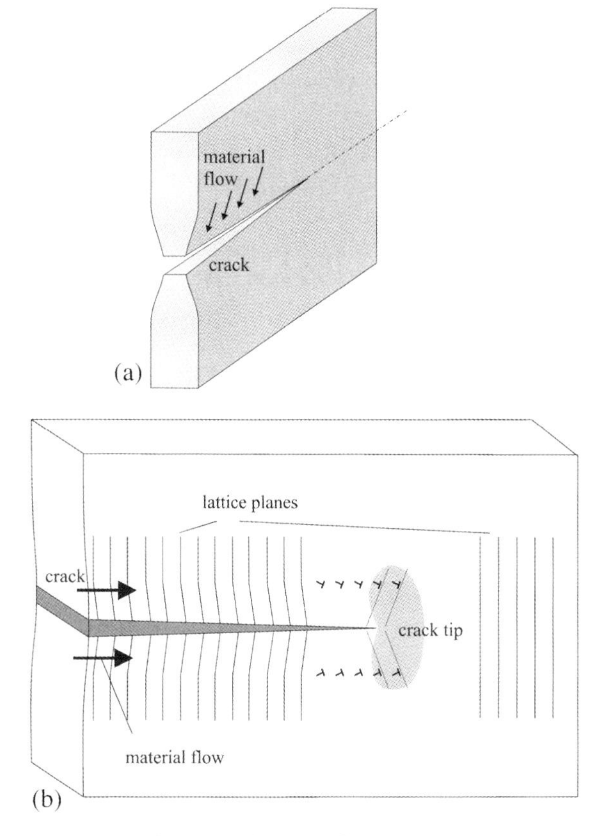

Fig. 6.21 Mechanism and material flow of plasticity-induced crack closure (a) for plane stress conditions (thin-walled components) and (b) plane strain conditions.

propagation direction. Hence, the successive transport of lattice half planes (dislocations) causes a net material flow from the crack exit at the material surface into the crack (cf. [366]).

When plasticity-induced crack closure is determined by dislocation motion along activated slip systems ahead of the crack tip, there must be a relationship with the crystallographic orientation of the respective grains, at least in the case of short cracks. Gall et al. [367] investigated such kinds of relationships between activated slip systems and crack-closure effects by means of finite-element simulations.

It is not necessarily required that the "additional wedge" material in the crack is exclusively due to plasticity at the crack tip. For real fatigue cracks, crack-closure effects by "additional" material can be attributed to crack deflection and plasticity-induced incompatibility of the crack faces (see Fig. 6.22a). Such a crack-closure phenomenon has been observed by Pippan et al. [368] by crack profile measurements. Furthermore, plasticity effects during crack branching might cause premature closure of fatigue cracks. This mechanism is sketched in Fig. 6.22b.To distinguish such cases from plasticity-induced crack closure after Elber, one may term it "plasticity-supported roughness-induced crack closure".

The development of plasticity-induced crack closure during fatigue-crack propagation is promoted by high stress amplitudes and fully reversed cycling ($R = -1$), as shown by finite-element method calculations carried out by Nicholas et al. [369] following the work of Newman et al. [370, 371] who used a ΔK_{eff} approach (FASTRAN) to simulate crack-closure effects for short and long fatigue cracks. In the latter work, the development of a plastic wake is accounted for by subdividing the plastic zone at the crack tip into discrete elements. Results of a yield strip model developed by Daniewicz and Bloom [372] support the assumption that crack closure is intensified for negative stress ratios R. Here, it is emphasized that the extent of crack closure is substantially influenced by the specimen or component geometry.

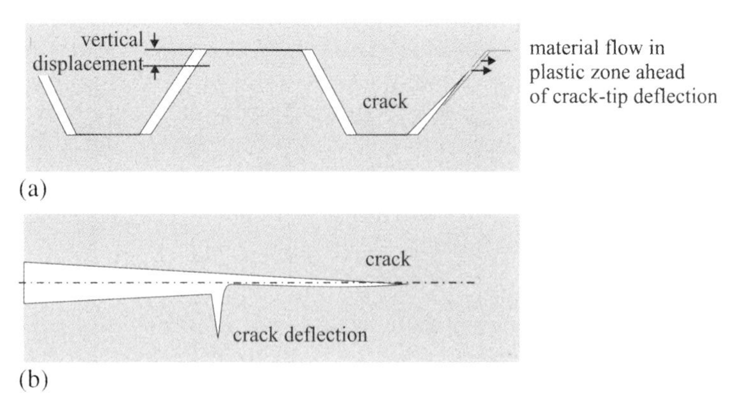

Fig. 6.22 Plasticity-supported roughness-induced crack closure by (a) crack deflection in combination with plastic deformation of the crack profile and (b) crack branching (after [368]).

This is in agreement with results of Xu et al. [373], which revealed, by comparing strain measurements immediately at the crack tip and in front of the notch of a compact tension (CT) specimen, that the occurrence of crack-closure effects varies across the specimen thickness, being particularly pronounced in areas where plane-stress conditions prevail.

The role of the stress ratio R during the development of fatigue-crack closure became particularly obvious through an experiment carried out by Ritchie et al. [374]. Here, the cyclic loading amplitude applied to CT specimens ($R = 0.5$ and 0.05) was stepwise reduced, until a situation was reached where crack advance stopped, i.e., $\Delta K_0 < \Delta K_{th}$. Afterwards, they applied a single overload in compression and continued cycling at ΔK_0. Since the crack faces were flattened by the compressive overload, the stress-intensity factor, where the crack closes, K_{cl}, was decreased, and hence crack propagation for $R = 0.05$ resumed, since $\Delta K_0 > \Delta K_{eff}$. Only after a plastic wake was developed once more, leading to an increase in the crack-closure stress, K_{cl}, was crack propagation stopped again, as shown in Fig. 6.23. At a stress ratio $R = 0.5$ the extrinsic threshold value corresponds generally to the intrinsic or effective threshold value of the range of the stress intensity factor, i.e., $\Delta K_{th} = \Delta K_{th,eff}$. Since the crack is always loaded by a remote stress in tension, it is always open. Thus, irrespective of the development of the plastic wake, crack-closure effects are not occurring. Certainly, this situation is not changed by any kind of compression overload. This is shown in the crack length vs. number of cycles curve in Fig. 6.23.

The strong dependence of plasticity-induced crack-closure effects on the stress ratio has been reported in various studies for a variety of materials. This is well documented in the review article by Liaw [358]. For instance in the case of the titanium alloy Ti-6Al-V4, crack-closure effects determine the fatigue crack-propagation behavior for stress ratios of $R = 0.02$ and 0.25, while they do not play any role at a stress ratio of $R = 0.5$.

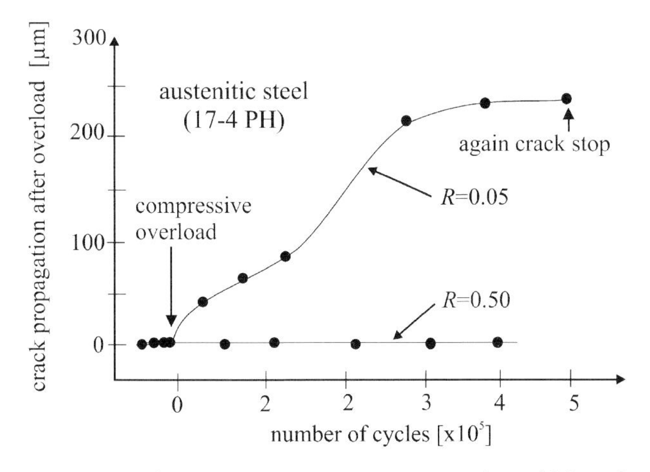

Fig. 6.23 Crack propagation in CT specimens at $R = 0.5$ and 0.05 and $\Delta K < \Delta K_{th}$ after single compressive overload (after [374]).

The frequently mentioned decrease in the extrinsic threshold value ΔK_{th}, when the stress ratio increases, seems to be at least partially correlated with environmental effects. As stated by Vasudevan et al. [361] and Liaw et al. [375], this decrease is not very significant in vacuum or inert gas atmospheres. Although it may be reasonable to attribute this observation to the absence of oxidation-induced crack closure, it is used by Vasudevan et al. [361] as an argument for their $\Delta K^{*}/\Delta K^{*}_{max}$ two-parameter approach that is introduced in Section 6.3.6.

6.3.3 Influence of Overloads in Plasticity-Induced Crack Closure

The experiment described in the preceding section (Fig. 6.23), where a nonpropagating crack was made propagating solely by a single compressive overload without changing the applied range of the stress-intensity factor ΔK, demonstrates that crack-closure effects and fatigue propagation in general depend substantially on the load history (see [182]). In the case of compressive overloads, flattening of the roughness peaks within the crack leads to a sudden increase in the effective range of the stress-intensity factor ΔK_{eff} [278].

The consequence of single and multiple tensile overloads is a sudden increase in the crack-propagation rate as well. However, this increase is followed by a decrease in the crack-propagation rate below its original value (see Lang and Marci [376]). According to Bichler and Pippan [133] this behavior can be attributed to an additional contribution to crack-tip blunting (residual CTOD) during the overload. Since the value of ΔCTOD correlates with the current crack-propagation rate, the intermediate increase in da/dN can be attributed to the residual CTOD. Even a 10% tensile overload causes the crack to stay open at complete unloading. The decrease in the crack-propagation rate after approximately 20–50 cycles is due to the temporary elevated plastic deformation at the crack tip, which causes an increase in the plastic wake, and hence an increase in plasticity-induced crack closure. Eventually, the crack-propagation rate approaches asymptotically its original value (at constant ΔK). The crack-propagation behavior as a consequence of tensile overloads is represented schematically in Fig. 6.24. Finite-element analyses by Ellyin and Wu [377] hint that the transient behavior as a result of compressive overloads is less pronounced than that resulting from tensile overloads.

A quantitative analysis of the influence of single and multiple overloads was carried out by Lang [378, 379]. He attributes the deceleration in crack propagation due to tensile overloads to the generation of compressive residual stresses in the plastic zone ahead of the crack tip. Only when the crack has passed though the compression zone does it return to its original propagation rate. Lang's analysis is based on a strict distinction between intrinsic influence factors, being relevant ahead of the crack tip, and extrinsic influence factors (crack-closure effects), which may only alter the loading conditions at the crack tip. Following this argument, plasticity-induced crack closure is not considered as a material-specific phenomenon, and hence an alternative treatment of fatigue crack propagation based on both the maximum value and the range of the applied stress-intensity factor might be useful, and this is introduced in Section 6.3.6.

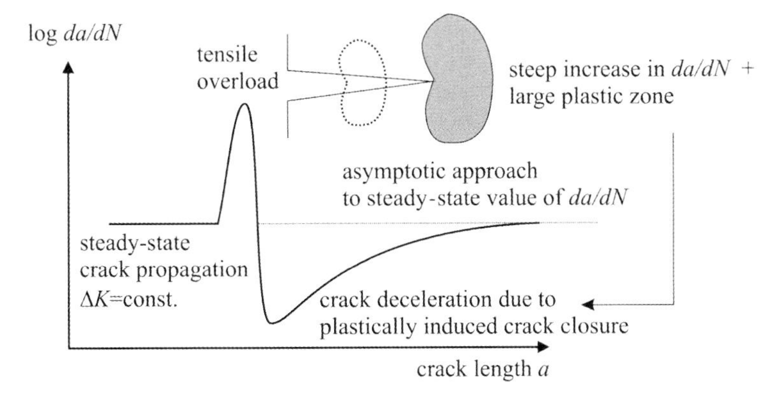

Fig. 6.24 Response of the crack-propagation rate to tensile overloads (schematic).

6.3.4
Roughness-Induced Crack Closure

Plastic deformation at the crack tip leads not only to the development of a plastic wake but also to a mutual displacement of the two crack faces. This mismatch causes locally premature contact of the crack faces before the sample/component is completely unloaded from remote tension.

Although a certain displacement of the crack faces occurs generally during plastic deformation in the plastic zone ahead of the crack tip, roughness-induced crack closure is particularly pronounced during crystallographic mode II crack propagation (cf. [380, 381]). In these cases, the crack-closure stress-intensity factor K_{cl} and, hence, the significance of roughness-induced crack closure depends on the alloy's grain size and the crystallographic misorientation relationships, when planar slip prevails. These microstructure parameters determine the roughness profile of the crack faces [340, 382]. With increasing grain size the extent of roughness-induced crack closure; i.e., the value of K_{cl}, increases and the effective range of the stress-intensity factor ΔK_{eff} decreases [340, 358, 382]. This is of particular significance for small or negative values of the stress ratio R, small ranges of the stress-intensity factor ΔK (HCF regime) and relatively low temperatures, or simply for conditions where plastic deformation at the crack tip is dominated by mode II shear displacement [358].

Wang and Müller [381] derived a relationship for roughness-induced crack closure in a Ti-2.5Cu alloy for small ΔK values, where K_{cl} is correlated empirically with the maximum value of the stress-intensity factor K_{max}, the yield strength σ_Y and geometrical parameters of a section through the crack-face profile (crack-deflection angle, height distribution). In a similar approach by Jung and Antolovich [383] K_{cl} is correlated with the height h of the roughness peaks within the fracture surface. For large values of h a relationship between roughness-induced crack closure and the crack length was found.

As discussed in the preceding section for plasticity-induced crack closure, also roughness-induced crack closure should be strongly influenced by compressive overloads, since they flatten the surface roughness of the crack faces. Consequently, K_{cl} decreases, and ΔK_{eff} and the crack-propagation rate increase [278, 374].

Modeling of roughness-induced crack closure and implementation in approaches to predict the fatigue-crack-propagation rate was carried out and published in various papers, e.g., by Suresh [384], Suresh and Ritchie [385], or Pokluda et al. [366, 386], taking grain-size statistics into account. Since roughness-induced crack closure is also relevant during propagation of microstructurally short fatigue cracks (see Section 6.3.7), it is accounted for in the numerical model that is introduced in Chapter 7.

6.3.5
Oxide- and Transformation-Induced Crack Closure

Just as Elber's plasticity-induced crack closure creates a plastic wake, which acts as a wedge between the two crack faces, oxidation or phase transformation products may cause a volume increase in the vicinity of the crack tip or in the wake leading to premature contact of the crack faces after unloading from remote tension [358].

Quantification of oxide- and transformation-induced crack-closure effects is difficult, since they depend on a variety of mutually interacting influence factors. Generally, the significance of oxide-induced crack closure increases with increasing temperature or oxygen partial pressure in the atmosphere, both leading to an acceleration in oxidation of the crack faces [358, 380, 387]. This is particularly relevant for the formation of massive thermal oxide scales at high temperatures, e.g., in nickel-based superalloy IN783, as shown by Ma et al. [388]. At the same time, the roughness of the crack faces and hence the extent of roughness-induced crack closure decreases. Consequently, the threshold value ΔK_{th} as a function of the temperature passes through a minimum (see Fig. 6.25) [358].

The relationship between the loading conditions and oxide-/transformation-induced crack-closure effects is controversially discussed in the literature, probably because this relationship depends on various additional influence factors, e.g., environmental effects. Low or negative stress ratios R cause an increase in frictional corrosion and hence an increase in oxide-induced crack closure [358]. On the other hand, at high temperatures admission of the atmosphere to the crack tip is eased by positive mean stresses ($R > 0$). Also, a material's strength is relevant for oxide-induced crack-closure effects. According to studies on copper using Auger electron spectroscopy (AES) by Liaw et al. [358, 389], the thickness of the oxide scale formed at the crack faces decreases with increasing strength. This can be attributed to a smaller extent of plasticity-induced crack closure (less plastic deformation at the crack tip of the stronger material), leading to a decrease in the contribution of frictional corrosion.

Some materials exhibit deformation-induced phase transformations. This effect, which is extensively used in the case of shape memory alloys, often leads to an increase in the specific volume as compared with the undeformed material. When deformation-induced phase transformation is restricted to the crack tip, the additional volume again causes premature contact of the crack faces when unloading

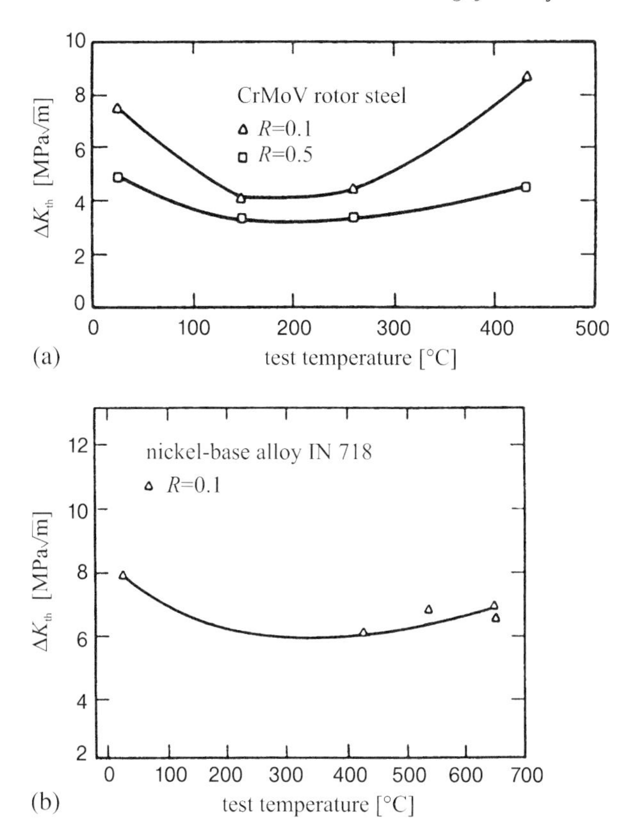

Fig. 6.25 Effect of the temperature on the threshold stress-intensity range ΔK_{th} for (a) A470 CrMoV rotor steel and (b) Ni-based alloy 718 (according to [358]).

from tension. This mechanism has been termed "transformation-induced crack closure" and is technically important for structural applications of metastable austenitic steels or TRIP steels [390], the latter being frequently used for the design of high-strength parts of autobodies.

The effect of transformation-induced crack closure during fatigue of metastable austenitic steel was studied by Hornbogen [391] (model alloy) and Mayer et al. [392] (metastable austenitic bainitic cast iron GGG100B), who found for a 30% deformation-induced transformation from fcc austenite via hcp ε martensite to bct α′ martensite a volume increase of 0.6% (cf. also [393]).

6.3.6
$\Delta K^*/K^*_{max}$ Thresholds: An Alternative to the Crack-Closure Concept

The fundamental concept of plasticity-induced crack closure distinguishes between an extrinsic threshold ΔK_{th} and an intrinsic threshold $\Delta K_{th,eff}$ for the range of the stress-intensity factor. While the extrinsic threshold accounts for shielding of the crack tip, the intrinsic threshold is a measure for the fatigue-crack-driving situation immediately at the crack tip. However, besides the plasticity-induced crack-closure concept alternative approaches have been introduced in the literature. These approaches consider simultaneous overcoming of two thresholds: a threshold ΔK^* of the range of the stress-intensity factor and a threshold K^*_{max} for the maximum value of the stress intensity during the fatigue loading cycles. Vasudevan et al. [361, 394] and Sadananda et al. [395] assume plasticity-induced crack closure as a material-specific phenomenon to be negligible and the influence of roughness-induced crack closure to be small. According to [394, 395], the microstructural development and the residual-stress state at the crack tip is of substantially higher significance than plasticity-induced crack closure. The frequently mentioned effects of the stress ratio on fatigue-crack propagation is accounted for by means of a two-parameter approach in Eq. (6.3), where the crack-propagation rate depends on both the range of the stress intensity factor ΔK and its maximum value K_{max}. The constants C, n and m have the same meaning as the constants in the Paris law (Eq. 2.5):

$$\frac{da}{dN} = C(\Delta K)^n (K_{max})^m \tag{6.3}$$

The condition for crack propagation is fulfilled when both thresholds, ΔK^* and K^*_{max}, are exceeded [361].

Krenn and Morris [396] compared two fundamental models, where crack propagation is substantially governed by (1) plasticity-induced crack closure and (2) the residual-stress state ahead of the crack tip, coming to the conclusion that extrinsic influence factors, like crack closure, should be studied based on the relevant microstructural mechanism rather than based on theoretical mechanical hypotheses. Some of the approaches introduced in Chapter 7 follow this suggestion in the framework of implementing microstructural short fatigue cracks in methods for fatigue-life assessment.

6.3.7 Development of Crack Closure in the Short Crack Regime

While the impact of crack-closure effects on the propagation of *long* and *physically short* fatigue cracks is fairly well understood with respect to the most relevant mechanisms and technical implications, knowledge about crack-closure effects during microcrack propagation is only marginal.

By definition, the plastic wake, being the key factor of plasticity-induced crack closure, must be small in the case of short cracks, and hence is of negligible significance [301, 356]. With increasing crack length, the size of the plastic zone becomes larger, manifesting itself in a larger plastic wake and a decrease in the crack-propagation rate da/dN. It can be concluded that plasticity-induced crack closure depends on the crack length [121, 269, 397, 398], and determines the transient behavior of short fatigue cracks within a distance of approximately 5 to 13 grain diameters (according to Larsen et al. [398]). Early crack propagation starts initially without crack closure at a high propagation rate da/dN, which quickly decreases due to a gradual development of plasticity-induced crack closure. Subsequently, the crack-propagation rate increases again. This is because the range of the stress-intensity factor ΔK increases with increasing crack length, once a steady state condition of crack closure is reached ($K_{cl} \approx$ constant) [358, 398] (cf. Fig. 6.32). James, Sharpe, and Graz [399, 400] classify the development of crack closure as an intrinsic material property, which can be correlated with the material strength. According to Jira et al. [401], the establishment of steady-state crack closure requires a material resistance in the wake that must be sufficient to act as an efficient wedge that unloads the crack tip. For the length d, within which crack closure is developed, they suggest the following empirical relationship depending on the crack length a, the maximum stress intensity factor K and the yield strength σ_Y:

$$\frac{d}{a} = -0.0005x + 0.5038x^2 - 0.0077x^3 - 0.1139x^4 + 0.0334x^5 \quad x = K/\sigma_Y \tag{6.4}$$

However, a closer inspection reveals that for short cracks such a general simplified description does not represent the actually active mechanisms. In particular in the case of microstructurally short fatigue cracks, crack-closure effects are determined by microstructural parameters, like the grain size and crystallographic orientation relationships.

To describe propagation results, obtained by *in situ* observations of short cracks in a 2219 Al alloy, Morris [214] derived a relationship between the crack-closure stress σ_{cl} related to the maximum stress σ_{max} and the distance to the next grain boundary, z, related to the surface crack length $2c$:

$$\frac{\sigma_{cl}}{\sigma_{max}} = \alpha \frac{z}{2c} \tag{6.5}$$

with α being a material parameter depending on the relative humidity ($\alpha \approx 1.2$ in dry air). Equation (6.5), being valid for $z/2c > 1$, characterizes the relationship between crack closure/propagation and the position of the crack tip relative to the next grain boundary. If the crack tip is close to a grain boundary, then the plastic zone limited by the grain boundary is small, and hence the contribution of crack closure is negligible and the crack-propagation rate should be high. It should be pointed out here that this conclusion is different from that which can be drawn according to the numerical short-crack model introduced in Chapter 7. In the case of a microstructurally short fatigue crack it is important to distinguish between single- and multiple-slip-driven crack propagation. In both cases, the plastic zone can be blocked by the grain boundary (cf. Fig. 6.15a), but the implications for the development of crack closure are different.

The crystallographic growth of microstructurally short fatigue cracks along slip bands (mode II crack propagation) is substantially influenced by roughness-induced crack closure [215, 402]. Due to plastic displacement within the slip plane (crack-tip-slide displacement, CTSD), matching between the crack faces is perturbed as soon as the crack has overcome the first grain boundary [403]. The extent of the corresponding positive crack-closure stress σ_{cl} is determined by the crack-de-

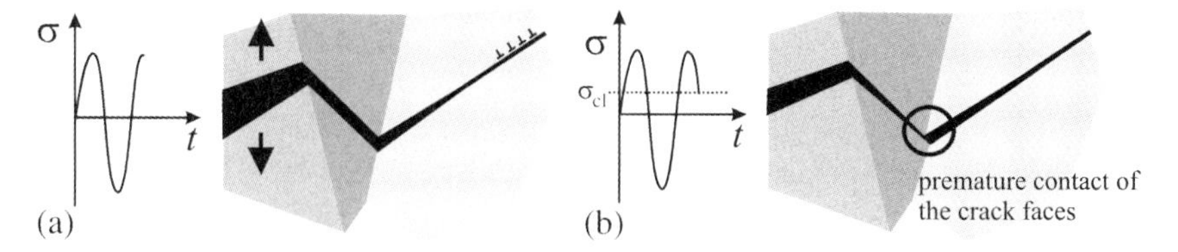

Fig. 6.26 Schematic representation of roughness-induced crack closure as a consequence of crystallographic microcrack propagation: (a) CTSD at maximum tension; (b) premature contact of the respective crack faces.

flection lengths [340], the misorientation of the slip systems of the neighboring grains [367] as well as by the degree of cyclic slip irreversibility at the crack tip [358]. These relationships, which are also considered in the numerical short-crack model introduced in Chapter 7, are represented schematically in Fig. 6.26.

By applying the interferometric strain/displacement gauge (ISDG) system, which is described in Section 3.1.3, crack-closure effects during early propagation of short fatigue cracks were quantitatively analyzed [227, 334]. For this purpose, Vickers microhardness indentations were placed above and below a propagating microcrack and illuminated by laser light. The shift of the resulting interference pattern yielded the crack-mouth-opening displacement (CMOD) at the surface, which was correlated with the applied remote stress. As an example, Fig. 6.27 shows two stress vs. CMOD hysteresis curves for microcracks in the β-titanium alloy LCB fatigued at $\Delta\sigma/2 = 400$ and 600 MPa, both at a stress ratio at $R = -1$ and a surface crack length of $2c = 300$ µm.

The different slopes of the hysteresis curves can be correlated with the following situations: (1) specimen with open crack and reduced stiffness (upper branch of

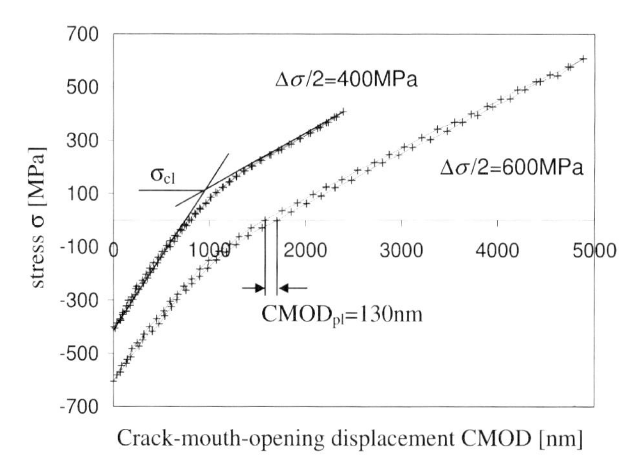

Fig. 6.27 Determination of crack-closure stress σ_{cl} by means of ISDG evaluation of short surface microcracks in the β-titanium ally LCB: remote stress vs. CMOD.

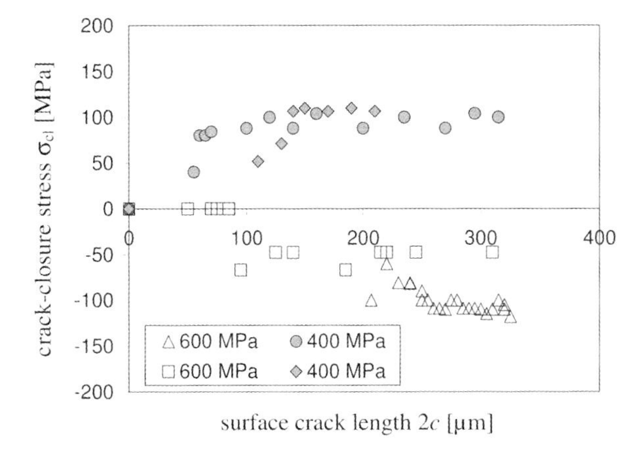

Fig. 6.28 Crack-closure stress σ_{cl} vs. surface crack length $2c$, obtained from stress vs. CMOD hysteresis loops (see Fig. 6.27; for details see text).

the curve) and (2) specimen with closed crack exhibiting a stiffness close to the compact material (lower branch of the curve). According to [215, 227, 404] the crack-closure stress σ_{cl} can be estimated by the intersection point of two tangents, applied to the two distinct slopes of the unloading branches of the hysteresis curves (as shown in Fig. 6.27). The development of crack closure during the propagation of microstructurally short cracks can be obtained by carrying out the procedure mentioned above for stress vs. CMOD hysteresis curves for various crack lengths. Such an evaluation is shown in Fig. 6.28, again for the β-titanium alloy loaded according to Fig. 6.27.

The results allow one to draw the following conclusions. At low stress amplitudes, crystallographic crack propagation along slip bands prevails. The mode II displacement at the crack tip leads, analogous to the mechanism sketched in Fig. 6.26, to roughness-induced crack closure and a positive crack-closure stress, as observed indeed during crack propagation at low remote stress ($\Delta\sigma/2 = 400$ MPa, Fig. 6.28). At higher stress amplitudes, the resolved shear stress acting on the slip planes increases, eventually leading to a situation where a change from single to multiple slip and crack-tip blunting occurs. According to the schematic representation in Fig. 6.29a and the micrograph in Fig. 6.29b, the resulting mismatch of the crack faces is reduced and the crack stays open, even when completely unloaded from maximum tension to zero (Fig. 6.29b). This also explains the increase in the plastic contribution to crack-mouth opening ($CMOD_{pl}$, see Fig. 6.27 and [405]).

The reduced displacement of the crack faces against each other and the residual compressive stresses ahead of the crack tip shift the crack-closure stress σ_{cl} towards negative values. Negative crack-closure stress values were also observed during microcrack propagation in an AlMgSi aluminum alloy (DIN1725, Al6082) by Lenczowski [94] (Fig. 6.30a) and in A533 carbon steel by James and Sharpe [399] (Fig. 6.30b), for two different grain sizes and a constant stress ratio of $R = -1$.

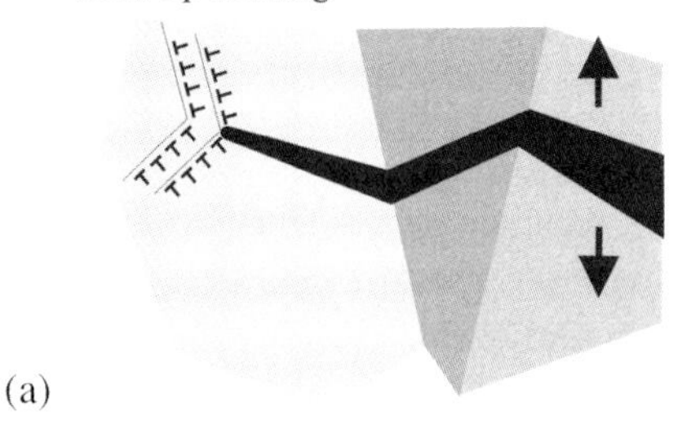

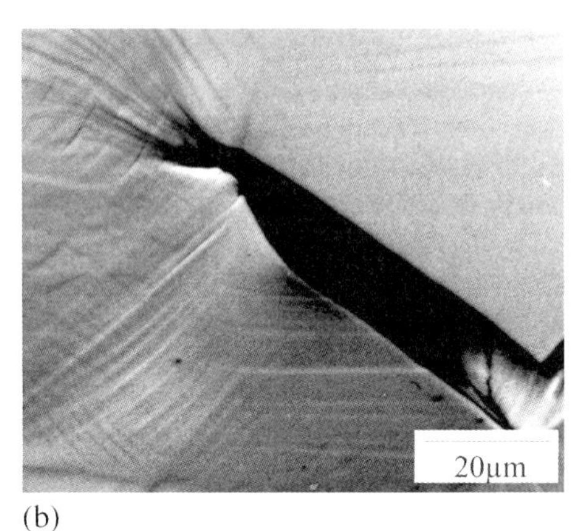

Fig. 6.29 Microcrack propagation under simultaneous activation of several slip systems: (a) schematic representation, (b) surface crack that remained open after complete unloading from $\Delta\sigma/2 = 600$ MPa (β-titanium alloy LCB).

For high stress amplitudes, the development of crack closure is characterized by a quick drop down to negative crack-closure stresses σ_{cl}, being equivalent to delayed crack closure, a relationship that was also pointed out by McDowell [406] for fatigue-crack growth in 1045 plain carbon steel. High stress amplitudes yield situations where crack-closure effects vanish; i.e., the ratio between the crack-opening stress (approximately of same value as that of the crack-closure stress) and the maximum stress is equal to $\sigma_{op}/\sigma_{max} = -1$. During ongoing fatigue loading, the ratio σ_{op}/σ_{max} increases until it reaches a steady-state value for plasticity-induced crack closure after Elber of $\sigma_{op}/\sigma_{max} = 0.3$–$0.4$. For low stress amplitudes roughness-induced crack closure due to single-slip displacement at the crack tip prevails during microcrack propagation.

Figure 6.31 shows the development of crack-closure stress for microcracks initiated during fatigue of AISI F51 duplex steel at a stress amplitude of $\Delta\sigma/2 = 370$ MPa (Fig. 6.31b) and $\Delta\sigma/2 = 400$ MPa (Fig. 6.31b) at a stress ratio of $R = -1$ [318]. ISDG measurements revealed a positive crack-closure stress σ_{cl} due to roughness-induced crack closure corresponding to single-slip crack propagation. After applying a 50% compression overload, the crack-closure stress dropped to negative values (Fig. 6.31a). In the case of the $\Delta\sigma/2 = 400$ MPa experiment, initial crack closure occurred at $\sigma = 0$ MPa. A substantial decrease to negative values was observed when applying 50% tensile overloads (twice, Fig. 6.31b). These results support the

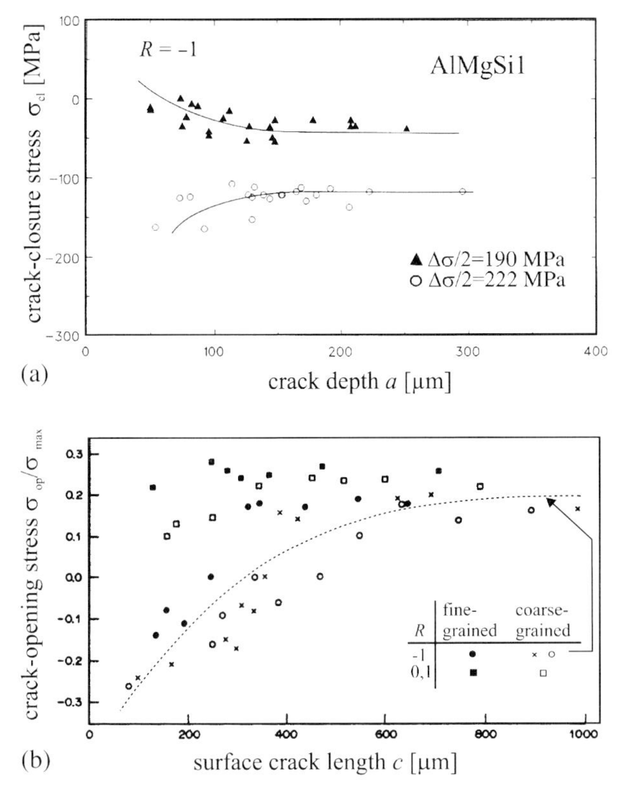

Fig. 6.30 Crack-closure stress σ_{cl} vs. surface crack length 2c, revealing negative crack-closure stress values, for an aluminum alloy (after [94]) and carbon steel (after [399]).

hypothesis that overloads promote the transition from single-slip to multiple-slip crack propagation, i.e., roughness-induced crack closure is replaced by residual CTOD at zero stress, which has to be overcome by a compressive crack-closure stress σ_{cl}.

In general, the transient regime of crack closure, owing to the development in microplasticity at the crack tip on the one hand and the increase in crack length on the other hand, can be considered as an explanation for the abnormal propagation behavior of microstructurally short fatigue cracks as compared to the behavior of long cracks (see [282, 357, 398, 407]). This is illustrated by the schematic representation in Fig. 3.32. Assuming crack initiation occurring within one single grain, one would expect crack-closure effects to be ineffective, i.e., $\sigma_{cl} = 0$. As discussed in Section 6.2.1, early crack propagation occurs mostly in a crystalline manner. The corresponding displacement of the crack faces leads to premature crack closure and crack-tip shielding at $\sigma_{cl} > 0$. Once the crack grows by operation of multiple slip systems, the plastic zone ahead of the crack tip becomes larger during loading in tension. This tensile deformation needs to be overcome by compression; the

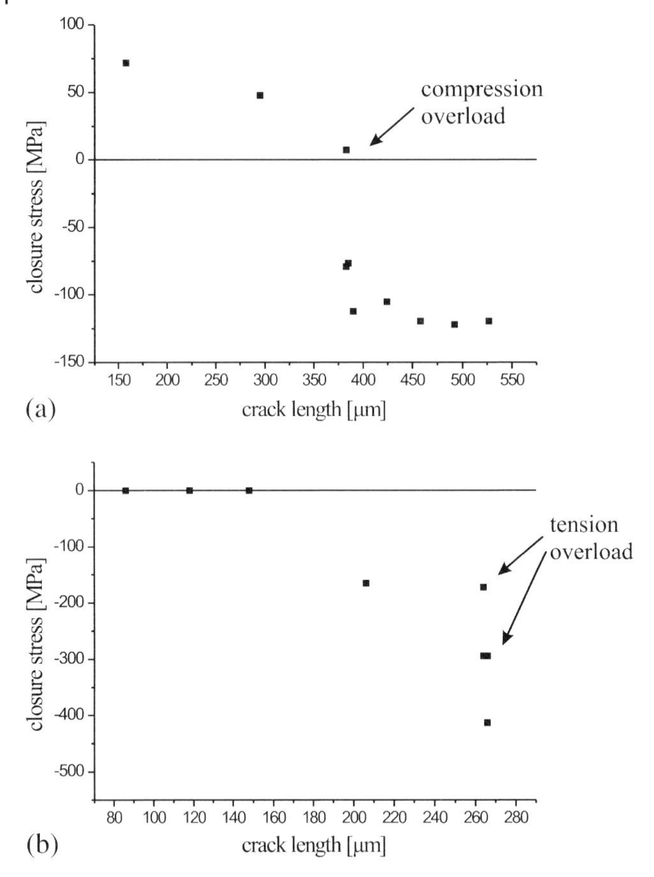

Fig. 6.31 Evolution of crack-closure stress during microcrack propagation in AISI F51 duplex steel: (a) $\Delta\sigma/2 = 370$ MPa and compressive overload, (b) $\Delta\sigma/2 = 400$ MPa and tensile overload (after [318]).

crack-closure stress becomes negative ($\sigma_{cl} < 0$). This leads to a situation for fully reversed loading ($R = -1$), where the complete applied range of the stress intensity factor can be effective [201] ($\Delta K = \Delta K_{eff}$, regime I in Fig. 6.32 with the effective range of the stress intensity factor ΔK_{eff} being dark gray), and microstructurally short cracks may grow for conditions that would not allow even the initiation of a technical fatigue crack. According to Liaw [358], the influence of the plastic wake increases faster than the applied range of ΔK. As a consequence, ΔK_{eff} passes through a minimum (transient regime II in Fig. 3.32), which might cause crack propagation to stop, if ΔK_{eff} falls below the local threshold. A similar mechanism is reported by Tanaka and Nakai [408] to explain crack propagation from notches.

After passing through the minimum, the crack-closure stress intensity K_{cl} gradually approaches its steady-state value and the crack-propagation rate increases because of the continuous increase in the applied range of the stress-intensity factor

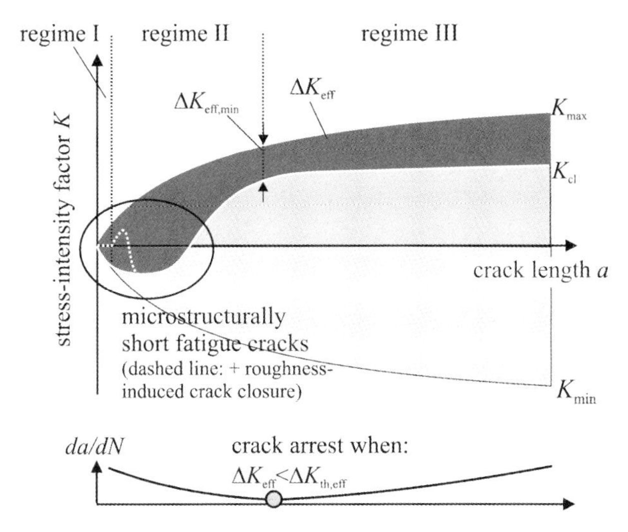

Fig. 6.32 Schematic representation of the evolution of crack-closure effects (crack closure stress intensity K_{cl}) during microcrack propagation (crack length a) – for microstructurally short cracks only a small plastic wake is present leading to a large contribution of ΔK to ΔK_{eff}.

ΔK, depending only on the crack length and geometry (regime III in Fig. 6.32, cf. James and Sharpe [399], for $\Delta\sigma_a$ = constant).

It should be pointed out that, strictly speaking, the application of the K concept to microstructurally short cracks is not useful, since their propagation does not follow the concepts of classical continuum mechanics. It is used in the preceding section only for the sake of illustration of the relationship between short- and long-crack propagation.

When comparing the effect of overloads on the propagation behavior of short and long fatigue cracks, it is worth mentioning that according to [409] tensile overloads lead to a decrease in the crack-closure stress σ_{cl} and an increase in the effective range for microstructurally short cracks, while in the case of long fatigue cracks σ_{cl} experiences an intermediate increase.

6.4
Short and Long Fatigue Cracks: The Transition from Mode II to Mode I Crack Propagation

Immediately after crack initiation, early fatigue-crack propagation is crystalline and predominately controlled by the resolved shear stresses (mode II) on the respective slip bands. With increasing three-dimensional crack advance, the influence of superimposed normal stresses (mode I) increases and the crack-propagation direction changes from close to 45° (maximum shear stress) to 90° with respect to the remote normal stress. This transition from mode I to mode II crack propagation, which is commonly considered as the transition criterion from short (stage I) to long (*K*-controlled stage II) fatigue-crack propagation, is schematically represented in Fig. 6.33. This concept was also postulated by Cheng and Laird [410] to model the stage I to stage II transition based on the presence of either single slip or multiple slip.

At which number of cycles the mode II to mode I transition (stage I to stage II transition) occurs depends on the loading conditions, the stress amplitude, the material's strength and its crystal structure, e.g., high strength bcc alloys tend to crystallographic single-slip microcrack propagation while soft fcc alloys tend to multiple-slip crack propagation, even in the early stages of the fatigue damage process (stage Ib crack propagation, cf. Section 6.2.2). At high load levels in the low-cycle fatigue (LCF) regime, microstructurally short cracks are not of high significance since the plastic zone ahead of the crack tip extends quickly to several grains often into an area of full plastification, either of a whole specimen or a notch area in a technical component. Under such circumstances, crack initiation may be followed immediately by physically short- and long-fatigue-crack propagation leaving out the stage of microcracks.

In the literature various semi-quantitative criteria for the transition form short to long crack behavior (stage I to stage II) are given. According to Tokaji and Ogawa [303], the transition occurs when the crack length has exceeded eight grain diameters. Taylor and Knott [304] correlate the transition from mechanically short to physically short fatigue cracks with a crack length of ten grain diameters. A somewhat more general formulation can be found in the work of Yoder et al. [411] or Rodopolous and de los Rios [305], the latter referring to the original model of Navarro and de los Rios [324] (see Section 7.2). Hence, crack propagation becomes independent of the microstructure, when the plastic zone ahead of the crack tip

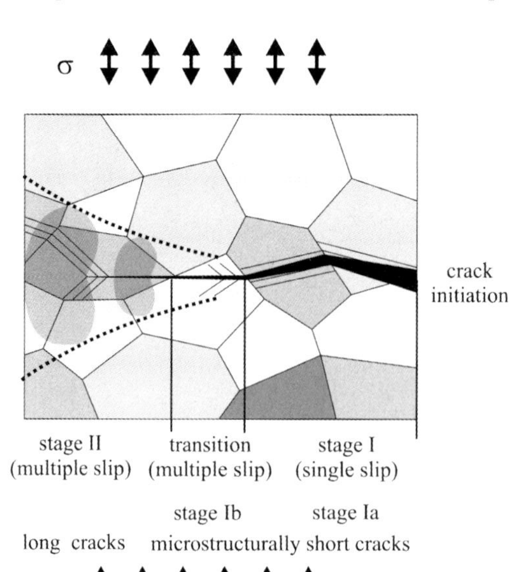

Fig. 6.33 Transition from stage I (stage Ia: shear-stress-controlled; stage Ib: normal-stress-controlled) to stage II crack propagation (normal-stress-controlled).

exceeds one grain diameter [411] or, as a more conservative assumption, exceeds two grain diameters [305].

A fundamental problem of implementing microstructurally short cracks in fatigue-life assessment is the common two-dimensional simplification of a problem, which is actually of pronounced three-dimensional nature. While long cracks can be considered as semicircular or semi-elliptical with an almost constant ratio between the crack depth a and half of the surface crack length $2c$ (aspect ratio a/c), the three-dimensional geometry of microcracks seems to be determined by local microstructural features, leading to a variation in the aspect ratio along the crack front (cf. Section 6.4.1). Since in the microcrack regime, slip in the plastic zone ahead of the crack tip is occurring along individual slip bands, deformation of the grains is constrained by the compatibility of the activated slip systems of neighboring grains, becoming increasingly complex as more grains are involved (Fig. 6.34). As a consequence, the driving force to activate multiple slip increases exponentially and the crack-propagation behavior becomes gradually homogeneous approaching a steady-state crack propagation rate (for ΔK = constant). This is in agreement with observations of Carlson [412], who found a decreasing scatter (standard deviation) in the short-crack-propagation rate with increasing crack length (or number of grain intersections). This is shown in Fig. 6.34a.

Besides the development in the aspect ratio, crack coalescence effects may play a key role during the transition from short- to long-crack propagation [413]. Crack coalescence effects are the subject of Section 6.4.2.

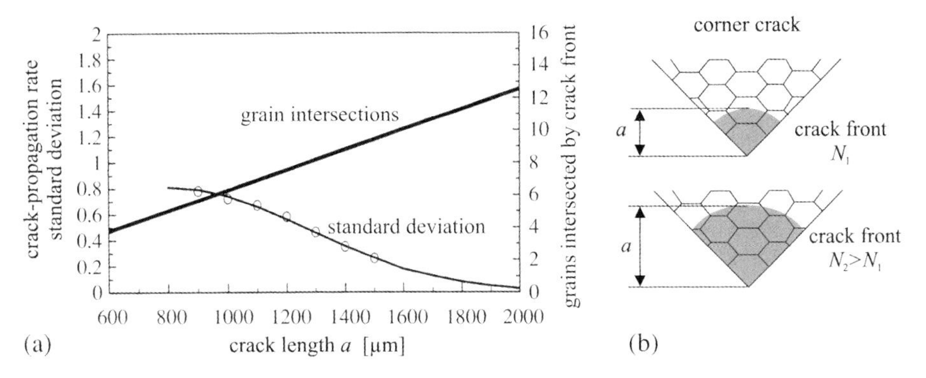

Fig. 6.34 (a) Increase in the number of grains involved and corresponding decrease in the standard deviation of da/dN during the propagation of (b) short corner cracks (according to [412]).

6.4.1
Development of the Crack Aspect Ratio a/c

For many materials it can be observed that cracks, originally initiated under very different conditions, like deep or shallow notches, inclusions or grain boundaries, converge when growing longer towards a semicircular contour with a constant aspect ratio of $a/c = 1$ [414], e.g., Ti-4Al and Ti-8Al, for surface crack lengths of $2c$ between 50 μm and 2 mm [398].

During early propagation of short fatigue cracks, the interactions with local microstructural features, which are discussed in the preceding sections, prevail. The variation in the strength of microstructural barriers leads to a certain band width of local aspect ratios. As a consequence of the strong constraints to single-slip deformation of the grains in the interior of the material (experiencing plane-strain conditions), microstructurally short cracks tend to propagate in a shallow manner (for surface grains the constraint is limited due to plane-stress conditions). In such cases, the aspect ratio is given by $a/c < 1$, as observed, for example, for a 7075 T6 aluminum alloy [202]. With increasing crack length the aspect ratio approaches the stable value of the semicircular shape $a/c \approx 1$ [371]. According to Ravichandran [414], the stress intensity factor K becomes homogeneous along the crack front for $a/c = 1$. It should be added here that shallow surface cracks tend to stop, e.g., at a microstructural barrier, while deep cracks, $a/c >> 1$, propagate preferentially, since they have already overcome the strong plane-strain constraint, which is present underneath the layer of surface grains.

A similar tendency for short cracks to operate with $a/c < 1$ was also observed for the β-titanium alloy LCB. To obtain the required three-dimensional data, fatigued specimens were stepwise polished down (15–30μm steps) after carefully analyzing microcrack propagation at the surface for various numbers of cycles. Figure 6.35a shows the procedure to obtain longitudinal sections of, for example, the microcrack shown in the surface micrograph in Fig. 6.35b. The course of the absolute crack depth as a function of the absolute crack length is given in Fig. 6.35c, referring to the marked measurement points in Fig. 6.35b. The respective crack aspect ratio is $a/c = 0.68$.

In addition to the observation that also crack propagation into the depth depends on the local microstructure, it was found that the crack segment marked in Fig. 6.35b, originally initiated along a grain boundary, propagated transgranularly into the bulk, as revealed by the longitudinal section in Fig. 6.35d. Obviously, only crack initiation was intercrystalline, probably due to stress concentration caused by elastic anisotropy. Further crack propagation occurred along a slip band, inclined at 45° with respect to the loading axis, driven by the maximum resolved shear stress on the respective slip planes. Prior to reaching the next boundary, the crack changed its propagation direction towards mode I, supporting the idea that mode II microcrack propagation is restricted to only a very shallow area of the specimen/component. A further example for the three-dimensional representation of microcracks, where the transition to multiple slip occurs underneath the surface, is provided by Fig. 6.36, showing a fatigue sample of AISI F51 duplex steel.

Similar results were obtained by Blochwitz and Tirschler [257] for the behavior of microcracks in 316 austenitic steel. They found crack initiation preferentially at twin boundaries, owing to stress concentration caused by elastic anisotropy (cf. Section 5.3.4). However, this kind of interfacial damage was restricted to the near-surface area only. Crack propagation into the bulk was governed by the maximum resolved shear stress acting on the slip systems of the respective grains.

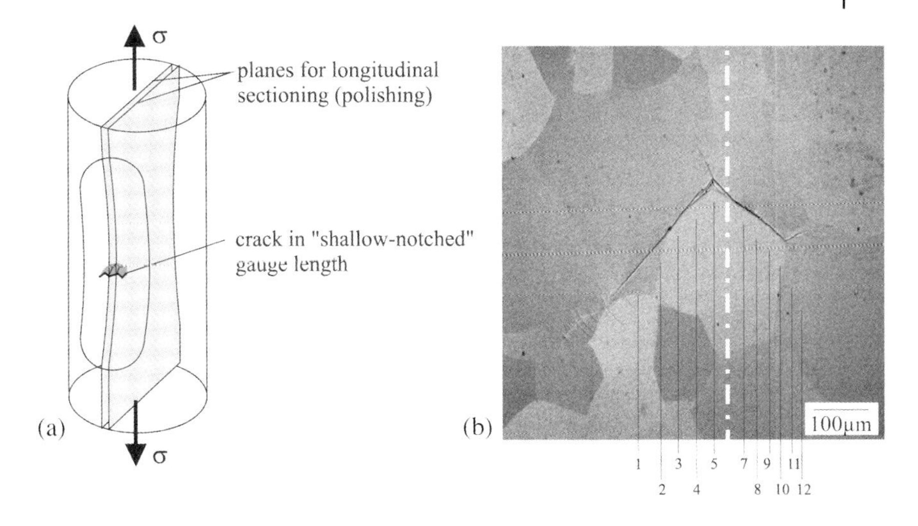

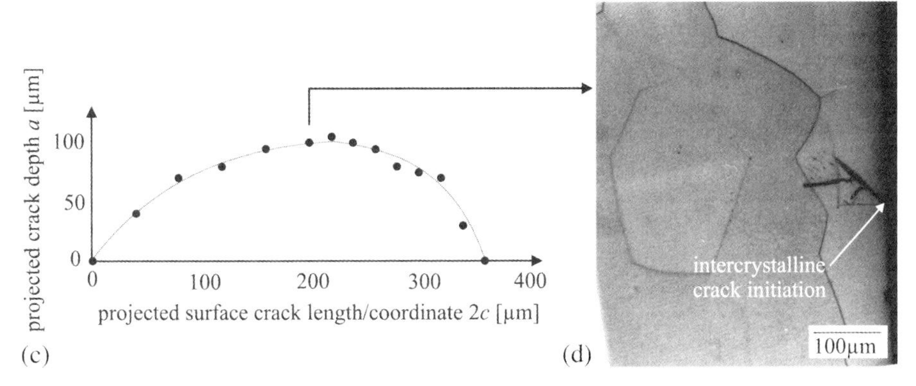

Fig. 6.35 (a) Schematic representation of three-dimensional crack-course analysis by sectional polishing; (b) surface crack that has been analyzed with respect to the corresponding crack depth profile (c, d) (β-titanium alloy LCB, $\Delta\sigma/2 = 600$ MPa, $R = -1$, $N = 3000$ cycles).

Besides the variation in the crack aspect ratio a/c, three-dimensional propagation of microstructurally short fatigue cracks becomes even more complex due to crack branching at microstructural barriers. An example of crack branching is given in Fig. 6.37, showing a cross section (parallel to the shallow-notched specimen surface) through a microcrack in a fatigued sample of the β-titanium alloy LCB, which has just overcome a grain boundary.

In spite of the local variation in the depth of microstructurally short fatigue cracks, the development of the crack aspect ratio vs. the surface crack length, again for a fatigue specimen of the β-titanium alloy LCB, was estimated by a compliance method (see below) and heat tinting (oxidation of the crack faces by locally heating the specimen). It was shown that the aspect ratio approaches a stable value for semi-elliptical cracks of $a/c = 0.9$ [90].

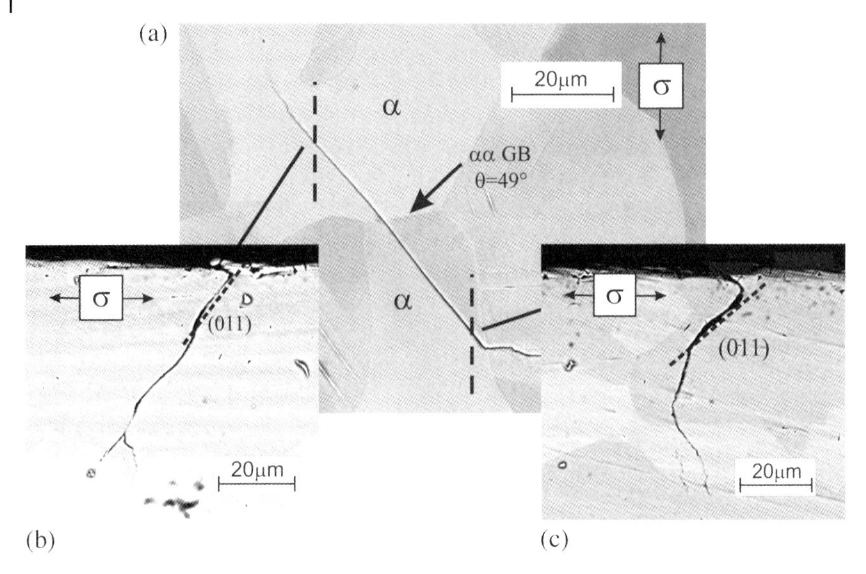

Fig. 6.36 Microcrack propagation in AISI F51 duplex steel at $\Delta\sigma/2 = 350$ MPa, $R = -1$, $N = 1.1 \times 10^6$ cycles: crack path at (a) the surface and (b, c) in the bulk according to the dashed lines in (a).

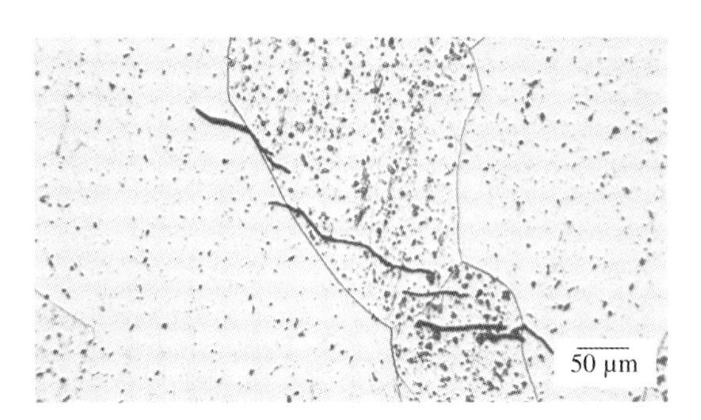

Fig. 6.37 Crack branching after crossing a grain boundary: etched longitudinal section parallel to the surface of a fatigued specimen of the β-titanium alloy LCB ($\Delta\sigma/2 = 600$ MPa, $R = -1$, loading axis: ⇕).

A similar development of the aspect ratio was observed by Ravichandran and Larsen [414] and Ravichandran [97] for the titanium alloy Ti-6A-2Sn-4Zr-6Mo. In the same way as the aspect ratio was obtained by Floer [90] for the titanium alloy LCB, they used an approach originally derived by Fett [415] that correlates the compliance $C(0, 0)$ in the center of a surface crack ($\rho = 0$, $\Phi = 0$, according to the geometrical arrangement in Fig. 6.38a) with the crack aspect ratio a/c. Following the solution of Newman and Raju [320] for the distribution of the stress-intensity factor along the front of three-dimensional surfaces and the work of Ravichandran and Li [417], half of the CMOD $u(\rho, \Phi) = 0.5 \times$ CMOD can be expressed as

$$u(\rho,\Phi)=\sum_{n=1}^{\infty}C_n\left(1-\frac{\rho}{r}\right)^{n+0.5} \tag{6.6}$$

where C_n are constants to be determined form boundary conditions [414]. By cutting the series in Eq. (6.6) for $n = 2$, one obtains after a few simplifications [414] the compliance $C(0, 0)$ as the ratio between CMOD and the applied remote stress σ:

$$C(0,0)=\frac{\text{CMOD}}{\sigma}=\frac{\sqrt{8}}{\sqrt{\pi}E}c\left(1-v^2\right)\lambda Ff_w g \tag{6.7}$$

with F, f_w and g being geometrical functions [414, 415], which depend on the surface crack length $2c$ and the crack depth a, as well as on the specimen thickness t and width w. For aspect rations $a/c \leq 1$ the parameter λ is equal to a/c, otherwise $\lambda=(a/c)^{0.5}$ is valid. To determine CMOD experimentally, Ravichandran und Larsen [414] used the ISDG technique (cf. Section 3.1.3). To account for the fact that microstructurally short fatigue cracks do not grow symmetrically with respect to their initiation site (cf. Section 7.3), they applied a correction that relates the measured CMOD, CMOD(ρ/r, 0), at an arbitrary point along the crack course at the surface with CMOD(0, 0) in the crack center ($\rho = 0$, $\Phi = 0$):

$$\text{CMOD}(0,0)=\frac{0{,}8\cdot\text{CMOD}(\rho/r,0)}{\left(1-\rho/r\right)^{0.5}-0{,}2\cdot\left(1-\rho/r\right)^{2.5}} \tag{6.8}$$

Using the surface crack length $2c$, which can easily be measured, e.g., by optical microscopy, the actual aspect ratio a/c is obtained by iterative variation of a/c until the calculated compliance coincides with the measured values (ISDG). Figure 6.38b and c represent examples for aspect ratios of short cracks emanating from a shallow and a deep notch [414] and for a naturally initiated short crack in the β-titanium alloy LCB [90].

It should be pointed out that there is a strong scatter in the development of the aspect ratios of short fatigue cracks, as was observed, for example, during fatigue of the β-titanium alloy LCB. According to Ravichandran et al. [416], this scattering can be attributed to the variation in the propagation rate in correlation with local microstructural features. This is supported by experimental results on the development of the microcrack aspect ratio in LCB, where the stable value for long fatigue cracks $a/c \approx 0.9$ was approached from above $a/c > 0.9$ and from below $a/c < 0.9$.

Ravichandran and Larsen [414] drew the conclusion that the deviations from the propagation rate dc/dN of short surface cracks from the estimated propagation rate da/dN following the long-crack scheme can be correlated with the development of the aspect ratio. Accordingly, the crack-propagation rate can be determined taking the considerations discussed above into account [414]:

$$\frac{da}{dN}=c\frac{d\left(\frac{a}{c}\right)}{dN}+\left(\frac{a}{c}\right)\frac{dc}{dN} \tag{6.9}$$

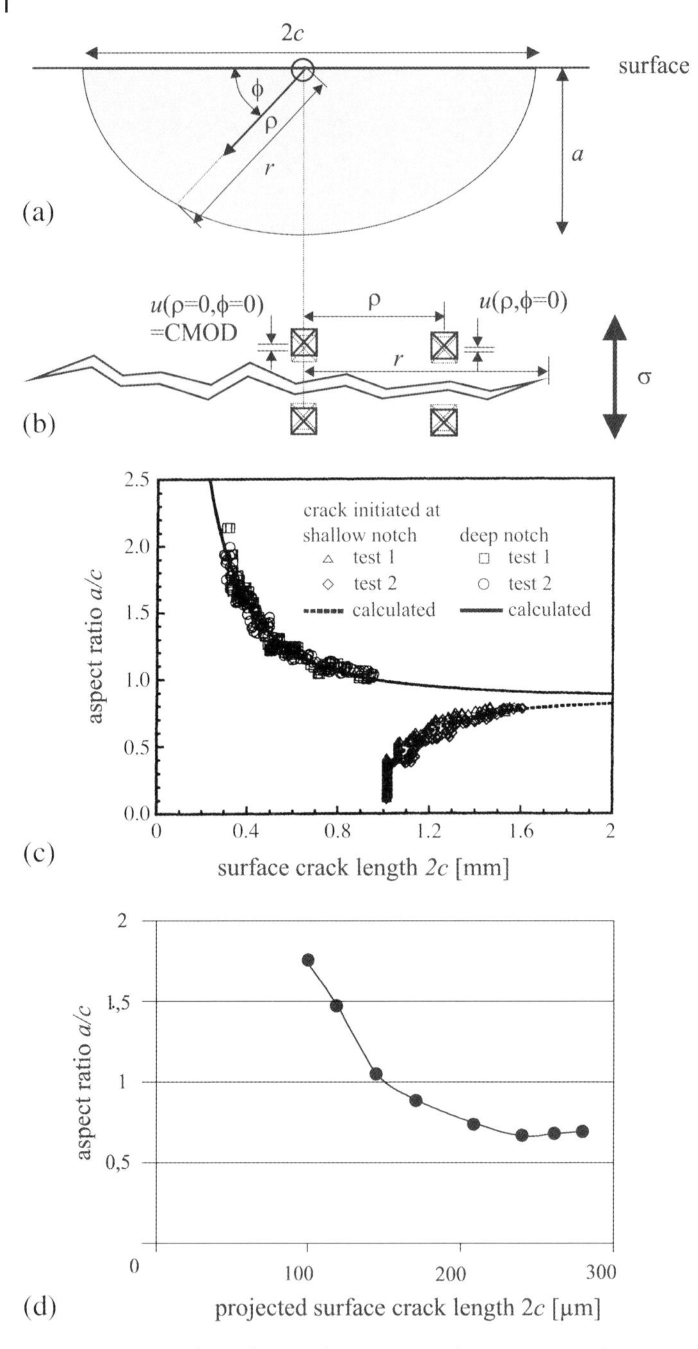

Fig. 6.38 Semicircular surface crack representing the parameters of Eqs. (6.6) to (6.9): (a) cross section, (b) surface; and, based on ISDG data, calculated aspect ratios for the Ti alloys (c) Ti-6Al-2Sn-4Zr6Mo (after [414]) and (d) LCB (with a stable aspect ratio of $a/c \approx 0.9$ for long cracks, after [90]).

Generally, the Newman–Raju concept [320] for assessment of the K distribution along a crack front implies the applicability of LEFM. Hence, it can serve only as an estimative approach for deriving the aspect ratio of microstructurally short fatigue cracks (where the conditions of LEFM are not fulfilled). Ravichandran and Li [417] improved the model, introduced in the preceding paragraphs, by a numerical calculation of local K values along the front of microstructural short cracks. In spite of the limited applicability, their model can be considered as one of the very few approaches to implement three-dimensional considerations in the theoretical analysis of short fatigue cracks.

6.4.2
Coalescence of Short Cracks

Crack coalescence effects are closely related to the development of the aspect ratio of short cracks. Varvani and Topper [278] found during biaxial tension/internal-pressure fatigue loading of a 1045 steel tube that failure occurred by rapid coalescence of the deep cracks, while the surface crack length remained almost constant. The relationship between crack propagation into the bulk, coalescence effects and failure is supported by the work of Stolarz and Kurzydloswki [413], who applied a stereological analysis to fatigue samples of the zirconium alloy Zircaloy4. According to Bayley and Bell [418] the propagation rate of coalescing cracks is mainly determined by the propagation into the bulk a_{coal}. Increasing of the surface crack length carries on only when the two coalescing cracks have reached the new equilibrium crack depth a, as shown in Fig. 6.39.

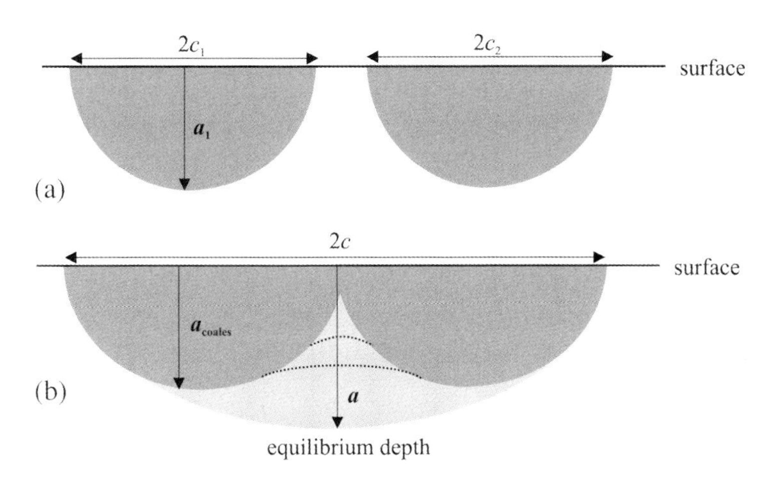

Fig. 6.39 Crack coalescence: two coalescing cracks (a) resume growing at the surface only when the new crack (b) has reached its equilibrium shape in depth a.

Depending on the material and in particular for HCF loading conditions, the early stage of fatigue life (approximately 20–30% of N_B) is determined by the initiation of many individual microstructurally short cracks. Only after exceeding about 50% of the total fatigue life do crack propagation and crack coalescence predominate, and hence the crack density may decrease. This was observed, for example, for α-Fe [199] (Fig. 6.40b) [199]. Alternatively, the crack density may reach a saturation value, when the initial microcrack density is small, as shown for the example of an Al–Cu–Mg alloy in Fig. 6.40a [199]. Normally, this is the case for very low plastic-strain amplitudes in the HCF regime. Consequently, the frequency of crack coalescence events is rather small. According to [300], the formation and propagation of new short cracks in the later stages of the fatigue life can be considered as a kind of background noise.

A strong influence of crack coalescence was observed during fatigue of the β-titanium alloy LCB. Figure 6.41a shows a replica, representing a variety of microstructurally short cracks, which later in fatigue life had grown together and formed the final crack (Fig. 6.41b), eventually leading to fracture.

Crack coalescence is strongly influenced by the stress–strain state at the respective crack tips. Meyer et al. [419] suggest an interaction function that is based on the ratio of the J integrals of mutually approaching crack tips with respect to the J integral for the corresponding isolated cracks. Hence, the angle between a straight line connecting the crack tips and the load axis is a key factor influencing crack coalescence. According to Ochi et al. [420], two microcrack coalesce when the distance between their tips falls below a critical value $l \leq 0.11(c_1 + c_2)$. In the case of the β-titanium alloy LCB a strong plastic interaction between two mutually approaching crack was observed for a distance between the tips of about l = 30 μm [254]. As can be seen in Fig. 6.41c, the crack tips pass by each other until the tips turn towards each other by operating respective slip systems. This is in accordance with

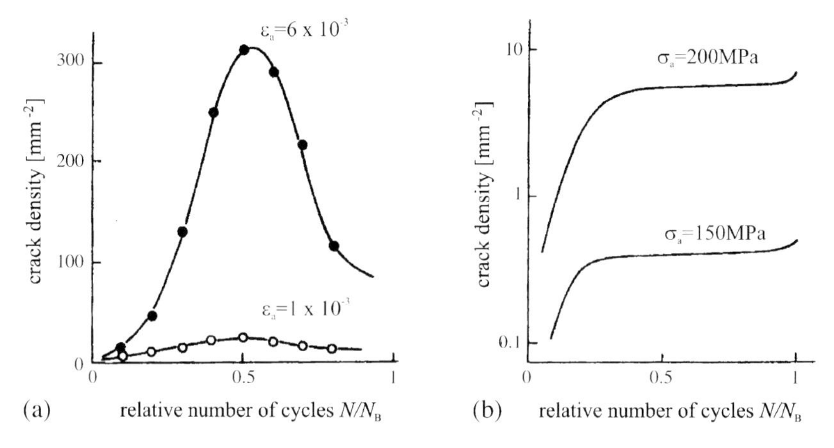

Fig. 6.40 Development of crack density during fatigue life: (a) for an Al–Cu-Mg alloy at two stress amplitudes $\Delta\sigma/2$, (b) for α-Fe at two different total-strain amplitudes $\Delta\varepsilon/2$ (both after [199]).

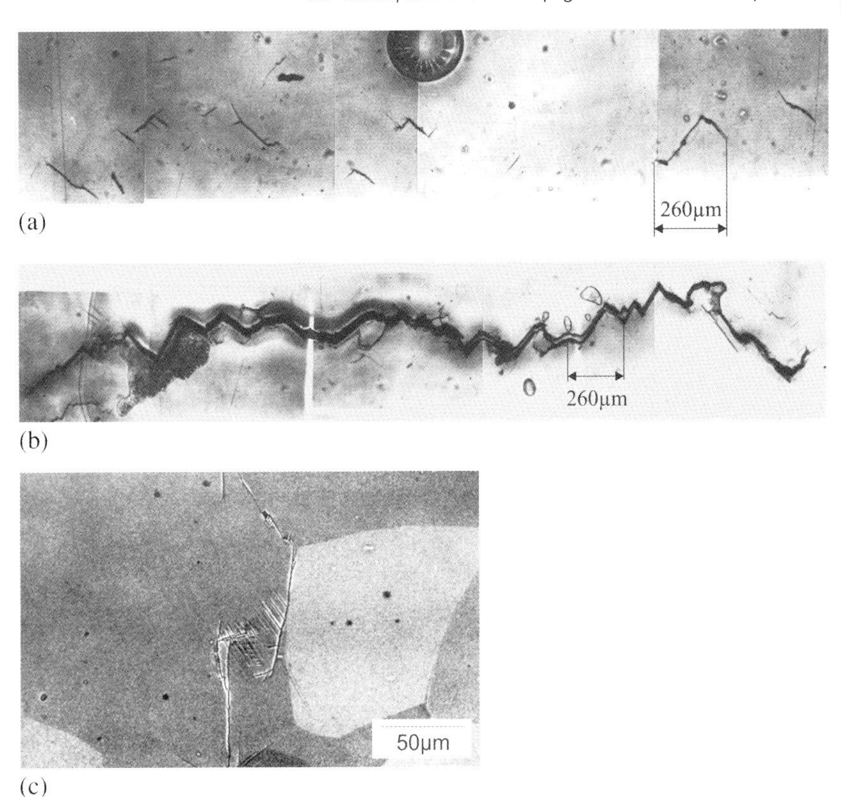

Fig. 6.41 Replica micrographs of the surface of a fatigued specimen of the β-titanium alloy LCB ($\Delta\sigma/2 = 600$ MPa, $R = -1$, loading axis: ⇕): (a) after 2500 cycles, (b) after 4700 cycles, (c) detail of two coalescing crack tips (loading axis: ⇔).

theoretical considerations of the stress field interactions between two tips, as was shown by Künkler [421] for coalescence of two normal-stress-controlled fatigue cracks.

6.5 Intercrystalline Crack Propagation at Elevated Temperatures: The Mechanism of Dynamic Embrittlement

6.5.1 Environmentally Assisted Intercrystalline Crack Propagation in Nickel-Based Superalloys: Possible Mechanisms

With increasing temperature, thermally activated or thermally influenced processes, like solid-state diffusion or dislocation climb, become increasingly significant.

As a consequence, the yield strength σ_Y of most materials drops, and creep and high-temperature corrosion become the key damage mechanisms for high-temperature applications. To improve the performance of highly stressed components, e.g., gas turbine blades and discs at temperatures up to 1200 °C, optimized nickel-based superalloys are used. These alloys are strengthened by precipitation of high fractions of ordered phases, such as the γ' phase ($Ni_3(Al,Ti)$), in precision-cast single-crystalline alloys up to a volume fraction $V_{\gamma'}$ of 0.7 and in polycrystalline wrought alloys up to $V_{\gamma'} = 0.3$, and in the case of alloy 718, the most-used superalloy in aeroengine design, by the additional γ'' phase ($Ni_3(Nb,Al,Ti)$). By this kind of strengthening (see [422] for details) the materials exhibit excellent creep resistance, and alloying with chromium and aluminum provides high-temperature corrosion resistance from the formation of protective and adherent Cr_2O_3 and/or Al_2O_3 surface layers (see [273, 423] for details).

In the case of high-strength polycrystalline superalloys, e.g., alloy 718, high elastic stresses at the crack tip allow the ingress of elements from the gas atmosphere. Particularly in the case of slow cycles or long dwell times at maximum tension during fatigue loading, the high elastic stresses may cause a dramatic increase in the crack-propagation rate da/dN in combination with a transition from transgranular crack propagation driven by plastic blunting to intercrystalline cracking. This is illustrated by the da/dN vs. cycle-duration diagram in Fig. 6.42 for the nickel-based superalloy IN718 cycled at various temperatures in air and under vacuum conditions. The detrimental effect of the environment becomes obvious when comparing the 650 °C/air results with the 650 °C/vacuum results, or those obtained at low temperature (400 °C). Since there is (1) no dwell-time effect at low temperature or in vacuum and (2) the crack-propagation rate increases with longer dwell times, the transition to fast intercrystalline crack propagation can be attributed to solid-state diffusion of the atmospheric constituents (mainly oxygen).

A further example of the dwell-time effect (sometimes also referred to as "hold-time cracking") is given in Fig. 6.43, again for IN718 fatigued at 650 °C at a total strain amplitude of $\Delta\varepsilon/2 = 0.7$ [424]. The LCF life of the specimen with no dwell time is higher by a factor of six than that of the specimen subjected to 300 s dwell time at maximum tension. This behavior correlates with the results of fracture-sur-

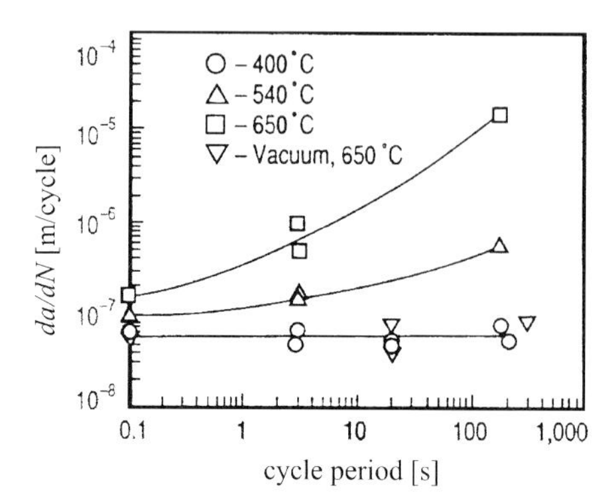

Fig. 6.42 Crack propagation rate da/dN vs. cycle period for fatigue-crack propagation at various temperatures in air and under vacuum at $\Delta K = 27.5$ MPa $m^{1/2}$ (after [425]).

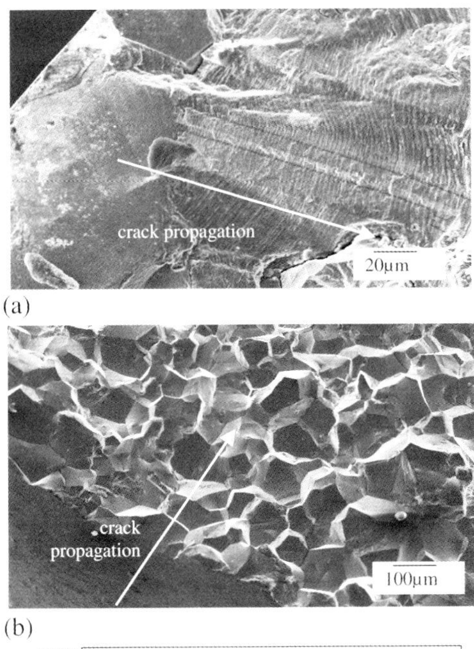

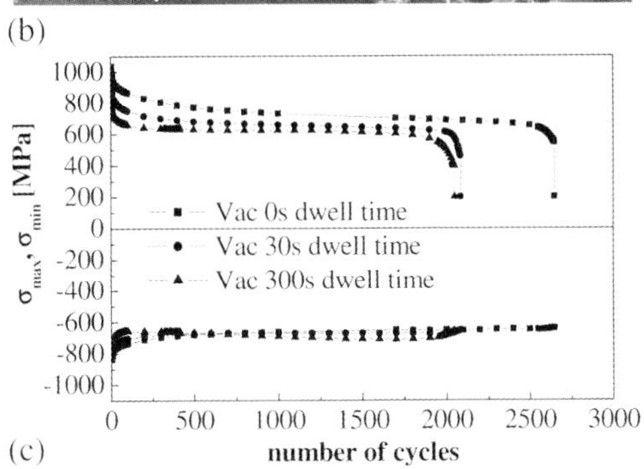

Fig. 6.43 Fracture-surface micrographs of alloy 718 LCF specimens after fatigue testing at $T = 650\,°C$ and $\Delta\varepsilon/2 = 0.7$ in air: (a) transgranular fracture surface with fatigue striations after testing with no dwell time; (b) fracture surface showing grain-boundary facets after testing with a 300 s dwell time at maximum tensile stress; (c) comparison of the respective numbers of cycles to failure in the cyclic deformation curve (representing σ_{max} and σ_{min} vs. the number of cycles) (after [424]).

face analysis: without a dwell time the fracture surface shows fatigue striations representing transgranular crack propagation driven by plastic blunting (Fig. 6.43b); with a 300 s dwell time the fracture surface is completely intercrystalline, showing deformation-less grain boundary facets (Fig. 6.43c).

Even though the dwell-time effect during high-temperature LCF loading of polycrystalline nickel-based superalloys has been extensively studied by many authors [388, 425–440], the relevant mechanisms are still the subject of discussion. For example, the question as to whether oxygen or water vapor, which are almost always present in normal air atmospheres, plays the key role in driving dwell-time crack propagation has not been definitely answered.

While Molins et al. [430] correlated the transition to fast time-dependent crack propagation with an oxygen partial pressure of $p(O_2) = 10^{-3}$ (see Fig. 6.44), Browning and Henry [441] as well as Hayes et al. [442] found that small concentrations of

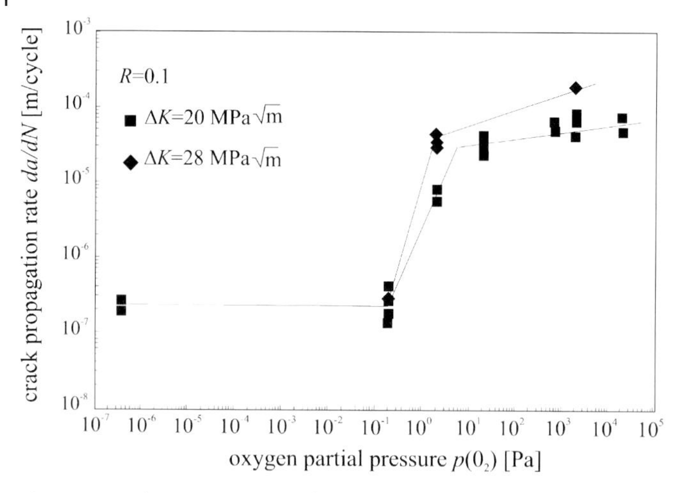

Fig. 6.44 Crack propagation rate *da*/*dN* vs. oxygen partial pressure for two different loading amplitudes (alloy 718) (after [430]).

water vapor affect the transition. According to Browning and Henry, who used $H_2/H_2O/Ar$ mixtures to fix both the oxygen and the water vapor concentrations, the transition occurs at $p(H_2O) = 10^{-2}$ mbar at 593 °C and $\Delta K = 30.5$ MPa $m^{1/2}$. Obviously, at the very crack tip, water vapor can be considered as a ready source for oxygen.

In general, an environmentally assisted increase in the fatigue-crack-propagation rate at elevated temperatures is correlated with the preferential oxidation attack of the grain boundaries, in a way analogous to stress-corrosion cracking. The effect of dwell-time cracking has been attributed to the formation of thermal oxides, occurring preferentially at the grain boundaries; hence, it has been termed stress-assisted grain-boundary oxidation (SAGBO) [438, 443, 444]. Wei et al. [445] found, besides a large fraction of niobium carbides, an enrichment of the γ'' phase (Ni_3Nb) in the vicinity of the grain boundaries of alloy 718, which they believed responsible for the formation of niobium oxides (Nb_2O_5) along the grain boundaries (compare also the work of Miller et al. [446, 447]). Eventually, the niobium oxides at the grain boundaries were thought to crack in a brittle manner, as it shown schematically in Fig. 6.45a. This idea was supported by a study of Chen et al. [448], who found, for a Ni-18Cr-18Fe alloy without niobium, no indication of a crack-accelerating effect of oxygen. Similar results were obtained by Huang et al. [449] for several powder metallurgy nickel-based superalloys of composition similar to the alloy IN100. With the addition of 2.5 and 5 wt.% niobium the environmentally assisted crack-propagation rate was considerably higher than for the alloy without niobium. However, the results of Molins et al. [430] are in contradiction to the proposed niobium mechanism. They found a transition to fast time-dependent crack propagation for Ni–Cr alloys without any niobium and attributed the dwell-time effect to the formation of epitactic NiO ahead of the growing crack. Since NiO grows by out-

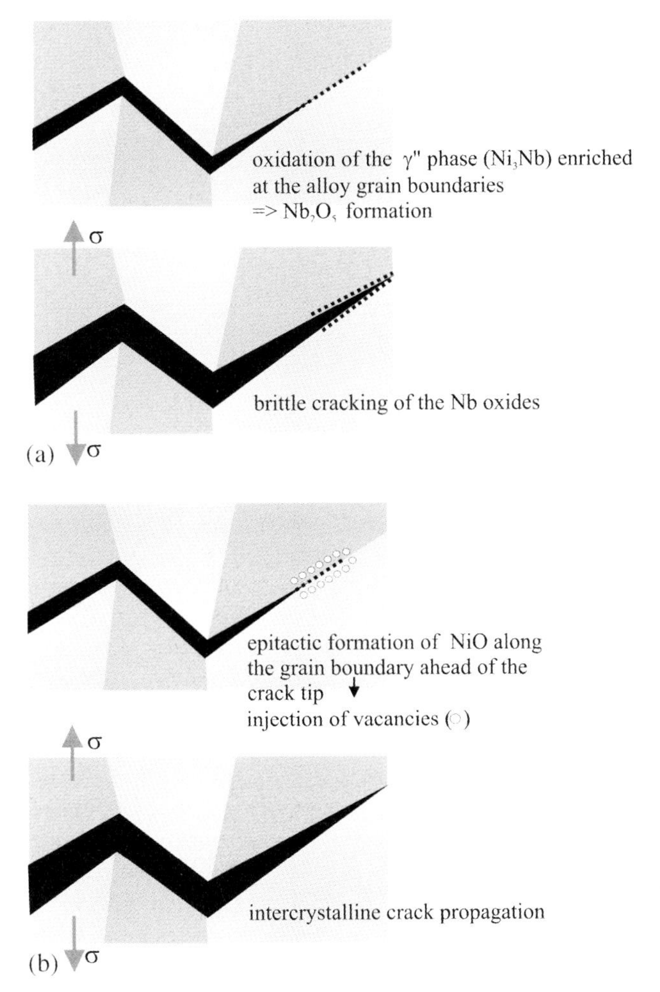

Fig. 6.45 Schematic representation of possible mechanisms for quasi-brittle intercrystalline crack propagation in alloy 718: (a) grain-boundary oxidation, as proposed by Wei et al. [445]; (b) epitactic NiO formation, as proposed by Molins et al. [430].

ward diffusive nickel-cation transport through the oxide scale, vacancies remain at the scale–substrate interface that could condense and form pores [450]. Eventually, these pores could weaken the grain boundary and promote intercrystalline crack advance. The mechanism proposed by Molins et al. [430] is represented schematically in Fig. 6.45b.

Oxidation studies of the niobium-containing alloy 718 carried out by Krupp et al. [451] did not confirm any preferential oxidation of the grain boundaries – neither Cr_2O_3 nor Nb_2O_5 formation – at the moderate temperature of 650 °C, which corresponds approximately to the in-service temperature of gas turbine discs.

The very slow overall oxidation process with a parabolic oxidation rate constant of $k''_p \approx 5.8 \times 10^{-15}\ g^2\ cm^{-4}\ s^{-1}$ (corresponding to the parabolic rate law for the square root of the area-normalized mass increase vs. exposure time $(\Delta m/A)^{1/2} = k''_p t$; cf. [423]) is mainly determined by the oxidation of the aligned niobium carbides (not along the grain boundaries) to Nb_2O_5 (see Fig. 6.46a and b).

Indeed, according to Connolley et al. [452], the niobium carbides may act as fatigue-crack-initiation sites due to their higher specific volume as compared to that of the substrate (cf. Fig. 6.46b). For the transition to fast intercrystalline crack propagation they are obviously not relevant. As shown in Fig. 6.46c, growth of the niobium oxides appears to take place only after the crack front has passed by. Otherwise, one would expect a higher fraction of broken oxide particles in the fracture surface in Fig. 6.46c.

Only at considerably higher temperatures can preferential oxidation of the alloy grain boundaries be observed. This is shown in Fig. 6.47, again for alloy 718, but now after exposure to air at 850 °C. Cr_2O_3 formation was found at the surface (Fig. 6.47a) and along the grain boundaries (Fig. 6.47b). Measured oxidation kinetics at $T = 850$ °C with a parabolic rate constant of $k''_p \approx 3.4 \times 10^{-13}\ g^2\ cm^{-4}\ s^{-1}$ are faster by two orders of magnitude than at $T = 650$ °C.

As an alternative to oxidation effects, which obviously cannot explain the dwell-time effect in a plausible way, the mechanism of dynamic embrittlement was suggested to account for quasi-brittle intercrystalline crack propagation at elevated temperatures [453–455]. According to the schematic representation in Fig. 6.48,

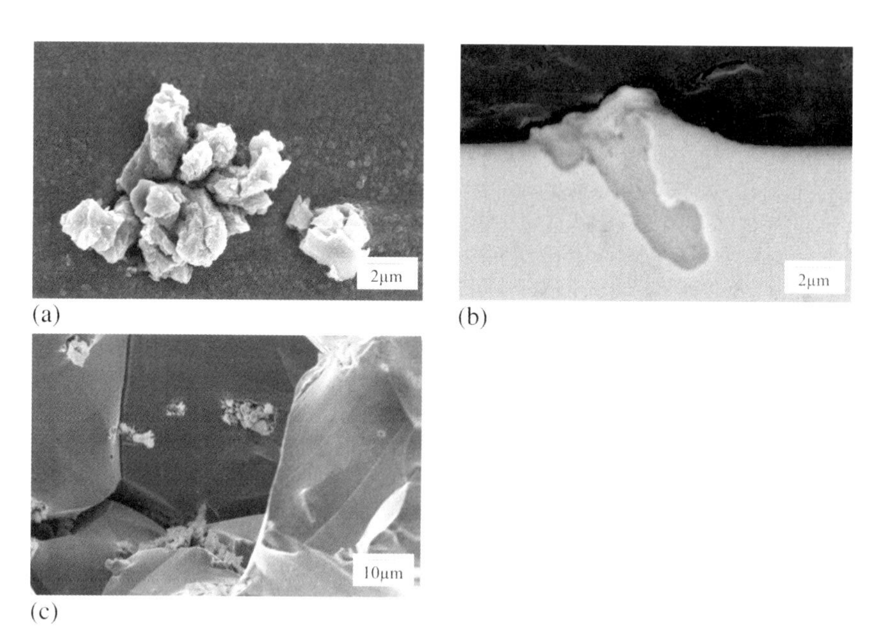

Fig. 6.46 Oxidation of alloy 718 at $T = 650$ °C (100 h exposure), oxidized niobium carbides: (a) surface and (b) cross section; (c) oxidized niobium carbides along the grain-boundary facets, after passage of the crack front.

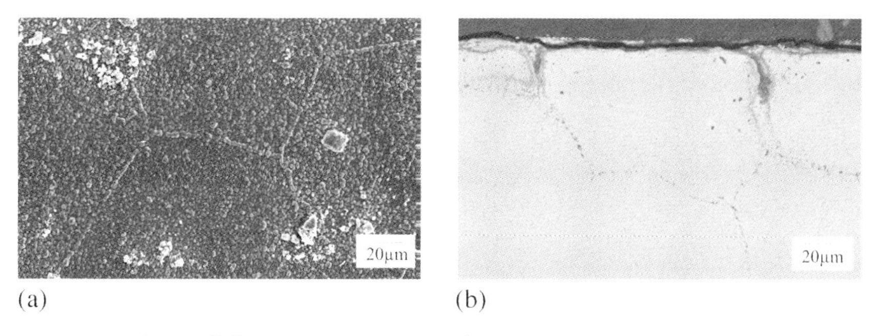

Fig. 6.47 Oxidation of alloy 718 at $T = 850\,°C$ (100 h exposure): (a) Cr_2O_3 formation at the surface, (b) intercrystalline oxidation.

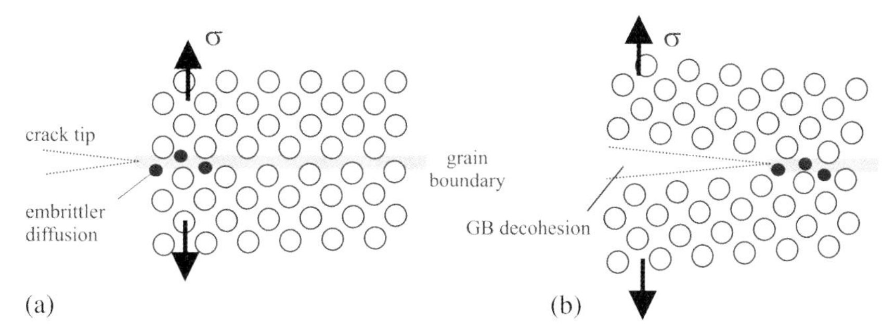

Fig. 6.48 Schematic representation of the dynamic embrittlement mechanism: (a) grain-boundary diffusion of an embrittling element in the elastically stretched cohesive zone ahead the crack tip, followed by (b) local decohesion.

high tensile stresses in the vicinity of an interfacial crack can drive grain-boundary diffusion of atomic oxygen. Since oxygen decreases the cohesion of the grain boundary, crack advance occurs.

Before discussing the mechanism of dynamic embrittlement in detail in Section 6.5.3, the following section provides an overview of technical phenomena that can be related to the generic failure mechanism known as dynamic embrittlement.

6.5.2
Mechanism of Dynamic Embrittlement as a Generic Phenomenon: Examples

The term "dynamic embrittlement" was used first by Liu and White [456], and later independently by Bika [453]. It has been used to define a generic phenomenon, in which the penetration of an embrittling species into a grain boundary is accomplished by stress-induced grain-boundary diffusion. As shown in Fig. 6.48a, a diffusional flux is induced by a tensile stress field in the cohesive zone ahead of the crack tip. The enrichment of the embrittling element in the grain boundary lowers the cohesive forces until interfacial separation occurs and the process zone is shifted forward (cf. Section 7.4). Since fundamentally this embrittlement mechanism requires the application of a tensile force (*dynamos* in Greek), it is referred to as being dynamic. As shown later, dynamic embrittlement does not imply necessarily any kind of dynamic loading, i.e., fatigue. It is active at sustained tension (or slow cycles) and is the major contribution to hold-time/dwell-time cracking.

Generally, several brittle or quasi-brittle crack-propagation phenomena can be explained by the dynamic embrittlement mechanism. As a phenomenon of great technical significance, certainly hydrogen embrittlement of high-strength steels can be considered. This has been intensively studied over several decades and reviewed, e.g., by Birnbaum and Sofronis [457] or McMahon [458]. Even at room temperature, the diffusivity of hydrogen is extremely high; according to [459] the effective diffusion coefficient reaches values of $D = 10^{-7}$ to 10^{-4} m^2/s^{-1}. Hence, the environmentally induced enrichment in hydrogen is not restricted to the alloy grain boundaries, it is generally concentrated to regions of high hydrostatic tensile stresses and correspondingly high lattice strains. In particular, the stress-induced enrichment of hydrogen in the vicinity of existing dormant microcracks may cause resumption of crack propagation [460]. Even though hydrogen can easily penetrate an alloy along dislocations terminating at the free surface and via interstitial sites, the fracture path is mostly observed to be intercrystalline along prior austenite grain boundaries. According to McMahon [458], this can be attributed particularly to pre-existing manganese and silicon segregation at the grain boundaries, the decohesive effect of which is enhanced by additional hydrogen. Eventually, the combination of hydrogen, Mn/Si segregation, and high tensile stress causes intercrystalline cracking. There is no general agreement in the literature about the details of the damage potential of hydrogen in steels; e.g., besides reduction of cohesion hydrogen has a tendency to be trapped at dislocation cores, influencing the dislocation mobility and promoting slip planarity (hydrogen-enhanced localized plasticity, HELP, see, e.g., [457]). The increase in dislocation mobility can also be attributed to a decrease in the dislocation-line energy by hydrogen segregation [461]. According to [459], the hydrogen concentration that must be considered as critical for damage by hydrogen embrittlement can only be given in combination with the mechanical strength of the material. In the case of high-strength martensitic steels, which are particularly endangered by hydrogen, an estimated value for the critical hydrogen concentration can be given as $c_{\mathrm{Hcrit}} = 0.9$ wt.-ppm for a yield strength of $\sigma_Y > 1000$ MPa (cf. [459]).

As a further intercrystalline crack-propagation mechanism, liquid-metal embrittlement (LME) or solid-metal embrittlement (SME) may be considered as a type of dynamic embrittlement. Diffusive penetration of a liquid metal in direct contact with the surface of a solid metal exposed to high mechanical tension can locally lower the cohesion of the grain boundaries and cause crack propagation [462, 463], e.g., in the case of liquid zinc in austenitic steel or liquid bismuth in solid nickel at elevated temperatures [464]. It should be pointed out that the mechanism of liquid-/solid-metal embrittlement is still a matter of controversy. For example, Lynch et al. [465, 466] have proposed a mechanism wherein the adsorption of the embrittling element at the crack tip affects atomic cohesion in the cohesive zone in such a way that dislocations are emitted, a mechanism that has been termed adsorption-induced dislocation emission (AIDE). The crack extension is then seen as a plastic process rather than as a result of the decrease in intergranular cohesion.

Bika [453] was the first to use the term dynamic embrittlement to describe stress-relief cracking in steels. Finely dispersed sulfides (FeS) that are precipitated along the grain boundaries as a consequence of atomic sulfur diffusion during high-temperature treatment (e.g., in the heat affected zone, HAZ, during welding) act as nuclei for the formation of pores when the material undergoes mechanical deformation at elevated temperatures or is just subjected to a stress-relief heat treatment [467]. According to Shin and McMahon [468], the surfaces of pores formed during stress relief or creep become covered by atomic sulfur. As a consequence of local tensile stress, either due to remote mechanical loading or residual stresses, sulfur diffuses into the grain boundaries, where a combination of critical sulfur concentration and stress leads to decohesion. This mechanism of stress-relief cracking can be understood to be similar to the Hull–Rimmer model for the growth of creep pores [469] (see Section 7.4 for a more detailed description of the relationship between stress, embrittler concentration and stress-assisted grain-boundary diffusion).

For the example of A508 steel, Fig. 6.49 shows sulfur-induced stress-relief cracking governed by the dynamic embrittlement mechanism for a CT specimen under sustained tension of $F = 2.67$ kN ($K_{start} = 14$ MPa m$^{1/2}$), at $T = 540\,°C$ and under

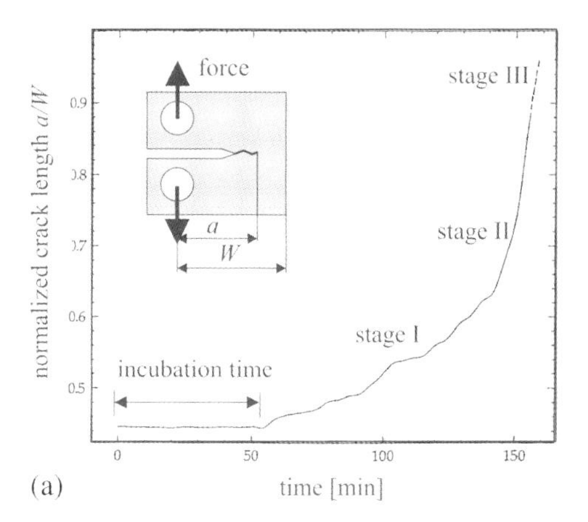

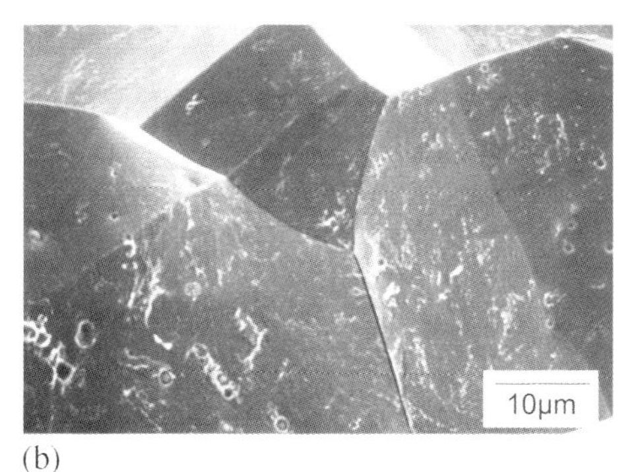

Fig. 6.49 Stress-relief cracking in A508 steel at $T = 540\,°C$ under ultrahigh-vacuum conditions ($p(O_2) = 10^{-7}$ mbar): (a) crack extension vs. time (measured by DCPD); (b) intercrystalline fracture surface of crack propagation regime II (after [453]).

high-vacuum conditions $p(O_2) = 10^{-7}$ mbar [453, 470]. Hence, considerable crack propagation sets in after a certain incubation time, during which the pores form and extend into intergranular cracks below the limit of detectability. According to Bika [453], one should distinguish between three regimes of failure by stress-relief cracking: (1) discontinuous intercrystalline crack propagation with oscillating crack-propagation velocity $da/dt = 0.1–1\ \mu m\ s^{-1}$ (regime I), (2) transitional crack propagation with an almost constant crack-propagation velocity of $da/dt = 15\ \mu m\ s^{-1}$ (regime II), during which the contribution of plastic deformation is continuously increasing, and eventually (3) unstable crack propagation, mainly governed by plastic deformation (ductile rupture, regime III).

The discontinuous, oscillating kind of stress-relief crack propagation, which has been observed also for sulfur-induced cracking of Cu–Cr alloy [471, 472], tin-induced cracking of Cu–Sn alloys [453, 473] and oxygen-induced cracking of alloy 718 [473, 474], can be attributed to a variable grain-boundary-diffusion process. Grain boundaries, which because of their structure are difficult to penetrate by an embrittling element, remain as unbroken ligaments behind the crack front and temporarily shield the crack [453, 473, 474]. Only when these ligaments break by slower dynamic embrittlement or rupture does local crack propagation resume (cf. Fig. 6.54).

Again, alternative mechanisms have been discussed in the literature to explain quantitatively the phenomenon of stress-relief cracking. Hippsly et al. [475] and Rauh et al. [476] assumed that volume diffusion of atomic sulfur is driven by the stress field in the vicinity of a crack tip in such a way that the smaller sulfur atoms migrate opposite to the hydrostatic stress gradient, i.e., towards areas of compressive stress. This was thought to lead to sulfur enrichment at the grain boundaries and, under the assumption of mixed-mode loading conditions, sulfur enrichment ahead of the crack tip. However, since the model implies volume diffusion of sulfur, it cannot explain in a reasonable way the high crack-propagation velocities that were observed by Bika [453] (cf. Fig. 6.48). This kind of sulfur-diffusion-controlled stress-relief cracking would require a diffusivity two orders of magnitude higher than the sulfur bulk diffusivity in steel [473].

The model of Chen [477] took sulfur grain-boundary diffusion into account, but in such a way that it would drive outward transport of iron atoms from the grain boundary into a growing cavity (= intercrystalline crack). The high sulfur concentration on the cavity surface would reduce the surface energy and, hence, would reduce the dihedral angle of the cavity tip and promote intercrystalline crack advance. Chen [477] obtained a good agreement with earlier experimental results by Shin and McMahon [468]; however, the stress and the high sulfur concentration required to plate out iron atoms into the grain boundary would be sufficient for decohesion of the grain boundary ahead of the cavity. Hence, the dynamic embrittlement mechanism would become active.

The preceding section can be summarized by the statement that there are several quasi-brittle, time-dependent intercrystalline cracking phenomena (cf. Fig. 6.50) that can be attributed to a combination of stress-assisted grain-boundary diffusion of an embrittling species followed by interfacial decohesion.

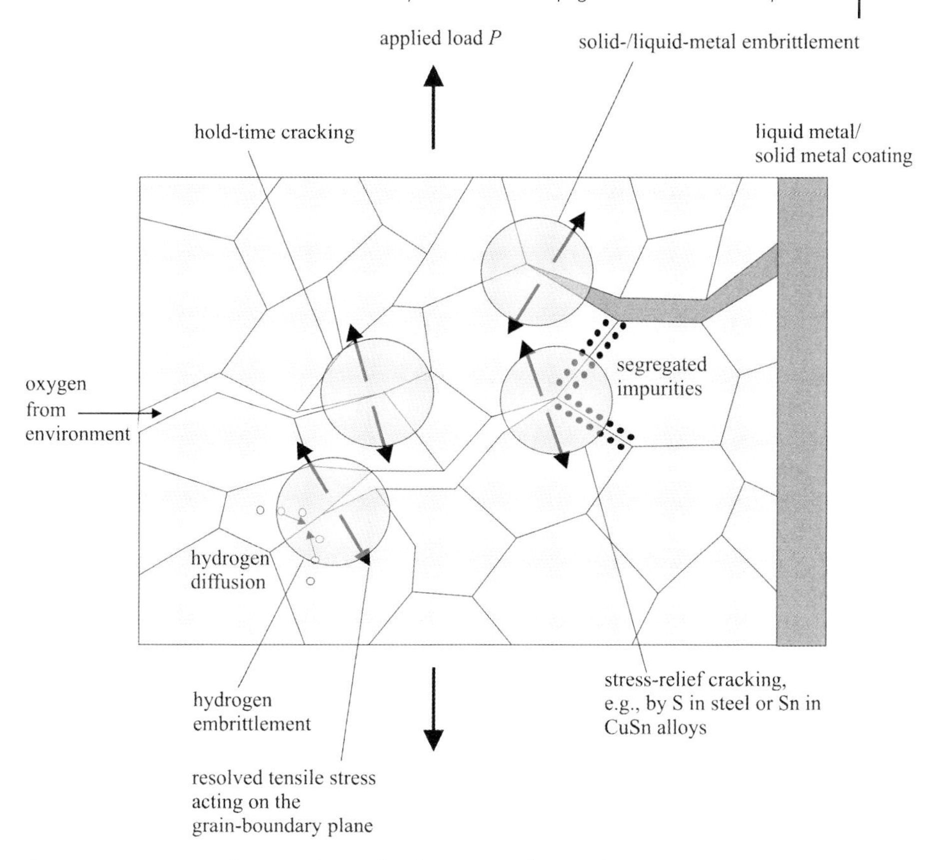

Fig. 6.50 Environmentally assisted quasi-brittle intercrystalline crack propagation mechanisms: hydrogen embrittlement, liquid-/solid-metal embrittlement, stress-relief cracking, dwell-/hold-time cracking.

The proposed dynamic embrittlement mechanism might be the rate-controlling step when the following conditions are fulfilled:

1. A low melting embrittling element coming either from the material itself or from the surrounding atmosphere adsorbs at the free surface or the surface of an internal cavity immediately in front of an advancing crack tip.

2. The temperature and the elastic tensile stress at the crack tip are high enough to ensure a sufficient concentration and (interfacial) diffusivity of the embrittling element into the grain boundary.

3. The embrittling element reduces the grain-boundary cohesion in such a way that a critical embrittler concentration by stress-assisted diffusion can be maintained to drive intercrystalline crack propagation.

In the following sections, dynamic embrittlement as the governing mechanism for dwell-time cracking of alloy 718 at elevated temperature ($T = 650\,°C$) is discussed further with the focus on some microstructural aspects (Section 6.5.3) and the grain-boundary-engineering concept (Section 6.5.4). A modeling concept for crack propagation by dynamic embrittlement is introduced in Section 7.4.

6.5.3
Oxygen-Induced Intercrystalline Crack Propagation: Dynamic Embrittlement of Alloy 718

As already mentioned in Section 6.5.1, the transition to fast intercrystalline crack propagation, which has been frequently observed for LCF loading of high-strength nickel-based superalloys at elevated temperatures (approximately in the range $500 < T < 750\,°C$) and low frequencies or with dwell times at maximum tension, can be attributed to the dynamic embrittlement mechanism. To focus on the dwell-time effect, static four-point bending tests at $T = 650\,°C$ and different oxygen partial pressures on IN718 specimens in an as-received condition were carried out, with the chemical composition and heat-treatment as given in Table 6.2 [473, 474]. The specimen geometry is given in Fig. 3.7 and the test setup in Fig. 3.3b.

Figure 6.51a shows load relaxation vs. time of a four-point bending experiment in pure oxygen that was started at an initial load of $F = 6600$ N [454]. The following load drop corresponds to an increase in compliance, i.e., crack propagation. Following a certain incubation time, during which the fatigue pre-crack transforms into intercrystalline separation along grain boundaries that were particularly susceptible to dynamic embrittlement, the crack propagation velocity increased, reaching a maximum value of $da/dt \approx 30\ \mu m\ s^{-1}$. By temporarily evacuating the test chamber to $p(O_2) = 10^{-5}$ mbar crack propagation was stopped immediately. When backfilling the chamber with oxygen again, crack propagation resumed within 10 s (time interval between two data points, see detail in Fig. 6.51b).

The relationship between the applied force and crack length was quantified by Pfaendtner [473], who measured the intercrystalline fraction of specimens that were broken open by impact after reaching different load-relaxation levels, and he correlated the results by means of a load vs. crack length calibration curve. The results corresponded with direct crack length measurements during test interruptions by means of an optical microscope. Using the calculated crack length and the maximum initial tensile stress at the notch tip

$$\sigma = \frac{3F}{W^2 B_{net}} \tag{6.10}$$

and using the dimensions according to the specimen geometry given in Fig. 3.6, the course of the stress intensity factor

$$K = \sigma \sqrt{\pi a}\, f(a/W) \tag{6.11}$$

with $f(a/W)$ being the geometry function of the single-edge notched bend (SENB) specimen [65], can be calculated and plotted vs. the crack-propagation velocity. This is shown in Fig. 6.52 for different oxygen partial pressures $p(O_2)$, which were ad-

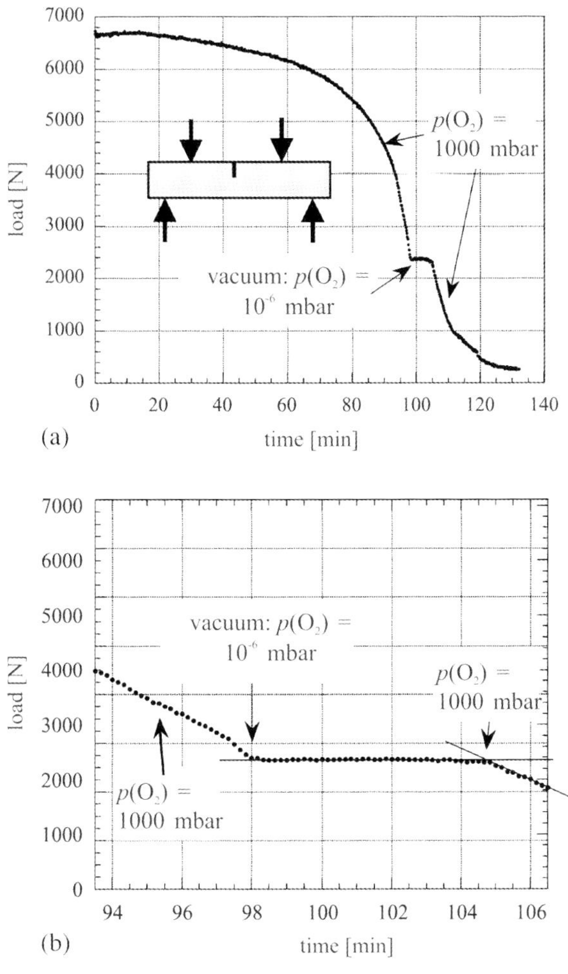

Fig. 6.51 (a) Load relaxation vs. time due to cracking of an IN718 four-point bending specimen at 650 °C in oxygen; cracking arrests when evacuating the chamber and resumes when backfilling the chamber with oxygen. (b) Detail of (a) (after [473]).

Table 6.2 Nominal chemical composition of IN718 and heat-treatment sequences (standard and direct aged).

Ni	Fe	Cr	Nb	Mo	Ti	Al	Co	Si	Mn	C	B
Balance	18.7	18.2	5.2	3.0	1.0	0.5	0.1	0.4	0.06	0.04	0.004

Solution annealing:	1050 °C (1 h) water-quenched
Aging:	720 °C (12 h) furnace-cooled 620 °C (12 h) air-cooled

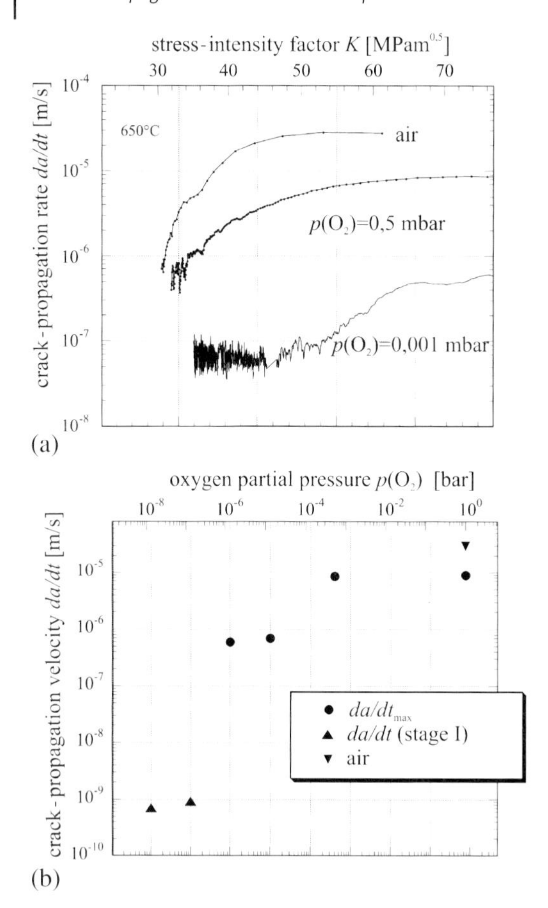

Fig. 6.52 Crack-propagation velocity (a) as a function of the stress intensity factor during four-point bending of IN718 specimens under constant displacement for different temperatures (cf. Fig. 6.51a); (b) as a function of the oxygen partial pressure at $T = 650\,°C$.

justed using a metering valve. The results reveal that the maximum crack propagation velocity decreases with decreasing $p(O_2)$ (Fig. 6.52b). A sudden drop was found for $p(O_2) < 10^{-3}$ mbar, which is in agreement with the observations from Molins et al. [430]. It must be pointed out that the *da*/*dt* data for the two lowest oxygen partial pressures in Fig. 6.52b correspond to the maximum crack propagation velocity during the first 100 h. Afterwards, the tests were terminated and the specimens were broken open by impact. Obviously, for $p(O_2) < 10^{-3}$ mbar crack propagation is governed by power-law creep only. Furthermore, it was found that the crack-propagation velocity in laboratory air was five times higher than in pure oxygen at the same pressure. Even though this effect was not studied systematically, it is an indication of the detrimental contribution of humidity to the quasi-brittle cracking process, as proposed by Browning and Henry [441]. Water vapor can act as a source for atomic oxygen at the crack tip promoting the dynamic embrittlement mechanism.

In particular, the very high crack-propagation velocities shown in Fig. 6.52, and the ability to switch crack propagation on and off together with the extremely low oxidation rates (cf. section 6.5.1), are indications that strongly support the hypoth-

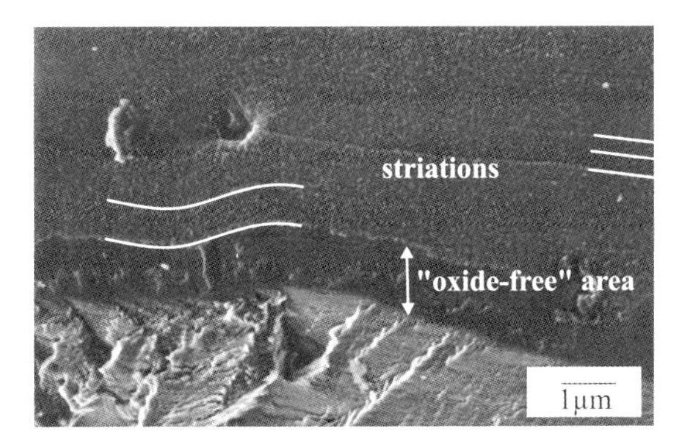

Fig. 6.53 Intergranular crack tip of a four-point bend specimen of IN718 after an interrupted fixed-displacement test at $T = 650\,°C$ (the specimen was broken open by impact) showing striation-like lines parallel to the crack front and an oxide-free area (arrow) (after [473]).

esis that dynamic embrittlement must be the governing mechanism for dwell-time cracking during fatigue of high-strength polycrystalline superalloys.

Analysis of the fracture surfaces of IN718 specimens that failed by oxygen-induced dynamic embrittlement (Figs. 6.53 and 6.54) revealed an obviously inhomogeneous cracking behavior. As proposed earlier by Bika [453], cracking by dynamic embrittlement seemed to be stepwise intermittent. Indeed, Pfaendtner and McMahon [454] found striation-like lines in the intercrystalline fracture surface that correspond to crack-propagation intervals. This is shown for a four-point bending specimen that was broken open by impact (Fig. 6.53). The interval closest to the crack tips remained nearly unoxidized and corresponds to the crack advance during the unloading process.

Also, macroscopic inspection of the fracture surface showed inhomogeneities. The area of the intercrystalline fracture surface shown in Fig. 6.54a, again for an interrupted test, where the SENB specimen was broken open by impact, reveals that the crack front proceeds like fingers into the material. The finding that dynamic embrittlement is inhomogeneous and depends on the local microstructure is supported by the observation of small patches/grain facets in the fracture surface, which showed some contribution of plastic deformation, e.g., slip traces or dim-

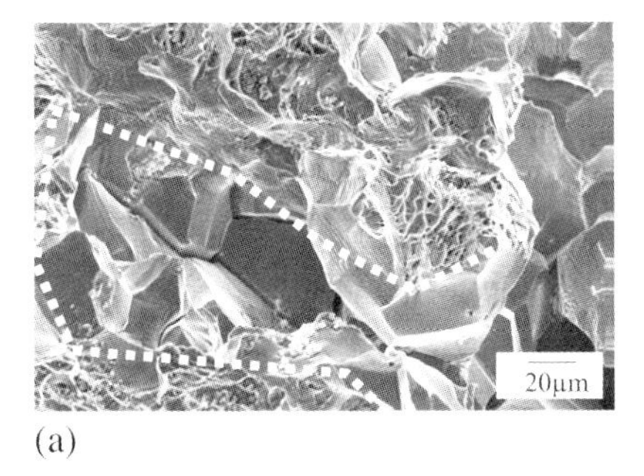

(a)

(b)

Fig. 6.54 (a) Irregular crack front (dashed line; interrupted test) and (b) grain-boundary facet with the termini of twin boundaries within the fracture surface of four-point bend specimens (IN718) after fixed-displacement tests at $T = 650\,°C$ (after [455]).

ples due to local plastic rupture or creep damage. The example in Fig. 6.54b shows a broken grain boundary with the terminus of twins. Obviously, a difference in crystallographic orientation alters the resistance to intercrystalline cracking by dynamic embrittlement.

Provided that dynamic embrittlement is governed mainly by stress-assisted diffusion of the embrittling element (here, oxygen), the influence of the local microstructure can be attributed to the following relationships (cf. Section 7.4):

1. High elastic tensile stresses in the cohesive zone ahead of the crack tip promote the diffusive penetration of oxygen into the grain boundary; i.e., the susceptibility to dynamic embrittlement increases with increasing yield strength.
2. A low grain-boundary cohesion increases the intercrystalline crack propagation velocity.
3. The concentration of oxygen in the grain boundary depends on the grain-boundary-diffusion coefficient for oxygen in the respective material (here, nickel-based superalloys).
4. The oxygen concentration profile within the grain boundary depends on the concentration immediately at the crack tip, which is determined by the oxygen solubility in the material and the oxygen partial pressure at the crack tip. Due to oxidation reactions at the crack tip, the oxygen concentration at the crack tip might deviate substantially from that in the environment.

During the initial time of dwell-time fatigue or sustained loading, intercrystalline crack propagation sets in along those grain boundaries that are particularly prone to dynamic embrittlement according to the criteria mentioned above (see Fig. 6.55). Temporarily, the remaining uncracked ligaments are able to counterbalance the increase in stress intensity. Only when the ligaments fail by ductile rupture or power-law creep (or slow oxygen-induced cracking), the stress at the several crack tips increases again, and crack propagation driven by interfacial oxygen diffusion resumes. This results in a discontinuous crack propagation on the microstructural length scale, as observed experimentally [453, 473].

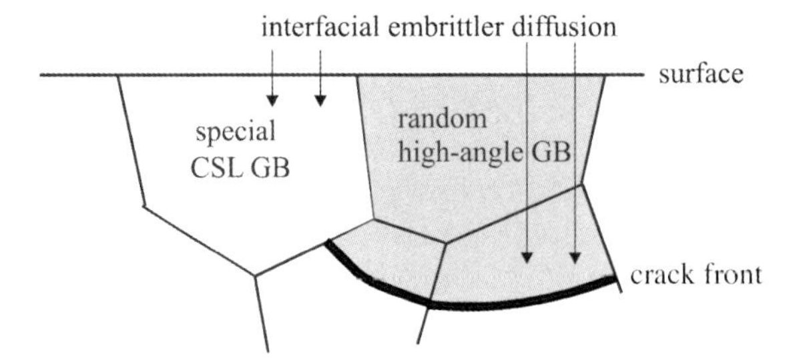

Fig. 6.55 Schematic representation of the composite effect during dynamic embrittlement; the resistance to dynamic embrittlement depends on local microstructural features that are responsible for a discontinuous intercrystalline crack-propagation velocity.

6.5.4
Increasing the Resistance to Intercrystalline Crack Propagation by Dynamic Embrittlement: Grain-Boundary Engineering

In the case of nickel-based superalloys, there are generally two ways to increase the resistance to environmentally and stress-assisted intercrystalline crack propagation: (1) alloying and (2) thermomechanical treatment. It was shown, for example, by Rösler and Müller [443] and White et al. [478], that adding small amounts of the elements boron, carbon, hafnium or zirconium might have a beneficial effect with respect to the susceptibility to intercrystalline cracking. This beneficial effect can probably be attributed to a decrease either in the grain-boundary energy or/and in the grain-boundary diffusivity of embrittling elements. The effect of the grain size is not fully understood yet. It is generally accepted that the creep resistance increases with increasing grain size. Intercrystalline cracking by dynamic embrittlement, however, seems to depend in a complex way on the alloy grain size. A fine-grained microstructure should therefore be beneficial, since the crack path becomes branched and tortuous, leading to a decrease in the peak stress at the crack tip [453]. Indeed, this relationship was observed for high-temperature fatigue crack propagation in alloy 718 by Osinkolu et al. [440]. The beneficial effect of a fine-grained microstructure was supported by sustained-load experiments on alloy 718 specimens taken from a forged gas-turbine ring in a "direct aged" heat-treatment condition; i.e., the component was solution heat treated below the solvus temperature of the δ phase to avoid excessive grain growth. The 5-μm grain-size SENB specimens revealed a superior resistance to dynamic embrittlement. Surprisingly, cracking changed from the radial to the axial direction (see Fig. 6.56). In accordance with [479], forged components show anisotropy with respect to their resistance to dynamic embrittlement.

Besides the grain size, the grain-boundary structure can also be modified by means of thermomechanical treatment. By applying a repeated sequence of cold work followed by an anneal (similar to recrystallization), the fraction of special CSL grain boundaries (coincident site lattice, Σ1–Σ29, cf. Section 3.3.5) can be increased. Such kinds of themomechanical-processing approaches have been termed grain-boundary engineering (GBE) or grain-boundary design [480] and have been extensively studied during the last 20 years. It has been shown in numerous papers and books that GBE-type processing has a promising potential to improve a material's performance with respect to intercrystalline damage mechanisms (see e.g. [152, 481]. For instance, the resistance to creep and fatigue damage [482–487] to intercrystalline corrosion and stress-corrosion cracking [488–491] or to intercrystalline oxidation [424, 492–494] of some nickel-based alloys could be increased by applying GBE. Figure 6.57 shows, for example, the decrease in intercrystalline corrosion vs. the fraction of special CSL grain boundaries that was increased by GBE.

From the literature [495], it is known that low-angle grain boundaries and special grain boundaries exhibit relative minima in the grain-boundary diffusivity of substitutional atoms, and also of interstitials. This can be attributed to the energy of formation of vacancies that is particularly high for special grain boundaries. Figure

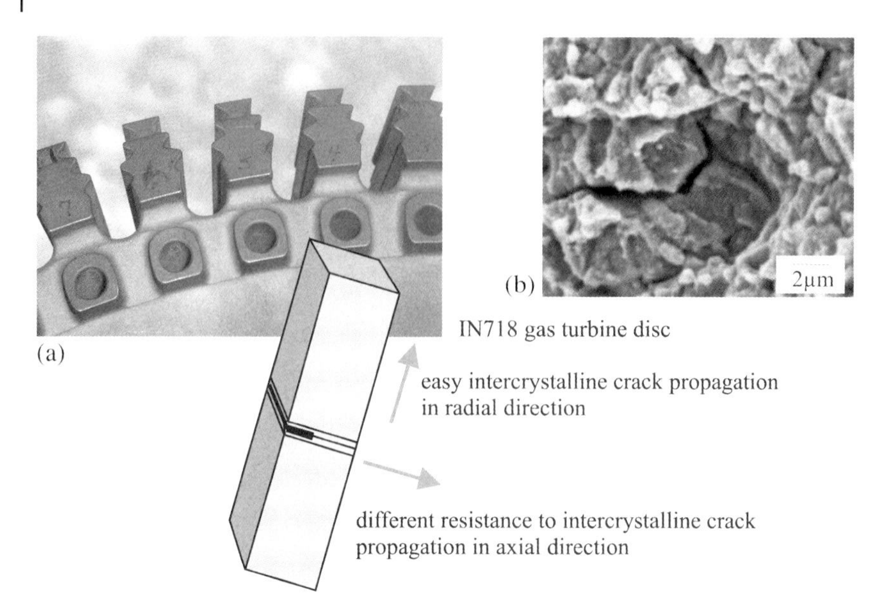

Fig. 6.56 Intercrystalline crack propagation in a four-point bend specimen taken from a forged gas-turbine ring (IN718) in the direct-aged heat-treatment condition: (a) position of the specimen with respect to the ring geometry; (b) detail of the fracture surface.

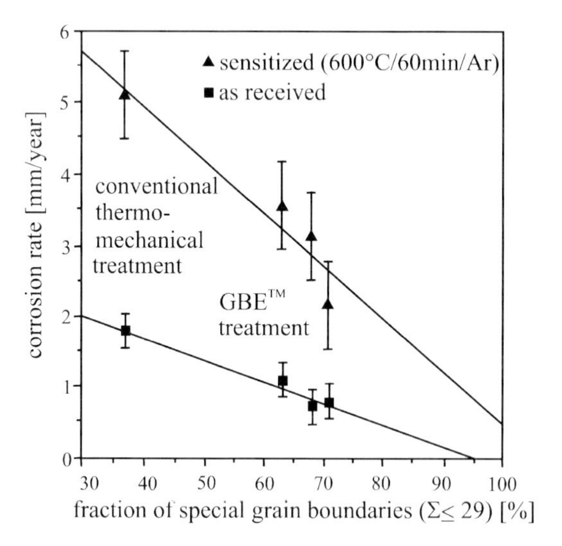

Fig. 6.57 Intercrystalline corrosion attack of the nickel-based alloy 600 in boiling sulfuric acid (50%, + 31.25 g L^{-1} FeS, according to ASTM G28) as a function of the fraction of special grain boundaries, obtained by conventional thermomechanical treatment (65% cold rolling + 5min anneal at $T = 1000\,°C$) and GBE™ treatment (after [489]).

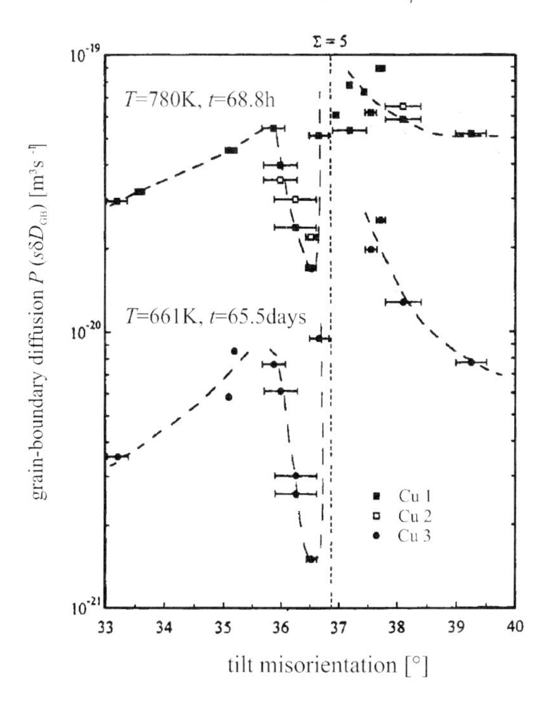

Fig. 6.58 Grain boundary diffusivity of gold in copper bicrystals vs. the tilt misorientation with a minimum close to the special Σ5 misorientation (from [275]).

6.58 shows as an example the grain-boundary diffusivity of gold in copper bicrystals vs. the tilt misorientation with a minimum close to the special Σ5 misorientation.

Provided that the relationship between grain-boundary diffusivity and misorientation is similar for the diffusion of the embrittling element governing the dynamic embrittlement process, and consequently assuming that CSL boundaries have a particularly high resistance to diffusion-controlled brittle cracking, then the resistance to dynamic embrittlement should also be improvable by GBE-type thermomechanical processing (TMP), according to the schematic representation in Fig. 6.59.

By means of one-time-, two-time- and four-time-repeated sequences of 20% cold rolling followed by a 1 h recrystallization heat treatment at 1050 °C, the fraction of special grain boundaries in specimens of IN718 was increased by a factor of two (four-times GBE, see Table 6.3 and Fig. 6.60a [275, 424]). The grain-boundary structures were quantified by means of automated electron backscattering diffrac-

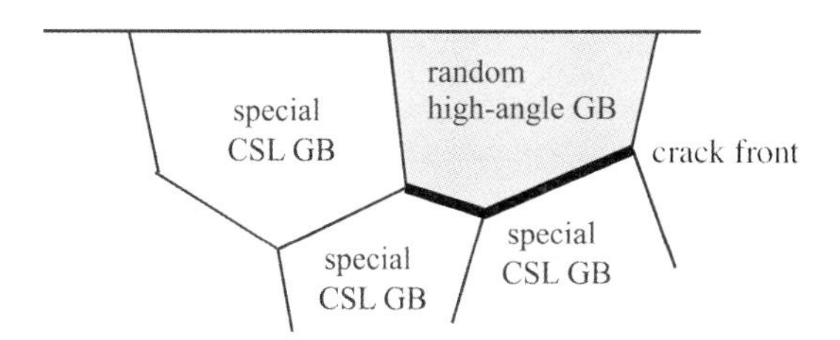

Fig. 6.59 Schematic representation of the effect of GBE-type processing on the susceptibility to crack propagation by dynamic embrittlement, governed by grain-boundary diffusion of the embrittling element.

tion on electropolished samples as CSL length fraction. Figure 6.60b shows an example of the grain orientation and special-grain-boundary distribution of an IN718 specimen after four-cycle TMP. It should be added that, due to the relatively low amount of deformation, recrystallization did not cause any increase in grain size, which remained constant at $d \approx 75$ μm during the whole TMP sequences.

Table 6.3 Fraction of special CSL grain boundaries (Σ3–Σ29) in IN718 in the as-received condition and after repeated application of GBE-type thermomechanical processing (TMP).

Σ value	Fraction of special CSL grain boundaries (length fraction, %)			
	As received	One-cycle TMP	Two-cycle TMP	Four-cycle TMP
3	12.2	13.7	25.1	25.4
5	0.2	0.4	0.4	0.9
7	0.3	0.1	0.5	0.5
9	0.7	0.8	1.0	1.3
11	0.3	0.4	0.8	0.8
13	0.2	0.2	0.4	0.2
15	0.2	0.2	0.3	0.4
17 a,b	0.1	0.2	0.2	0.4
19 a,b	0.2	0.2	0.4	0.3
21 a,b	0.2	0.1	0.2	0.1
23	0.1	0.1	0.1	0.3
25 a,b	0.1	0.1	0.1	0.3
27 a,b	0.2	0.3	0.1	0.4
29 a,b	0.1	0.4	0.3	0.4
Sum	14.9	17.3	30.0	31.7

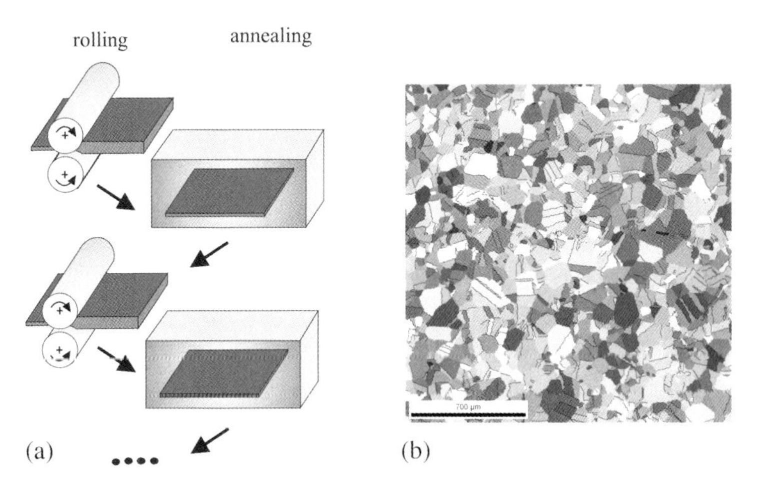

Fig. 6.60 (a) GBE-type TMP and (b) corresponding distribution of crystallographic orientations and CSL grain boundaries (four-cycle TMP, measured by electron backscattering diffraction, CSL boundaries marked as black lines).

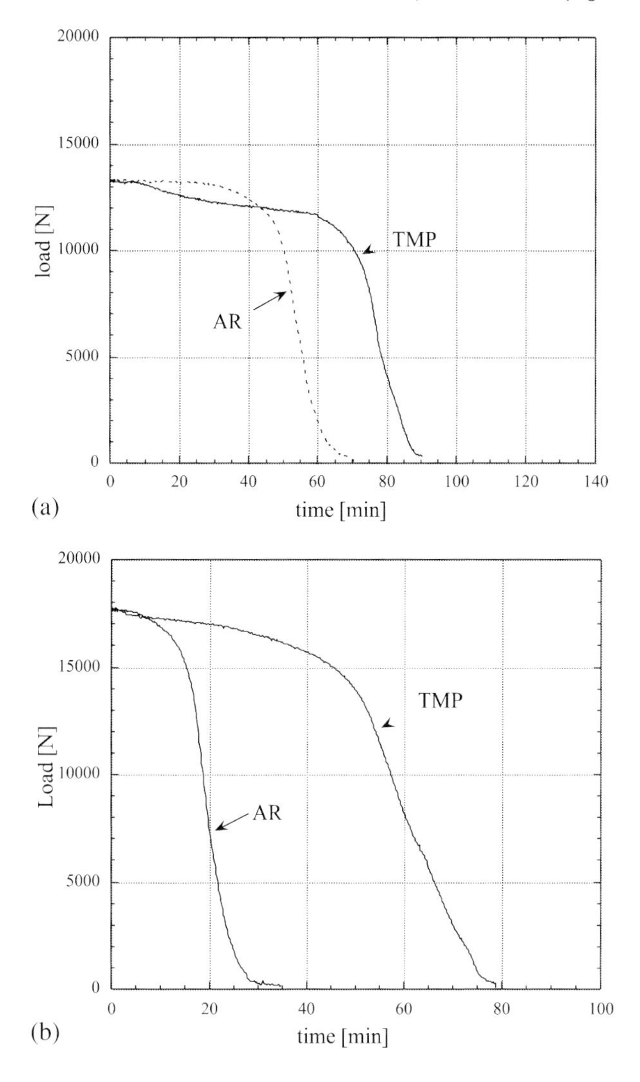

Fig. 6.61 Load relaxation due to intercrystalline cracking at $T = 650\,°C$ in four-cycle TMP specimens and as-received specimens of IN718 for two different initial load levels: (a) $F_0 = 13\,350$ N (=3000 lb), (b) $F_0 = 17\,800$ N (4000 lb) (four-point bending of notched specimens, after [474]).

Comparison of load-relaxation experiments carried out on as-received specimens (standard heat treatment according to Table 6.2) and on TMP specimens at 650 °C in air revealed that the higher fraction of special CSL grain boundaries indeed led to (1) a longer incubation time before intercrystalline fracture by dynamic embrittlement set in and (2) a decrease in crack-propagation velocity [474]. This is shown in Figs. 6.61 and 6.62 supporting the hypothesis that the susceptibility to dynamic embrittlement depends on local microstructural features.

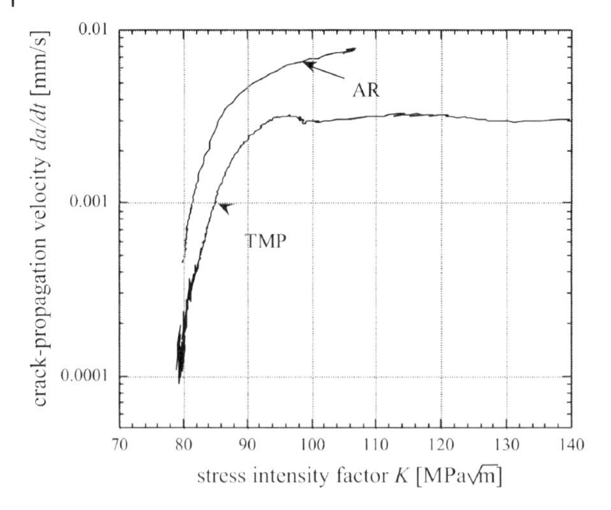

Fig. 6.62 Crack-propagation velocities vs. stress-intensity factor during four-point bending testing in a four-cycle TMP and an as-received specimen of IN718 under constant displacement (corresponding to the data in Fig. 6.61b) (after [474]).

Quantification of the fracture surfaces of the SENB specimens (Fig. 6.63) subjected to load-relaxation tests (cf. Fig. 6.61b) with respect to the plastically failed patches showed excellent agreement with the CSL grain boundary fraction. The increase in the CSL grain boundary fraction for the four-cycle TMP as compared to the as-received material was by a factor of two. Correspondingly, the fraction of plastic patches in the fracture surface of the four-cycle TMP material was twice as high as that in the as-received material, reflecting the remarkable effect of GBE-type processing.

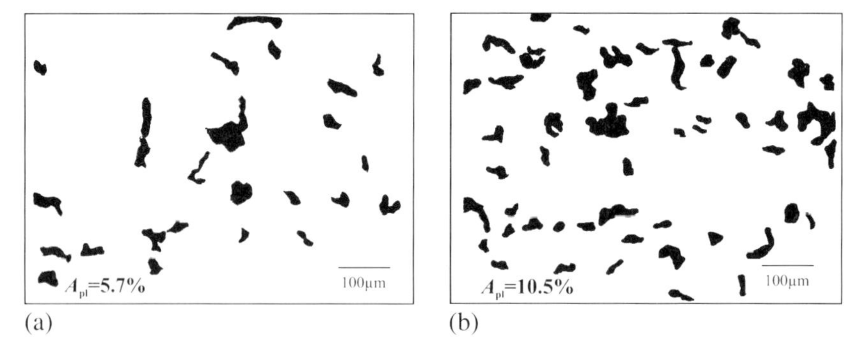

Fig. 6.63 Visual comparison of the ductile area fractions of IN718 four-point bend specimens in the (a) as-received and (b) four-cycle TMP condition (after [474]).

The possibility of improving a material's resistance to dynamic embrittlement by GBE-type processing implies qualitatively the relationship between the grain-boundary structure and grain-boundary diffusivity. To obtain a quantitative picture of the embrittlement process, further criteria have to be taken into account:

1. The local stress normal to the crack plane is not only determined by the geometrical arrangement of the grain boundaries (see [496]) with respect to the remote loading direction but also by elastic anisotropy (cf. Section 5.3).

2. Based on the assumption that crack propagation by dynamic embrittlement is restricted to nonspecial ($\Sigma > 29$) random, high-angle grain boundaries, the beneficial effect of increasing the fraction of special CSL grain boundaries by GBE-type processing can be considered to be indirect. In several studies [497–504] it has been pointed out that it is rather the effective interruption of the weak random-grain-boundary network that is responsible for the improvement in the brittle-cracking performance of metals and alloys. Palumbo et al. [505] and Kumar et al. [506] have suggested a stochastic approach wherein the maximum length of interconnected grain boundaries is considered to be the main measure of the resistance to intercrystalline damage [505]:

$$l_{max} \propto \frac{1}{\log(1-P)} \tag{6.12}$$

   with P being the probability at a triple point (in a two-dimensional section) that an interconnected grain-boundary cluster terminates.

3. CSL boundaries are normally defined for a certain misorientation interval, e.g., according to the Brandon criterion (Eq. 3.15) [154], owing to the fact that, within such an interval, the condition that every Σth lattice site is coincident for the adjacent grains is maintained by geometrical necessary dislocations. This corresponds to the general idea of low-angle grain boundaries, which are established just by an array of dislocations within a misorientaion angle up to $\Theta_0 = 15°$.

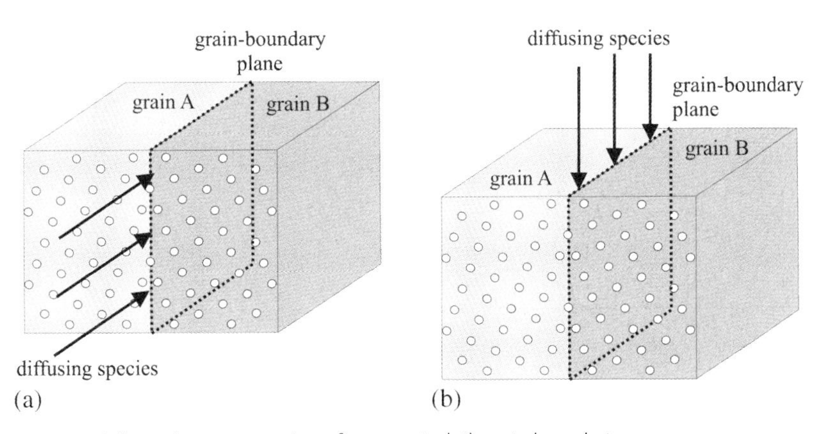

Fig. 6.64 Schematic representation of symmetrical-tilt grain boundaries aligned for (a) fast and (b) slow interfacial diffusion.

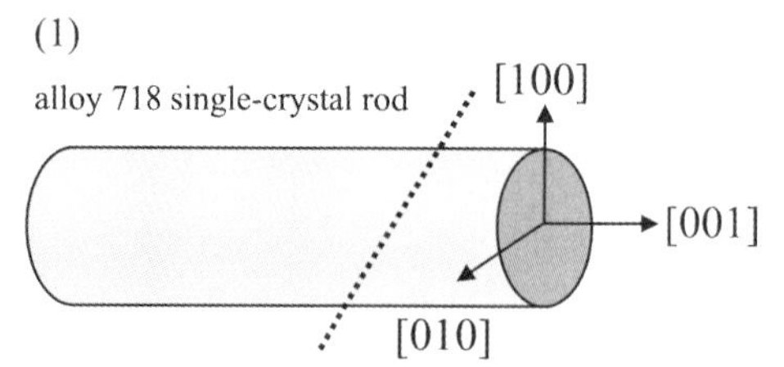

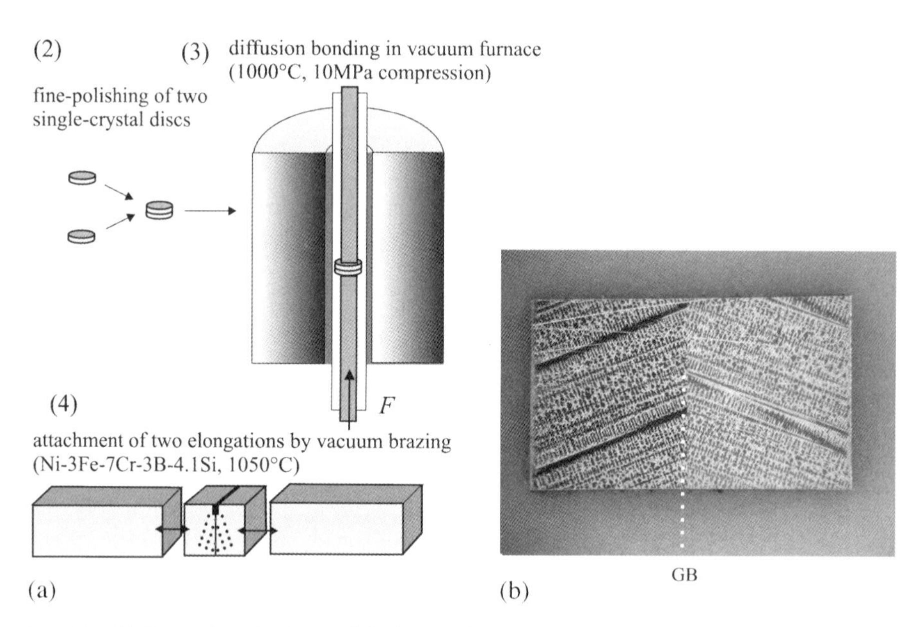

Fig. 6.65 (a) Bicrystal production by diffusion bonding; (b) etched cross section of a diffusion-bonded alloy 718 bicrystal with a symmetrical Σ5 (031)[001] tilt grain boundary showing the dendritic structure aligned parallel to the [001] direction.

According to Randle [507] and Palumbo and Aust [508], a further improvement in a material's grain-boundary strength can be obtained when the actual misorientation angle is as close as possible to the exact CSL misorientation. Instead of using the Brandon criterion, application of the Palumbo–Aust criterion [508] for the misorientation interval ΔΘ of CSL grain boundaries as a function of the Σ value is suggested:

$$\Delta\Theta \leq 15° \cdot \Sigma^{-5/6} \tag{6.13}$$

A direct and quantitative correlation of the intercrystalline crack propagation mechanism with the grain-boundary structure is possible by means of mechanical

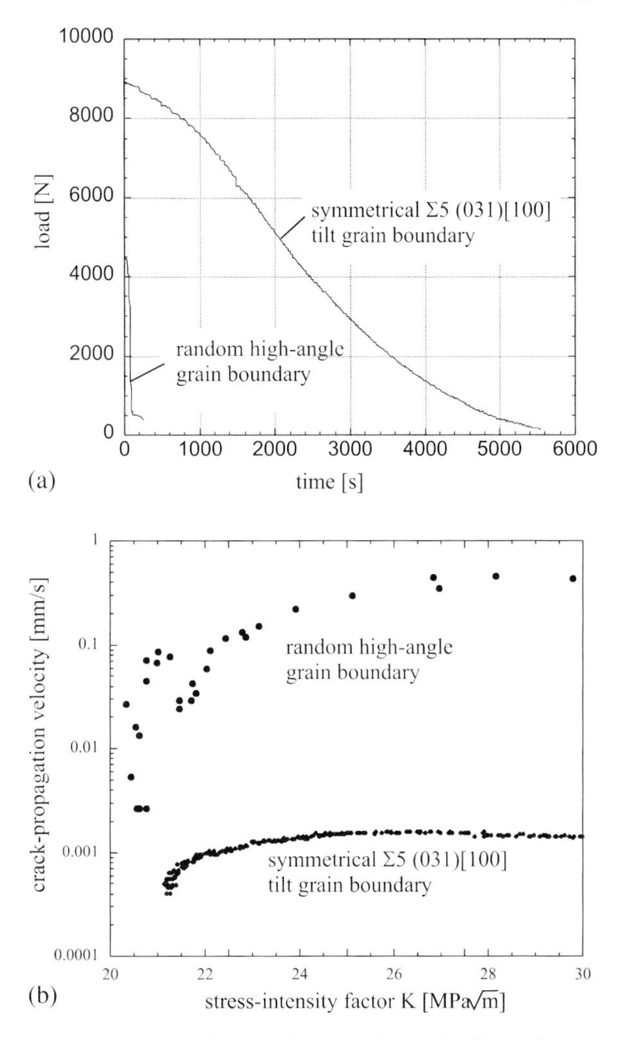

Fig. 6.66 (a) Load relaxation due to cracking within bicrystal specimens of alloy 718 (see Fig. 6.65) with a symmetrical Σ5 (031)[001] tilt grain boundary and a random high-angle grain boundary at $T = 650\,°C$; (b) corresponding crack-propagation rates vs. stress-intensity factor (after [455]).

experiments on bicrystals. Muthiah et al. [509] studied the mechanism of tin-induced dynamic embrittlement using bicrystals with a symmetrical Σ5 tilt grain boundary. They found that this grain boundary fails easily when the misorientation axis was oriented parallel to the crack-propagation direction (Fig. 6.64). On the other hand, specimens with the misorientation axis perpendicular to the cracking direction of the SENB specimen were shown to have a high resistance to dynamic embrittlement. Muthiah et al. [509] attributed this observation to the pipe effect for diffusion along the grain boundary when the tilt axis is parallel to the concentration gradient of the embrittling element.

Experiments on alloy 718 bicrystals, produced by diffusion bonding according to Fig. 6.65a, with either a symmetrical Σ5 <031>(001) tilt grain boundary (Fig. 6.65b) or a random high-angle grain boundary, respectively, supported the hypothesis that special CSL grain boundaries exhibit a substantially higher resistance to dynamic embrittlement than random grain boundaries [455, 496].

This is documented by the load relaxation vs. time and the respective crack propagation velocity vs. stress intensity factor curves in Fig. 6.66.

It is worth mentioning that during load relaxation of bicrystalline SENB specimens neither an incubation time nor discontinuous cracking were observed. Hence, the effects of damage accumulation prior to intercrystalline cracking and stepwise discontinuous crack advance in technical alloys can be solely attributed to the polycrystalline nature of the microstructure.

7 Modeling Crack Propagation Accounting for Microstructural Features

7.1 General Strategies of Fatigue Life Assessment

Generally, fatigue-life assessment can be carried out by means of (1) the nominal-stress concept, (2) the structure-stress concept, (3) the local-stress or notch approach or (4) the fatigue-crack-propagation approach. While the nominal-stress concept is based on analytical stress values for simplified geometries and/or experimental Wöhler (S/N) diagrams for component-like samples or load cases, application of the structure-stress concept requires either a detailed finite-element stress analysis of complex-shaped or welded geometries or suitable correction factors to be used together with the nominal-stress approach [40]. The local-stress approach places the focus towards the highest-stressed sites of a component accounting for local plastic deformation by using strain-based Wöhler (S/N) diagrams and cyclic stress–strain curves. With the help of suitable counting procedures to translate the real loading spectrum into groups of weighted loading contributions, the actual fatigue damage can be expressed as a damage sum according to Palmgreen [9] and Miner [10] (cf. Chapter 2). The basic ideas and differences of the fatigue life assessment methods are illustrated in Fig. 7.1 (cf. also Fig. 2.16).

Fatigue-life assessment based on crack-propagation concepts accounts for the residual life of components with cracks. As introduced in Section 2.2.3, Paris' law [26] and its modifications can be considered as an empirical but technically the most widely accepted approach. Here, the range of the stress-intensity factor ΔK (or the cyclic J integral, ΔJ or Z, resp.), combining the remote stress range with the actual crack length, defines the driving force for crack propagation. However, the Paris-type approach is limited to crack cases where the size of the plastic zone is negligible as compared to the crack length (linear elastic fracture mechanics, LEFM) or where the crack length is large as compared to the characteristic microstructure length (e.g., grain size, elastic–plastic fracture mechanics, EPFM). This restriction to long cracks and physically short cracks is fulfilled for most practical applications of fracture mechanics in fatigue-life assessment, since existing cracks smaller than approximately 500 μm cannot be detected by the typical methods of nondestructive materials testing, e.g., ultrasonic, eddy current or dye-penetrant inspection.

Fatigue Crack Propagation in Metals and Alloys: Microstructural Aspects and Modelling Concepts. Ulrich Krupp

ISBN: 978-3-527-31537-6

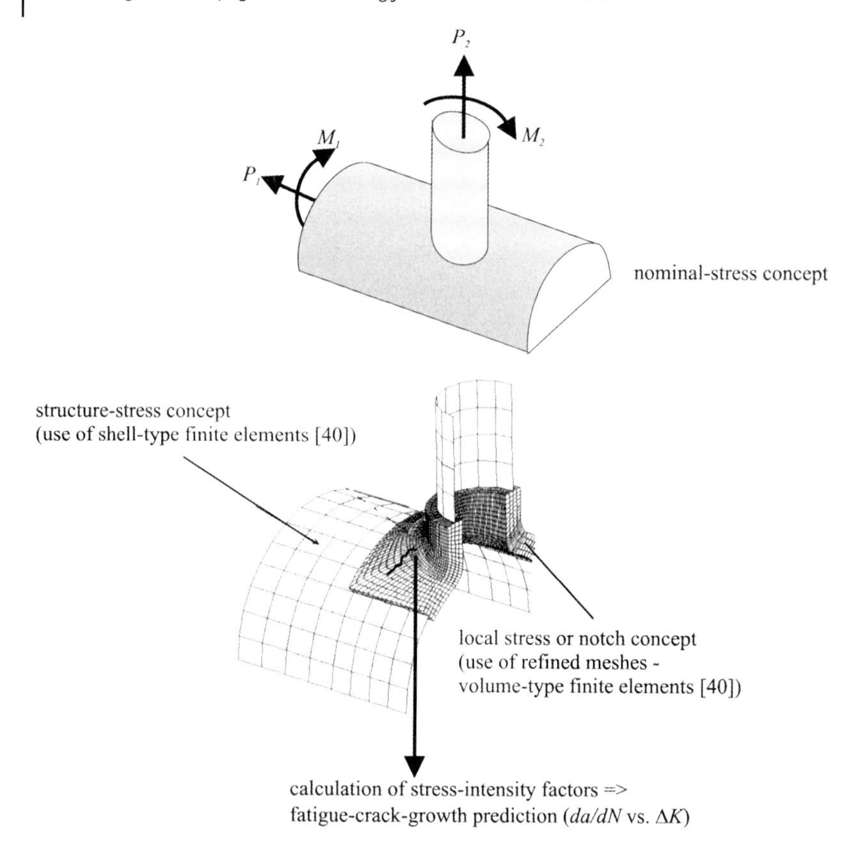

Fig. 7.1 Basic concepts of fatigue-life assessment in engineering design.

The situation is different for smooth structural components made of high-strength materials and subjected to high-cycle fatigue (HCF) or very-high-cycle fatigue (VHCF) loading conditions, e.g., polished shafts or contact surfaces of internal combustion engines. Here, the end of service life is almost reached when the crack length has exceeded the characteristic microstructural length scale. This is represented schematically in Fig. 7.2a. Hence, for microstructurally short cracks, application of conventional damage-tolerant approaches is not very useful, since the residual service life after occurrence of first technical cracks would be of less duration than reasonable inspection intervals. Frequently, such high-strength structural components are dynamically loaded for numbers of cycles above 10^6. Hence, total-life design must be in such a way that the maximum occurring stress amplitude is well below the material's fatigue limit. At this point, it must be mentioned that in the HCF/VHCF regime the values of measured fatigue life show a strong scatter, easily by a factor of two and more, owing to the increasing significance of microstrucural features like pores, inclusions or crystallographic orientation relationships. To improve structural integrity concepts for dynamically loaded components either (1) application of the local-strain concept should include such microstructural features and their interaction with early fatigue damage or (2) fatigue-crack propagation concepts should account for the mechanisms of microcrack propagation (cf. Chapter 6).

In general, the fatigue limit that can be obtained by Wöhler-type experiments, and the threshold for the growth of technical cracks can be brought together in the so-called Kitagawa–Takahashi diagram, which implies the condition for nonpropagation of fatigue cracks as a function of the applied stress range and the total crack length (see Fig. 7.2b). Kitagawa and Takahashi [510] assumed that below a certain crack length a_0 ($a_0 \approx 0.5$ mm) the nonpropagation criterion is determined by the fatigue limit, while above a_0 it is given by the fracture-mechanics threshold condition $\Delta K_{th} = (\Delta\sigma\sqrt{\pi a})_{th}$ (a_0 is obtained by the intersection of the fatigue limit with the line representing ΔK_{th}). Hence, the Kitagawa–Takahashi diagram represents a crack-length-dependent fatigue limit.

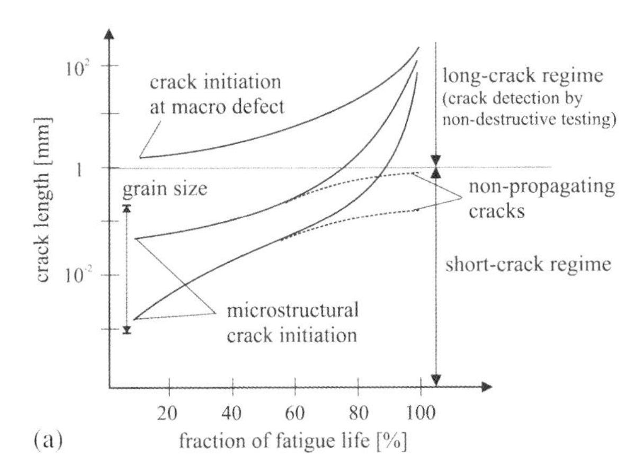

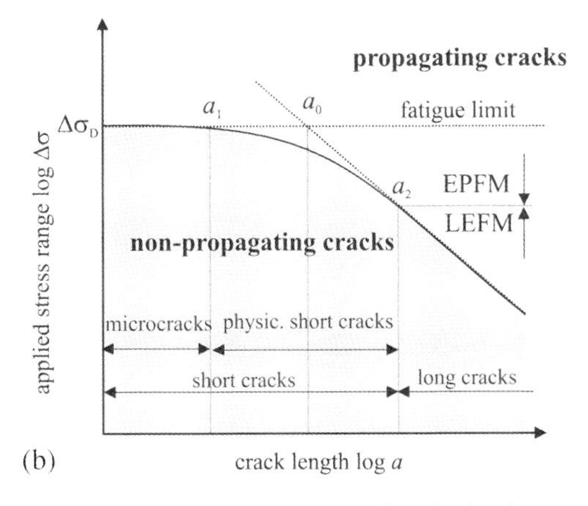

Fig. 7.2 Schematic representation of (a) the development of the crack length vs. the fraction of fatigue life for natural initiation of microcracks and crack initiation at large material defects and (b) the Kitagawa–Takahashi diagram.

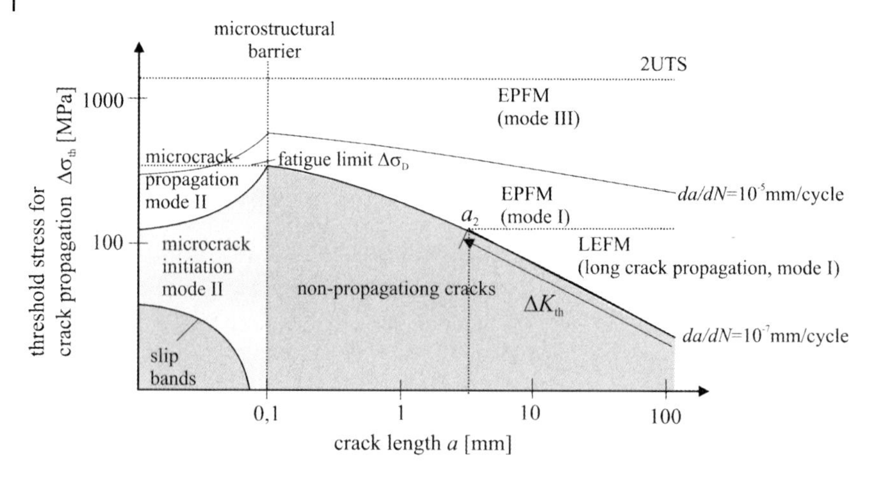

Fig. 7.3 Modified Kitagawa–Takahashi diagram (following a paper of Brown [511]).

In reality, the nonpropagation condition of fatigue cracks shows a smooth transition from the fatigue-limit-controlled constant stress value $\Delta\sigma_{FL}$ to the ΔK_{th}-controlled stress decrease (with a slope of –0.5 in the double-logarithmic representation) for the nonpropagation condition (Fig. 7.2b). The area below the intersection point a_o can be subject to nonconservative design, since cracks may propagate below the fatigue-limit condition as well as below the ΔK_{th} condition.

As experimentally proven (cf. Chapter 6) and shown in Fig. 7.3 (according to [511]), crack initiation and early propagation of microstructurally short cracks occurs well below the fatigue limit. With increasing crack length, the threshold stress range $\Delta\sigma$ for propagating cracks increases, owing to the fact that dislocations are paired or piled up along a slip band in front of a microstructural barrier (phase or grain boundary, see Section 6.2). Once the crack has overcome the boundary, the propagation condition corresponds to a much lower stress range. Hence, the fatigue limit is determined by the variation in microstructural barriers and the stress distribution and therefore being subject of strong scatter. Only when the crack length has reached a value of a_2 can the transient propagation regime, being coincident with a large plastic zone size ahead of the crack tip, be considered to be finished. This value is often defined as the threshold between short- and long-crack behavior [105]. Taylor [512] assumes as an estimate value for a_2 the higher value of either (1) ten times the grain size ($a_2 = 10d$) or (2) ten times the size of the cyclic plastic zone size ($a_2 = 10r_p$). Generally, the transition crack length a_2 is of higher values for ductile materials.

A simple equation for the smooth transition from the fatigue-limit-controlled to the ΔK_{th}-controlled regime of the Kitagawa–Takahashi diagram, depending on the length a_0, was given by El Haddad et al. [513]:

$$\Delta K_{th} = \Delta\sigma_{th}\sqrt{\pi(a + a_0)} \tag{7.1}$$

They understand the parameter a_0 as a measure for the missing constraint in the surface grains due to plane-stress conditions [514]. Although Eq. (7.1) describes quite well the gradual decrease in the stress range, defining the nonpropagation criterion between a_1 and a_2, it does not account for the actually relevant mechanisms of microcrack propagation. The interactions between the crack tip and the local microstructure cannot be represented adequately by the concepts of LEFM or EPFM. Application of Eq. (7.1) (for the crack-driving force ΔK) would yield a decreasing crack-propagation rate for a constant stress range $\Delta\sigma$ and a decreasing crack length a. From numerous studies (see Chapter 6) it is known that the opposite is the case. Immediately after initiation, microcracks grow at a very high rate that decreases when approaching the first microstructural barrier.

Sections 7.2 and 7.3 provide an overview, certainly incomplete, of the current state in modeling microstructurally short fatigue-crack propagation with the focus on the model of Navarro and de los Rios [324] and recent work on a numerical modification and extension [421, 515].

7.2 Modeling of Short-Crack Propagation

7.2.1 Short-Crack Models: An Overview

In several studies it was tried to transfer fracture-mechanical concepts, which are well established to predict long fatigue-crack propagation (ΔK concept, Paris law), to short-crack problems, e.g. [201, 406, 516–518]. Generally, it is possible to implement the transient behavior of short cracks with respect to the development of the plastic zone size and crack-closure effects, as is the case for the model of Vormwald [518] and its modification by Anthes [397] by means of using the effective ΔJ concept in combination with elastic–plastic material data obtained from stress–strain hysteresis curves. A further example of a fracture-mechanics-based short-crack model was proposed by McEvily et al. [517] (see also [302]), where the stress-concentration factor includes the stress concentration due to a fatigue crack tip of a radius ρ_e, to be obtained as an empirical parameter from experimental data. To account for the plastic zone size r_p, the crack length a is replaced by an effective crack length $a_{eff} = a + 0.5r_p$:

$$K = \left(\left(\sqrt{\frac{\pi\rho_e}{4}}\right) + Y\sqrt{\pi a_{eff}}\right)\sigma \tag{7.2}$$

To represent the transient development of crack closure within the propagation regime of short cracks of length l, the model of McEvily et al. [517] implements an exponential expression for the crack-opening stress-intensity factor K_{op} (cf. Section 6.3): $K_{op} = (1 - \exp(-\kappa a)\ K_{op\ max})$, with $K_{op,max}$ being the crack-opening stress-intensity factor for fully developed long-fatigue-crack closure and κ being a material parameter of units of mm^{-1}. With the material constant C the crack-propagation rate can be expressed as:

$$\frac{da}{dN} = C\left(\Delta K - \left(1 - \exp(-\kappa a)\right) K_{\mathrm{op,max}} - \Delta K_{\mathrm{eff,th}}\right)^2 \tag{7.3}$$

which yields together with an appropriate form of ΔK based on Eq. (7.2) a description for short and long cracks. The model has been applied successfully to experimental data for 2024-T3 aluminum alloy and 1045 plain carbon steel [519].

However, like most other fracture mechanics approaches to predict short-crack propagation, the model of McEvily et al. [517] does not take microstructural barriers into account, and hence it is not capable of accounting for the frequently observed oscillation in microcrack propagation.

There are various models where the variation in barrier strength of microstructural features is taken into consideration. For instance, Grabowski and King [520] described the oscillating crack propagation in nickel-based superalloys by a sequence of weak and hard barriers, representing crack deflection at the grain boundaries and within the grain, respectively. A similar approach was suggested by Bomas et al. [521, 522], where the short-crack-propagation rate varies within a maximum and a minimum value, accounting for interactions with pearlite islands and grain boundaries in 1017 low-carbon steel. The distances between the barriers in the model are generated by random numbers between 0 and 1, which are correlated with the Weibull distribution of the measured distances [522]. Figure 7.4 shows an example of calculated and measured crack-propagation rates for two different total-strain amplitudes.

A further empirical model, which takes local barrier effects during microcrack propagation in ferritic–pearlitic carbon steel into account, was proposed by Hobson et al. [523]. Here, propagation of short surface cracks of length $2c$ is described by two semi-empirical equations. The first one (Eq. 7.4) accounts for the early decrease in the surface-crack-propagation rate $d(2c)/dN$ as a function of the applied stress range $\Delta\sigma$ and $2c$. Intersection of the $d(2c)/dN$ vs. $2c$ curve (Eq. 7.4) with the abscissa ($d(2c)/dN = 0$) yields the respective barrier spacing d being equivalent to the mean distance between the ferrite–pearlite boundaries. The second equation (Eq. 7.5) represents propagation of physically short cracks in the form $d(2c)/dN = f(\Delta\varepsilon)$ minus a constant to account for the threshold behavior (ΔK_{th}).

$$\frac{d(2c)}{dN} = 1.64 \times 10^{-34} (\Delta\sigma)^{11.14} (d - 2c) \qquad \text{for } 2c < d \tag{7.4}$$

$$\frac{d(2c)}{dN} = 4.1 \times (\Delta\varepsilon)^{2.06}\, 2c - 4.24 \times 10^{-3} \qquad \text{for } 2c > d \tag{7.5}$$

In the semi-empirical Eqs. (7.4) and (7.5) the parameters must be given in units of μm and MPa. An example of the application of the model of Hobson et al. [523] is shown in Fig. 7.5a. The main disadvantages of the model are that (1) it is empirical and has to be adapted to the respective materials and (2) it generally considers transition from microstructure-dependent (stage I) to microstructure-independent

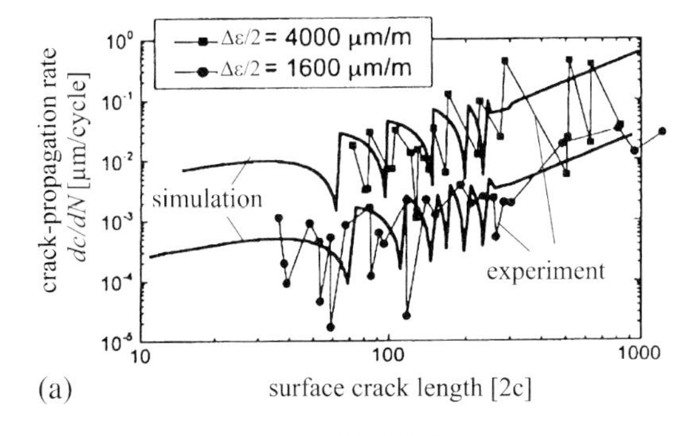

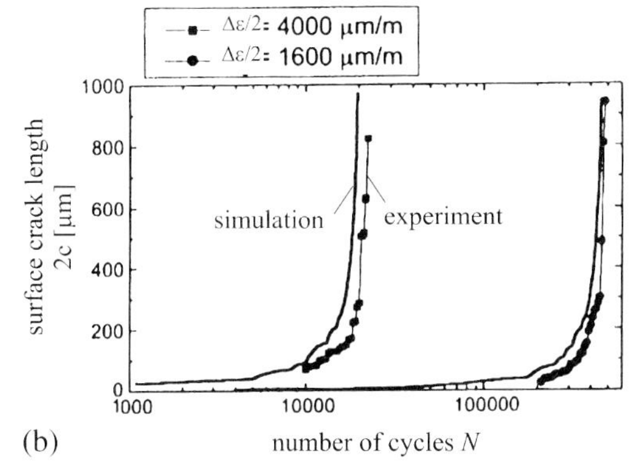

Fig. 7.4 Calculated and measured (a) crack propagation rates and (b) crack lengths vs. number of cycles for 1017 plain carbon steel (Cm15) fully reversed loaded ($R = -1$) at two different total strain amplitudes (after [522]).

(stage II) crack propagation always at the first barrier. In most real cases, however, microstructurally short cracks interact with several barriers of various strengths, until propagation becomes microstructure-independent. This restriction can be overcome by parameterizing the characteristic length d by d_i ($i = 1 \dots 4$). By this modification Murtaza and Akid [524] could apply the model of Hobson et al. [523] to cases where microcrack propagation is hindered subsequently by four different barriers until stage II long-crack propagation proceeds.

In a more quantitative way, the models of Chan und Lankford [525], and in a very similar way the model of de los Rios et al. [526], considers the barrier strength as a function of the resolved shear stresses τ_1 and τ_2 acting on the slip planes of two neighboring grains by means of the crystallographic misorientation function $\Phi(\Theta) = 1 - (\tau_1/\tau_2)$. The value of this function becomes zero when the shear stresses τ_1 and τ_2 are identical, i.e., the misorientation angle is equal to zero. The crack-propagation rate according to Chan and Lankford [525] can be expressed as

$$\frac{da}{dN} = C\Delta K^n \left(1 - \Phi(\Theta)\left(\frac{d-2x}{d}\right)^m\right)^2 \tag{7.6}$$

where C, n and m are constants in analogy to the Paris–Erdogan equation (Eq. (2.5), d is the mean spacing between the microstructural barriers, e.g., the mean grain size, and x the distance between the crack tip and the barrier (grain boundary). The effect of the misorientation function and the distance x form the grain boundary is shown in Fig. 7.5b. Hence, the model represents the decreasing crack-propagation rate when the crack tip approaches a barrier, i.e., the distance x becomes smaller.

Analytic models for initiation and propagation of microstructurally short fatigue cracks were proposed, for example, by Doquet [527], Tanaka and Mura [528], Chan [529] (providing an extension of Tanaka and Mura's model), Pippan [530], Tanaka et al. [531] and Navarro and de los Rios [324].

Doquet [527] uses a K concept for mode I crack propagation, where the effective stress-intensity factor K is reduced by crack-tip shielding due to an array of planar arranged dislocations ahead of the crack tip. By balancing the shear stresses acting on individual dislocations, the model is capable of predicting a threshold for fatigue-crack propagation also for local mode I loading conditions, which prevail during the propagation regime of microstructurally short cracks (cf. Wilkinson et al. [330, 532], Tong et al. [533] and Lin and Thomson [534]).

Fatigue-crack initiation due to planar-slip irreversibility has been modeled on the basis of continuously distributed dislocations by Tanaka and Mura [528]. They assume dislocation pileup at grain boundaries and consider crack initiation as to occur when the stored strain energy of the accumulated dislocations exceeds a certain critical value (cf. Fig. 7.6). As a possible way of crack initiation, the buildup of tensile stresses in the area of the slip band, where vacancy-type dislocation dipoles are present, can be considered.

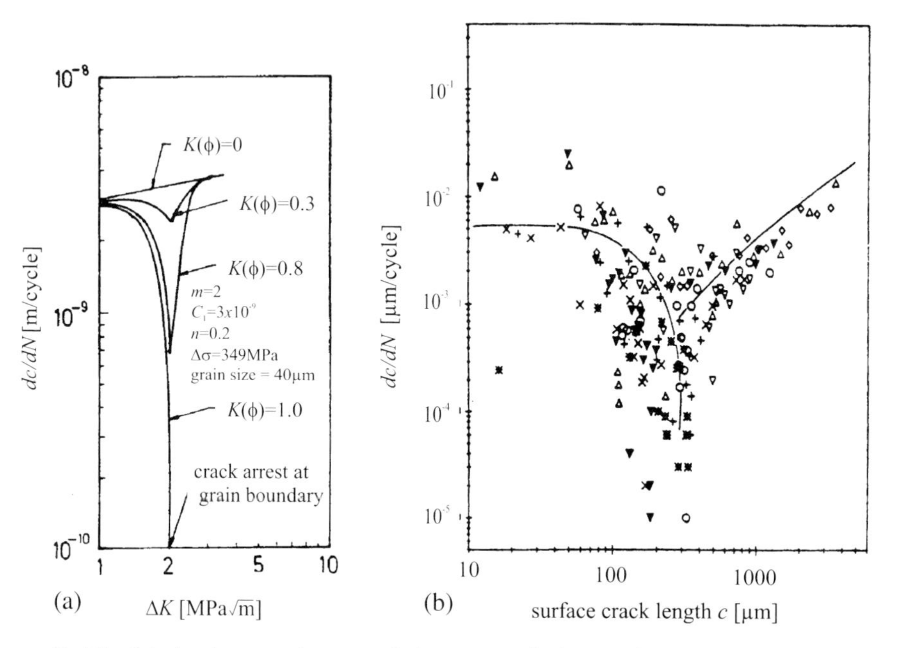

Fig. 7.5 Calculated propagation rates of microstructurally short cracks (a) according to the model of Hobson (after [302]) and (b) according to the model of Chan and Lankford [525].

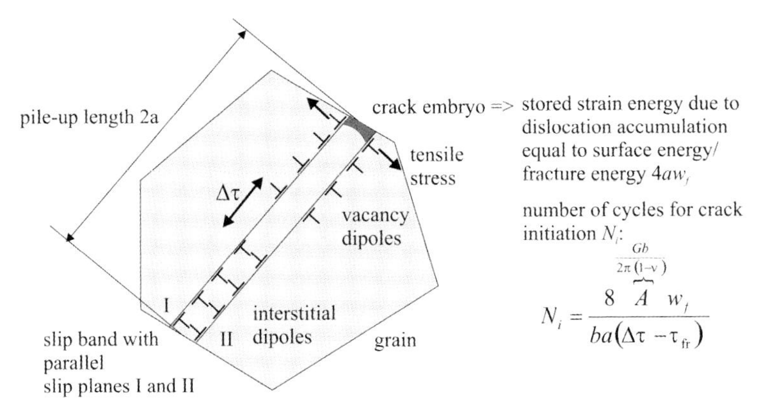

Fig. 7.6 Crack initiation due to dislocation pileup and formation of extrusions and intrusions in favorably oriented grains (after Tanaka and Mura [528]).

The general concept of describing a crack and its plastic zones by an array of distributed dislocations suggested by Bilby, Cottrell and Swinden [71] (BCS crack) has been modified to quantify the barrier efficiency of grain boundaries by Taira et al. [535]. This theoretical approach has been applied by Tanaka et al. [531, 536] to technical cases of fatigue-crack propagation, where in further studies also crack-closure effects [537] and statistical aspects [538] were taken into account. Basically, the model assumes a crack of length $2a$ emanating from the center of a grain of diameter d and growing along a single slip band or by alternate operating slip systems. The extension of the plastic zone is equal to $\omega = c - a$ (according to Fig. 7.7). Tanaka et al. [531, 536] distinguish between three cases: (1) the crack and the plastic zones at the crack tips are in equilibrium and located within the first grain (solution according to Bilby et al. [71] is valid), (2) the plastic zones are blocked at the first grain boundaries (solution after Taira et al. [535] is valid), and (3) the plastic zone has passed the first grain boundaries (solution according to Tanaka and Akiwa [539] is valid). Hence, the plastic zone (mode I, alternate-slip crack propagation) or the slip band (mode II, single-slip crack propagation) blocked by grain boundaries spreads into adjacent grains when a microstructural stress-intensity factor K^{m} exceeds a critical value $K^{\mathrm{m}}_{\mathrm{crit}}$. The microstructural stress intensity factor K^{m} depends on the applied normal stress σ, the crack length a and the plastic zone length c:

$$K^{\mathrm{m}} = \sigma\sqrt{\pi c}\left[1 - \frac{2\tau_{\mathrm{fr1}}}{\pi\sigma}\arccos\left(\frac{a}{c}\right)\right] > K^{\mathrm{m}}_{\mathrm{crit}} \tag{7.7}$$

Here, τ_{fr1} is the resolved friction stress in grain 1 that counteracts the resolved applied shear stress τ_{a} on the respective slip band. The barrier strength of a grain boundary depends on the amount of elastic strain energy that is required to induce plastic slip in the adjacent grain. According to Tanaka et al. [531], this can be expressed by the friction stress τ_{fr2}, accounting for the difference in the friction stresses between grains 1 and 2 (see Fig. 7.7, $\tau_{\mathrm{fr2}} = (\tau_{\mathrm{fr2}} - \tau_{\mathrm{fr1}}) + \tau_{\mathrm{fr1}}$). For spreading

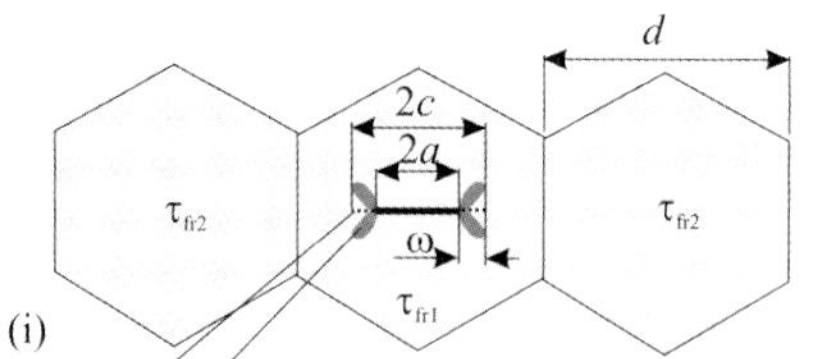

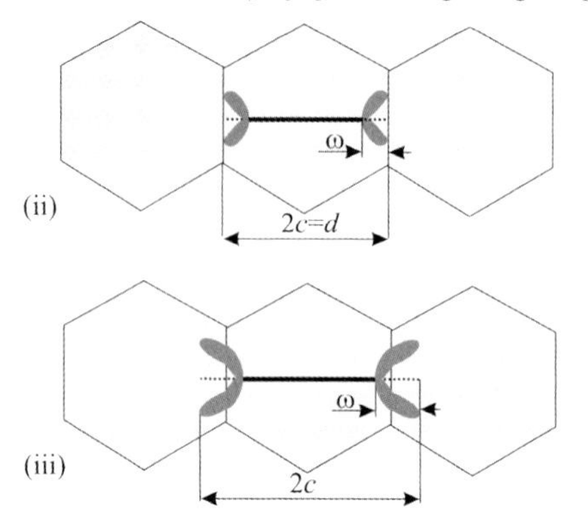

Fig. 7.7 Schematic representation of the plastic zone ahead of a crack tip interacting with a grain boundary according to the model of Tanaka et al. [531] (see text).

of the plastic zone $\omega = c - a$ across the first grain boundaries the following relationship is valid [531, 537]:

$$\frac{\pi}{2}\sigma_a - \tau_{\text{fr1}} \arccos\left(\frac{a}{c}\right) - \left(\tau_{\text{fr2}} - \tau_{\text{fr1}}\right) \arccos\left(\frac{d}{4c}\right) = 0 \tag{7.8}$$

Equation (7.8) corresponds to the original Dugdale approach [69] for the plastic zone size in Eq. (2.46), when it is assumed that the friction stress is of the same value for all the grains and is set equal to the global yield strength $\tau_{\text{fr}} = \sigma_{\text{Y}}$. Equation (7.8) can be applied to polycrystalline microstructures by summing up the differences in friction stress for n grains instead of the last term in Eq. (7.8) [538]. In the case of shear-controlled crack propagation σ_{a} might be multiplied with the local Schmid factor M_{S}.

Tanaka et al. [531] applied the solutions for the distributed dislocation theory (Bilby et al. [71] and Taira et al. [535]) for cases (1) to (3) introduced above to determination of crack-tip-opening displacement (CTOD, normal-stress-driven mode I crack propagation) and crack-tip-slide displacement (CTSD, shear-stress-driven mode II crack propagation). Eventually, fatigue crack propagation follows a simple exponential function:

$$\frac{da}{dN} = B\left(\Delta\text{CTOD}\right)^m \tag{7.9}$$

where the constant B accounts for the irreversible part of cyclic slip ΔCTOD or ΔCTSD, and the exponent m represents a possible relationship between ΔCTOD/ΔCTSD and the constant B [540]. Figure 7.8 shows an example of the application of the model [531].

Even though the analytical short-crack models, which are based on the theoretical work of Bilby et al. [71] and Taira et al. [535], describe crack propagation close to the fundamental mechanisms of fatigue damage, the statistical uncertainty of predicting the propagation rate of microstructurally short fatigue cracks, due to the distribution of grain size, geometry, crystallographic orientation, etc., remained a major problem.

In more recent work, it was attempted to take these statistics into account. For instance, Hoshide et al. [541] developed an algorithm that adapts regular hexagons to a grain size distribution by shifting their edges. The distribution in crystallographic orientation is represented by pairs of slip bands that are randomly assigned to each of these modified hexagons. Crack initiation in the synthetically generated microstructure is calculated based on the slip-blocking model according to Tanaka and Mura [528], while the crack-propagation model takes coalescence events into account, when the distance between two crack tips falls below a critical value. This approach was adopted by Olfe et al. [542], who also used modified hexagons to generate synthetic microstructures. These hexagons have the same grain size distribution as the material studied (Al-3Mg), and were used in combination with randomly distributed three-dimensional slip systems. Calculation of crack propagation is based on the slip-band-blocking model of de los Rios et al. [526] (similar to the model of Chan and Lankford [525], Eq. 7.6). Crack coalescence is considered by an influence zone ahead of the crack tip, the size of which increases with increasing crack length. Coalescence occurs when two influence zones encounter each other. Besides using hexagons [541], there are several further effective algorithms available to represent polycrystalline microstructures, e.g., the

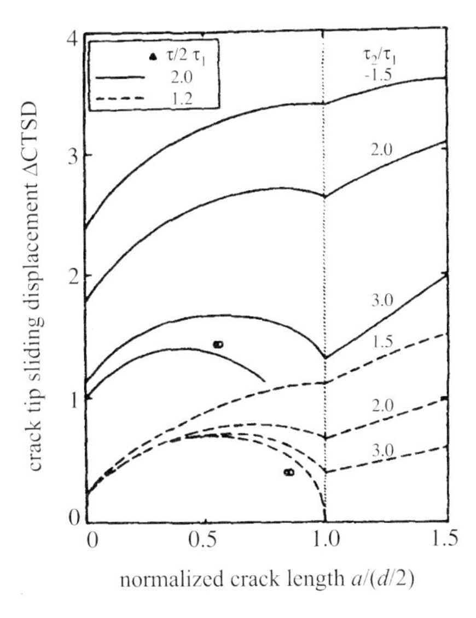

Fig. 7.8 Calculated crack-tip slide displacement vs. crack length normalized with the grain diameter d (after Tanaka et al. [531]).

Johnson–Mehl algorithm [543] or the Voronoi algorithm [544]. The latter, which is also used for the numerical short-crack model introduced in Section 7.3, has been applied by Meyer et al. [545] to model crack propagation in martensitic steel. Here, cracks growing along martensite laths are classified as one- and two-segmental cracks as well as zigzag cracks, the latter resulting from crack coalescence. Brückner-Foit and Huang [546] assumed that immediately after crack initiation according to the model of Tanaka and Mura [528] the respective grain is cracked completely. Accordingly, the stress distribution in the model microstructure is rearranged by a finite-element calculation. Early fatigue damage accumulation is then considered as an increase in crack density.

Bertolino et al. [547] use the finite-element method to describe the inhomogeneous stress distribution in a polycrystalline microstructure. The calculated shear stress between an existing crack and the closest grain boundary was used as input for a discrete-dislocation-dynamics (DDD) simulation of crack-tip plasticity and the corresponding crack growth rate. They found a qualitative agreement with the experimentally observed scatter in microcrack growth rates being responsible for the relationship between local texture and the variation in the fatigue limit.

It should be noted that the models introduced in the preceding sections and in Sections 7.2.2 and 7.3 are only a few representative examples of a steadily increasing number of short-crack models. Overviews are provided by Anthes [397] and Schick [515] (both in German) or by Hussain [302].

7.2.2
Model of Navarro and de los Rios

The model of Navarro and de los Rios, published in 1988 [324, 548], can be considered as a further development of the ideas of Dugdale [69], Bilby et al. [71] and Taira et al. [535]. It forms the theoretical basis of a numerical model that was originally developed by Schick [515] and that is introduced in Section 7.3.

The model of Navarro and de los Rios is capable of reproducing the decreasing efficiency of microstructural barriers with increasing crack length, accounting for the influence of crystallographic misorientation relationships by means of an orientation factor m^*. By manifold extensions and modifications, e.g., the implementation of slip-band hardening by Xin et al. [549], the consideration of variable load amplitudes by James and de los Rios [550], the application to composite materials by de los Rios et al. [551] and the use of variable crystallographic orientation factors by de los Rios et al. [552] or Wilkinson [553], the model of Navarro and de los Rios has become one of the most powerful and prominent concepts for fatigue-life assessment taking interactions with the material's microstructure into account.

Similar to the model of Tanaka et al. [531] (see previous section), the starting point of the model of Navarro and de los Rios is a crack of length $2a$ initiated in the center of a grain of diameter d. Ahead of the two symmetrical crack tips a plastic zone of total length $2c$ (including the crack length) is located, according to the schematic representation in Fig. 7.9. In the most simple case, this situation corresponds to Dugdale's yield strip model for a normal-stress-loaded crack in an infinite plate with the boundary conditions that the crack is stress free, $\sigma = 0$ for $|x| \leq a$,

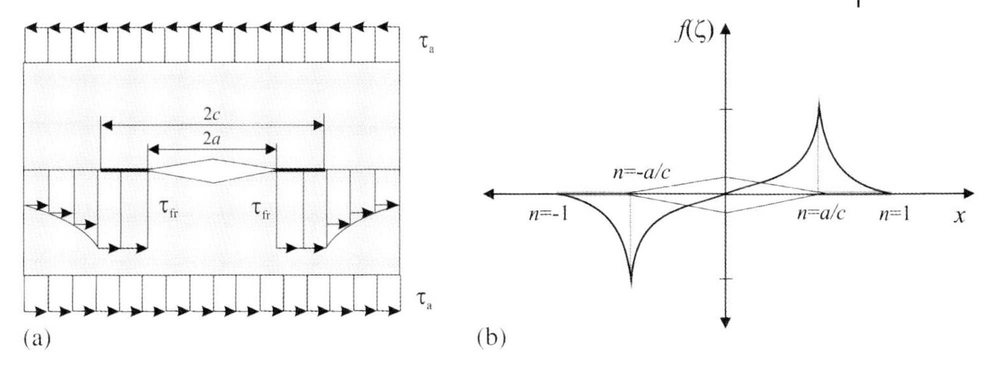

Fig. 7.9 Schematic representation of (a) a crack of length $2a$ with its plastic zone of length $2c$ loaded under mode II by a shear stress of τ_a and (b) corresponding dislocation-density distribution according to Bilby et al. [71].

and the stress within the yield strip is equal to the yield strength, $\sigma = \sigma_Y$ for $a < |x| \leq c$ (see [69] and Section 2.3.2). A more detailed picture provides the theory of continuously distributed dislocations by Bilby, Cottrell and Swinden [71] (BCS model). Here, the crack and its plastic zone are represented by a one-dimensional dislocation-distribution function along a slip band. It must be pointed out that the slip band and the dislocations are rather of mathematical character than based on physical interactions between dislocations, representing macroscopic plastic deformation. In analogy to the elastic theory of dislocations (cf. Section 4.2), the condition that the resolved shear stress acting on the dislocations is equal to the friction stress has to be fulfilled, in order to maintain equilibrium. For a mode II crack (similar for mode II or mode III cracks) loaded by τ_a the shear stress along the slip band as a function of the normalized coordinate $\zeta = x/c$ and the dislocation-density distribution $f(\zeta)$ can be expressed by the following integral equation:

$$\tau(\zeta) = A \cdot \int_{-1}^{1} \frac{f(\zeta')}{\zeta - \zeta'} d\zeta' + \tau_a \tag{7.10}$$

with A being a constant containing the shear modulus G, the Poisson ratio ν and the Burgers vector b; $A = Gb/[2\pi(1-\nu)]$ for edge dislocations and $A = Gb/2\pi$ for screw dislocations. With the boundary conditions mentioned above and the assumption that outside of the plastic zone the dislocation density is equal to zero, $f(|\zeta| > 1) = 0$, one obtains for the dislocation-density distribution (after Bilby et al. [71]):

$$f(\zeta) = \frac{\tau_{fr}}{\pi^2 \cdot A} \cdot \left[\operatorname{arcosh}\left(\left| \frac{1 - n \cdot \zeta}{n - \zeta} \right| \right) - \operatorname{arcosh}\left(\left| \frac{1 + n \cdot \zeta}{n + \zeta} \right| \right) \right] \tag{7.11}$$

where the parameter n denotes the ratio between the crack length and the length of the plastic zone, $n = a/c$; τ_{fr} is the friction stress, acting opposite the motion of dislocations on the slip band and being zero within the crack. Figure 7.9 shows schematically the dislocation-density distribution according to the BCS model [71].

When assuming the applied resolved shear stress τ_a to exceed the friction stress τ_{fr} at the slip band, the BCS model would predict a dislocation motion towards infinity. However, this is not the case when the grain boundaries act as effective barriers. Mathematically, the barrier effect of grain boundaries can be accounted for by a supplementary term in Eq. (7.11), leading to a singularity of the dislocation-density function at the barrier [324]. The corresponding dislocation-density function for the so-called unbounded solution (according to Taira et al. [535], see Fig. 7.10) is given by

$$f(\zeta) = \frac{\tau_{fr}}{\pi^2 \cdot A} \cdot \left[\operatorname{arcosh}\left(\left| \frac{1 - n \cdot \zeta}{n - \zeta} \right| \right) - \operatorname{arcosh}\left(\left| \frac{1 + n \cdot \zeta}{n + \zeta} \right| \right) \right] + \\ + \frac{\tau_{fr}}{\pi^2 \cdot A} \cdot \frac{\zeta}{\sqrt{1 - \zeta^2}} \left[2 \cdot \arcsin(n) + \pi \cdot \left(\frac{\tau_a}{\tau_{fr}} - 1 \right) \right] \tag{7.12}$$

The solution after Navarro and de los Rios (NR model) is identical with the BCS model when the second term of Eq. (7.12), representing the singularity at the end of the plastic zone, $|\zeta| = 1$, vanishes. This is the case only when the friction stress is larger than the actual stress acting on the slip plane, $\tau_{fr} > \tau_a$, and hence the dislocations are not able to reach the grain boundary. Accordingly, the second term becomes zero for

$$\left[2 \cdot \arcsin(n) + \pi \cdot \left(\frac{\tau_a}{\tau_{fr}} - 1 \right) \right] = 0 \tag{7.13}$$

With a few rearrangements, using the goniometric relationship $\sin(\alpha - \beta) = \sin\alpha \cos\beta - \cos\alpha \sin\beta$, the ratio $n = a/c$ for the BCS crack with freely expanding plastic zone (bounded solution) can be expressed as

$$n_{BCS} = \cos\left(\frac{\pi \tau_a}{2 \tau_{fr}} \right) \tag{7.14}$$

In the model of Navarro and de los Rios, spreading of the plastic zone is considered to occur stepwise grain by grain. When the stress acting on a slip system in a neighboring grain exceeds a critical value, the plastic zone size immediately increases by one grain diameter. Hence, the extension of the plastic zone c in the ith grain and the ratio n_{NR} can be given as (cf. Fig. 7.10)

$$c = i \frac{d}{2}, \quad i = 1, 3, 5, \ldots \qquad n_{NR} = \frac{2a}{id} \tag{7.15}$$

For a given crack length a the ratio n can be determined using Eq. (7.14) (BCS bounded solution) or Eq. (7.14) (NR unbounded solution). According to Wilkinson [553], the BCS solution (Eq. 7.11) should be used as long as $n_{BCS} < n_{NR}$ is valid.

Analogous to Eq. (7.9) (model of Tanaka et al. [531]), the model of Navarro and de los Rios considers an exponential relationship to calculate the crack-propagation rate da/dN as a function of the cyclic crack-tip-slide displacement:

$$\frac{da}{dN} = B(\Delta \text{CTSD})^m \tag{7.16}$$

where according to [550, 554] the factor B can be used to fit experimental data by means of a linear function of the applied shear stress: $B = B_1 + B_2\tau_a$.

The value of ΔCTSD for the solution of Navarro and de los Rios is obtained by integration of the dislocation-density function $f(\zeta)$ (Eq. 7.12) over the plastic zone from $\zeta = a/c$ to $\zeta = 1$, and multiplication with the absolute value b of the Burgers vector. The ratio n in the subsequent equations corresponds to n_{NR} according to Eq. (7.15):

$$\Delta \text{CTSD} = \frac{2b\tau_{\text{fr}}c}{\pi^2 A}\left[n \cdot \ln(1/n) + \sqrt{1-n^2}\left(\arcsin(n) + \frac{\pi}{2}\cdot\left(\frac{\Delta\tau_{\text{a}}}{\tau_{\text{fr}}} - 1\right)\right)\right] \tag{7.17}$$

To obtain the solution for the BCS crack the second term in Eq. (7.17) has to be omitted. Transfer of the plastic zone into the neighboring grain occurs when the stress increase S due to dislocation pileup at the grain boundary exceeds a critical value S_{crit} at a dislocation source a distance of r_0 ahead of the boundary (see Fig. 7.10). Solution of the integral in Eq. (7.10) for r_0, corresponding to $\zeta_0 = 1 + 2r_0/d$, yields the stress $S(\zeta_0)$ as follows [549, 554]:

$$\frac{S(\zeta_0)}{\beta\tau_{\text{fr}}} = \frac{\tau_{\text{a}}}{\tau_{\text{fr}}}\left[1 - \frac{2}{\pi}\arctan\left(\frac{n}{\beta\sqrt{(1-n^2)}}\right)\right] - \frac{2}{\pi}\arccos(n) \tag{7.18}$$

$$\text{with} \quad \beta = \frac{\zeta}{\sqrt{\zeta^2 - 1}}$$

The critical stress at r_0 can be considered as driving force being required to activate dislocation movement or multiplication (according to the Frank–Read mechanism; Section 4.2) in the respective grain. However, there are no precise data for the length r_0 in the literature available. One may correlate r_0 with the width of a slip band [324], or with the characteristic length of a Frank–Read source [548]. Accordingly, r_0 should lie between $r_0 = 0.1$ and 1 μm [324] or at approximately 1% of the slip band length [324, 548].

Certainly, slip activation in neighboring grains depends also on the crystallographic misorientation between the adjacent grains. In the original model of Navarro and de los Rios [324] this is considered by an orientation factor m^*. Hence, transfer of the plastic zone into the neighboring grain requires the fulfillment of the following condition:

$$S(\zeta_0) \geq m^* S_{\text{crit}} \tag{7.19}$$

Strictly speaking, the orientation factor m^* should vary from grain to grain in a real polycrystalline microstructure. Owing to the variation in crystallographic orientations, Wilkinson [553] used measured orientation data to derive the orientation factor for the ith grain boundary from the orientation of the neighboring grain as the reciprocal value of the minimal Schmid factor of the respective grain $M_{S,i}$:

$$m_i^* = \min \frac{1}{M_{S,i}} = \min \left(\frac{1}{(\vec{F}_e \vec{n}_{crys})(\vec{F}_e \vec{s})} \right) \tag{7.20}$$

Here, $\vec{F}_e$, $\vec{n}_{crys}$ and $\vec{s}$ are the unit vectors of the force, the slip-plane normal and the slip direction of the various slip systems in grain i. As a first approximation, Navarro and de los Rios [324] used for the orientation factor the mean of all Schmid factors for single slip, revealing for cubic crystals an orientation factor of $m^* = 2.24$ (corresponding to the Sachs factor M_{Sa}, cf. Section 4.2). Closer to reality is the assumption that for the first grain the orientation factor is small, having a value of $m_1^* = 1 \ldots 2$, since crack initiation can be expected to start at sites where the resolved shear stress is of maximum values. Initially, the orientation factor should depend strongly from the local crystallographic orientation. However, during further crack propagation it approaches, due to prevailing multiple slip, an orientation-independent value of approximately $m^* = 3.1$ (corresponds to the Taylor factor M_{Ta}, cf. [553]). Xin et al. [549] describe this development in the orientation factor by a simple logarithmic relationship:

$$\frac{m_i}{m_1} = 1 + 0.5 \cdot \ln(i) \tag{7.21}$$

between the orientation factor m_1 of the first grain (which should be low, see paragraph above) and the ones m_i of the subsequent grains i. The unknown critical stress S_{crit} in Eq. (7.18) can be eliminated by setting $n = 1$ in Eq. (7.18) [554]. This corresponds to the condition that the applied stress is just not sufficient to cause transfer of the plastic zone to the neighboring grain. In the case of the first grain this situation is equivalent to the fatigue limit $\Delta\sigma_{FL}$, for all subsequent grains it can be understood as a crack-length-dependent fatigue limit $\Delta\sigma_{FLi}$ (cf. Kitagawa–Takahashi diagram, Fig. 7.2b). Hence, Eq. (7.18) together with Eq. (7.19) can be simplified as follows:

$$\text{(a)}\ S(\zeta_{0,1}) = \beta_1 \Delta\sigma_{FL} = m_1^* S_{crit} \qquad \text{(b)}\ S(\zeta_{0,i}) = \beta_1 \Delta\sigma_{FLi} = m_i^* S_{crit} \tag{7.22}$$

Combination of Eqs. (7.22a) and (7.22b) and inserting in Eq. (7.18) yields the criterion for transfer of the plastic zone in the neighboring grains when the critical ratio n_i^c is exceeded, according to the derivation of de los Rios and Navarro [555]:

$$\frac{\pi}{2} \left[\frac{\Delta\tau_a}{\tau_{fr}} \left(1 - \frac{\beta_1 m_i \Delta\sigma_{FL}}{\beta_i m_1 \Delta\tau_a} \right) + \frac{1}{\beta_i} \left(1 - \frac{2}{\pi} \arctan \left(\frac{n_i^c}{\beta_i \sqrt{(1 - n_i^{c2})}} \right) \right) \right] - \arccos(n_i^c) \geq 0 \tag{7.23}$$

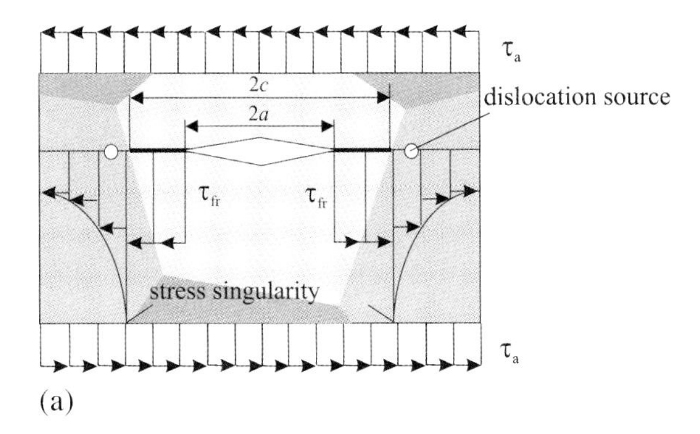

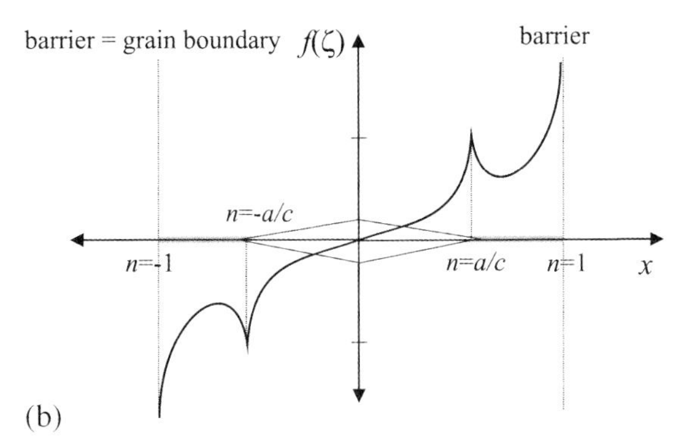

Fig. 7.10 Schematic representation of (a) a crack of length $2a$ with its plastic zone of length $2c$ loaded under mode II by a shear stress of τ_a and (b) corresponding dislocation-density distribution (according to Taira et al. [535] and Navarro and de los Rios [324]).

Schick [554] developed an algorithm to implement the short-crack-propagation model of Navarro and de los Rios [324, 548, 549] into the simulation environment MATLAB (cf. Fig. 7.11):

1. As first step, the ratios n_i^c and n_{i+2}^c need to be determined. This is done by applying a numerical approach to get zero points (MATLAB) to Eq. (7.23). When the critical value n_i^c is reached the plastic zone spreads into the two neighboring grains $i+2$ (left- and right-hand side) and is immediately increased by $2d$, while the crack length remains at first constant. The start value for the ongoing calculation changes to $n_{i+2}^s = n_i^c\ (i/i+2)$.

2. The actual crack-propagation rate da/dN vs. the crack length a (that varies with the ratio n) can be calculated by the crack-propagation law in Eq. (7.16) as a function of the cyclic ΔCTSD that is obtained by Eq. (7.17). Integration of Eq. (7.16) between n_i^s and n_i^c yields the number of cycles ΔN_i of the fatigue life of grain i. The crack length is represented by the actual value of the ratio n_i with the plastic zone being temporary constant, $c_i = id$. For the first grain the start ratio is $n_1^s = 0$.

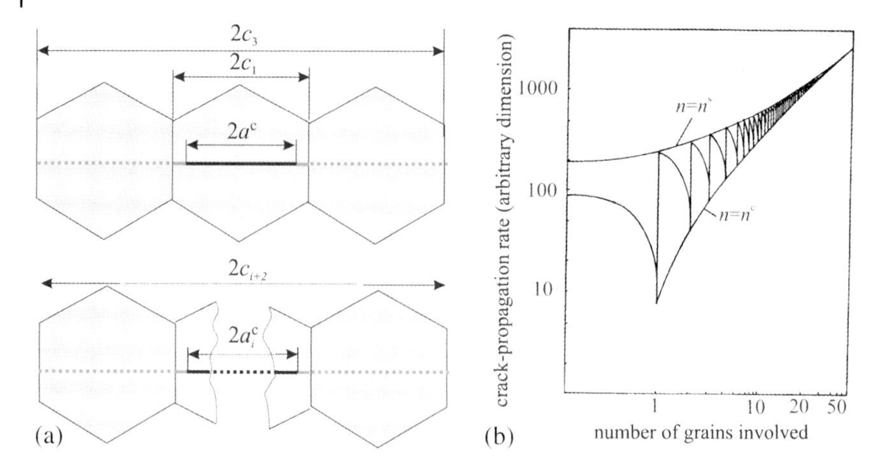

Fig. 7.11 (a) Spreading of the plastic zone into the neighboring grains after exceeding of a critical ratio $n_i^c = n_i^c/c_i$; (b) corresponding development of the crack-propagation rate (after [557]) (details in text).

3. The crack-propagation rate is governed by the parameter $n_i = a_i/c_i$, i.e., since the plastic-zone size always coincides with integer multiples of the grain size, it is governed by the distance of the crack tip from the adjacent grain boundary. Once this distance becomes very small, ΔCTSD and hence the (lower) crack-propagation rate da/dN decreases substantially. This is shown in Fig. 7.11a. As soon as the critical ratio n_i^c is reached, a dislocation source in the neighboring grain is considered to be activated and the plastic zone spreads into the two neighboring grains ahead of the two crack tips. Immediately the next start ratio n_{i+2}^s becomes by a factor $i/i + 2$ smaller than the preceding critical end ratio n_i^c. Correspondingly, the new (upper) crack-propagation rate jumps to a higher value. With increasing crack length and an increasing number i of participating grains, the upper and lower crack-propagation rate converges. This can be considered as the gradual transition from microstructure-dependent short-crack to microstructure-independent long-crack propagation [556, 557] (see Fig. 7.11b).

4. As termination condition, a critical crack length needs to be defined that either is sufficient to cause immediate failure of the respective component by rupture or that corresponds to the transition to long-crack behavior to be treated by a Paris-type equation for long-crack propagation (two-step model). The latter case can be assumed to occur when the extension of the plastic zone exceeds one grain diameter ahead of the crack tip. This is the case for $a_{i+2}^c < (i+2)\,(d/2)$.

Figure 7.12 shows examples for the application of the model of Navarro and de los Rios to predict fatigue life and crack-propagation rates. Schick [554] compared experimental data of the β-titanium alloy LCB in the solution-heat-treated condition (cf. Chapters 5 and 6) and the austenitic steel alloy 800 with the model predictions including work hardening according to the model extension by Xin et al. [549]:

$$\tau_{fr} = h\tau_{fr}^0 \tag{7.24}$$

with τ_{fr}^{0} being the initial friction stress and h a function of the slip band width, the applied resolved shear stress and the work-hardening coefficient (for details see [549, 554]). It must be added that the excellent agreement between the experimental and simulation results in Fig. 7.12a is partially due to the adaptation of the factor B (with the two constants B_1 und B_2) in Eq. (7.16). Figure 7.12b shows crack-propagation rates obtained by Wilkinson [553] for an Al–Li alloy and various orientation factors (application of Eq. 7.20). Choosing random orientation factor distributions results in a scattering of the crack-propagation rates, which is closer to reality than the continuous increase resulting from a mean orientation factor (Eq. 7.19) or an gradually increasing orientation factor (Eq. 7.21).

In general, the analytical model of Navarro and de los Rios is capable of representing the basic mechanism of early crystallographic fatigue-crack propagation, owing to microstructural parameters like the grain size or the crystallographic orientation and the gradual transition to long-fatigue-crack propagation. However, its restriction to one dimension allows only a very limited implementation of transient effects like roughness-induced crack closure or pronounced three-dimensional influence factors like the grain geometry, crack coalescence or crack branching.

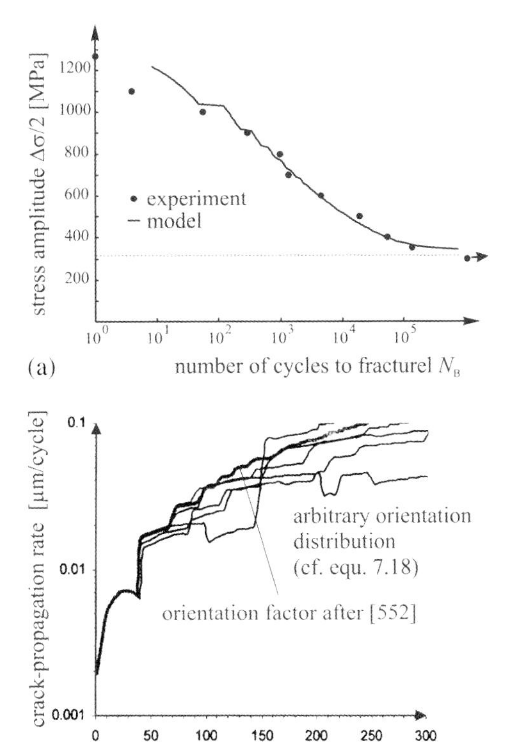

Fig. 7.12 Application of the model of Navarro and de los Rios: (a) comparison between measured and calculated numbers of cycles to fracture for the β-titanium alloy LCB at $R = -1$; (b) crack-propagation rates vs. crack length for arbitrary orientations (after [553], Al–Li alloy, $\Delta\sigma/2 = 110$ MPa, $R = 0$).

7.3
Numerical Modeling of Short-Crack Propagation by Means of a Boundary Element Approach

7.3.1
Basic Modeling Concept

The concept of the numerical short-crack model developed by Schick [515] and Künkler [421], which is introduced in this section, uses the principles of the boundary element method (BEM, cf. [558,559]) and is based on the fundamental concept of the model of Navarro and de los Rios, which is discussed in the preceding section. Again, ΔCTSD, either on a single slip band (mode II) or on two alternate operating slip systems (mode I), is the driving force for crack advance. When the crack tip approaches a grain boundary, crack propagation is hindered and the stress on a dislocation source representing the slip systems in the adjacent grains increases. Plastic deformation will transfer on the slip system in the adjacent grain, where at first the critical stress for activation of dislocation sources is exceeded. Extension of the plastic zone promotes ΔCTSD, and hence leads to an increase in the crack-propagation rate da/dN.

To represent the real distribution of the size, geometry and crystallographic orientation relationship of the surface grains of a specimen – data that can be obtained easily by means of electron backscattered diffraction (EBSD) measurements in a scanning electron microscope (see Section 3.3.2) – the crack and its plastic zones are meshed by boundary elements. This is shown schematically in Fig. 7.13a for a microstructurally short crack growing by single slip. The boundary elements are treated mechanically in the same way as paired edge dislocations – each element corresponds to a shear (slip band elements) or a shear and normal displacement (crack elements), described by mathematical dislocation dipoles. Contrary to the physical dislocations, the Burgers vectors of the boundary elements are not related to the crystal lattice. The Burgers vectors of the crack elements for normal and tangential displacement are given as (b_n^i, b_t^i); the Burgers vectors in the slip band are limited to tangential displacement, $(0, b_t^j)$.

In analogy to physical dislocations, mathematical dislocations cause elastic distortion (see Section 4.2). The stress field in a distance $r = \sqrt{x_0^2 + y_0^2}$ from the dislocation core (cf. Fig. 7.14) is given as follows [515, 560]:

$$\begin{bmatrix} \sigma_{xx}(x_0, y_0) \\ \sigma_{yy}(x_0, y_0) \\ \tau_{xy}(x_0, y_0) \end{bmatrix} = \frac{G}{2\pi(1-\nu)} \cdot \begin{bmatrix} -\frac{y_0}{r^4}\left(3x_0^2 + y_0^2\right) & \frac{x_0}{r^4}\left(x_0^2 - y_0^2\right) \\ \frac{y_0}{r^4}\left(x_0^2 - y_0^2\right) & \frac{x_0}{r^4}\left(x_0^2 + 3y_0^2\right) \\ \frac{x_0}{r^4}\left(x_0^2 - y_0^2\right) & \frac{y_0}{r^4}\left(x_0^2 - y_0^2\right) \end{bmatrix} \cdot \begin{bmatrix} b_x \\ b_y \end{bmatrix} \tag{7.25}$$

Since each of the boundary elements consists of two dislocations (dipoles), and hence the stress field is related to the element center ($x_0 - y_0$ coordinate system), a

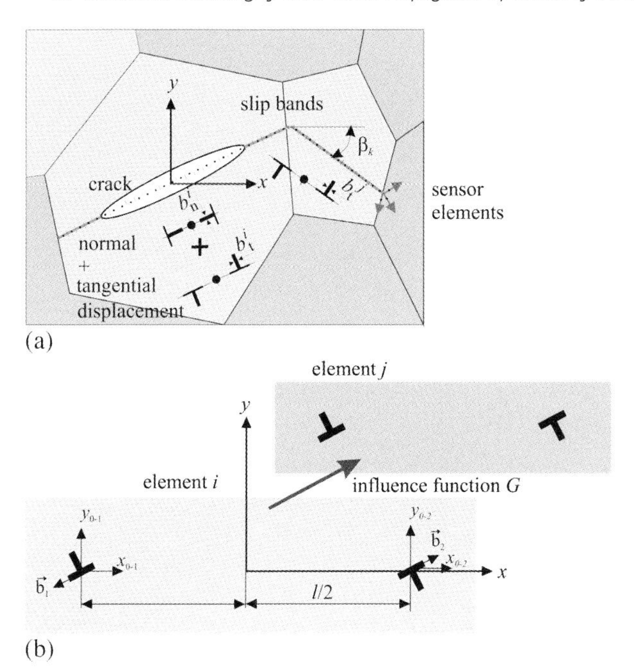

Fig. 7.13 Schematic representation of (a) a slip-band crack meshed by boundary elements and (b) one of these boundary elements consisting of two dislocations representing tangential and normal displacement.

transformation of the x coordinate of dislocation 1 from $x_{0,1}$ to $x - l/2$ and for dislocation 2 from x_0 to $x + l/2$ is required, according to the schematic representation in Fig. 7.13b.

The displacement of the two dislocations of an element j results in mechanical stress acting on all other elements i of the meshed slip band/crack line according to Eq. (7.25). Hence, the overall stress acting on an individual element i can be calculated as the sum of the stresses due to interactions with all other elements j plus the applied stress σ_a.

Since the analysis is focused on the slip band geometry within a two-dimensional polycrystalline microstructure it is useful to transform the global x–y coordinate system into the respective t–n slip-band coordinate system being inclined by an angle β^j with respect to the x axis (see Fig. 7.13). The transformation yields the displacement vector $\vec{b}^i_{t,n}$ (Burgers vector) for the $t-n$ coordinate system with respect to that in the x–y coordinate system as follows [515]:

$$\vec{b}^i_{t,n} = \underline{\underline{A}}\,\vec{b}^i_{x,y} = \begin{bmatrix} \cos\beta^j & \sin\beta^j \\ -\sin\beta^j & \cos\beta^j \end{bmatrix} \vec{b}^i_{x,y} \qquad (7.26)$$

Accordingly, this transformation needs to be applied also to the stress values.

The relationship between the normal stresses σ^i_{nn} (normal to the slip band after applying the coordinate transformation), the shear stresses τ^i_{tn} (tangential along the slip band) acting on a boundary element i and the displacements b^j of all other elements j can be expressed as

$$\sigma^i_{\mathrm{nn}} = \sum_{j=1}^{p} G^{ij}_{\mathrm{nn,n}} b^j_{\mathrm{n}} + \sum_{j=1}^{p+q} G^{ij}_{\mathrm{nn,t}} b^j_{\mathrm{t}} + \sigma^i_{\mathrm{nn,a}} \leq 0\,, \qquad \text{for } i = 1 \ldots p \tag{7.27}$$

$$\left|\tau^i_{\mathrm{tn}}\right| = \left|\sum_{j=1}^{p} G^{ij}_{\mathrm{tn,n}} b^j_{\mathrm{n}} + \sum_{j=1}^{p+q} G^{ij}_{\mathrm{tn,t}} b^j_{\mathrm{t}} + \tau^i_{\mathrm{tn}}\right| \begin{cases} = 0 & \text{for } i = 1 \ldots p \\ \leq \tau_{\mathrm{fr}} & \text{for } i = p+1 \ldots p+q \end{cases} \tag{7.28}$$

Here, G includes the geometry data and material constants, representing the matrix term and the constant in Eq. (7.25) (Green functions). $\sigma^i_{\mathrm{nn,a}}$ and $\tau^i_{\mathrm{tn,a}}$ are the resolved normal and shear stresses, resulting form the uniaxial applied stress, acting on the element i. Furthermore, the equation/inequality system (Eqs. 7.27 and 7.28) must fulfill the boundary conditions of the model:

1. Within the crack ($i = 1 \ldots p$, crack elements) no shear stresses can be present. Only compressive normal stresses are possible (Eq. 7.27) that cause geometry-induced crack-closure effects. Normal displacements within the crack must be positive, $b_{\mathrm{n}} \geq 0$, i.e., material penetration is not possible.
2. Within the plastic zone ahead of the crack tip ($i = p + 1 \ldots p + q$, slip-band elements) the shear stresses are limited by the friction stress τ_{fr} in compression as well as in tension.

Solution of the linear inequality system of Eqs. (7.27) and (7.28) yields the tangential and normal displacements of all elements of the meshed slip-band/crack line.

Since initially the extension of the plastic zone is unknown, Schick [515] developed a fast algorithm that firstly treats the crack fully elastic, considering only the p crack elements in Eq. (7.26) (first iteration; Fig. 7.14a). This leads to a linear system of equations that can be solved easily. In a second step, the resulting shear stresses within the meshed slip band are calculated. Those elements where the friction stress τ_{fr} is exceeded are considered as slip-band elements; the shear stress is set to τ_{fr}. This procedure is repeated in an iterative way, as illustrated schematically in Fig. 7.14a (after [515]), until the boundary conditions in Eq. (7.28) are fulfilled. In a similar way, the crack-closure condition that the normal displacements of the crack can only be positive, $b_{\mathrm{n}} \geq 0$, is treated. In a first calculation step the condition $b_{\mathrm{n}} \geq 0$ is ignored and interpenetration of the crack faces is allowed. The maximum value of the penetration is subdivided in intervals; all penetration values exceeding the first interval are set to the respective new maximum penetration value (see Fig. 7.14b). The iterative approach is required to fulfill the geometrical crack-closure condition, $b_{\mathrm{n}} \geq 0$, without obtaining misleading results. The latter would be the case when the complete crack-face penetration $b_{\mathrm{n}} < 0$ was set to zero [515].

Thus, the plastic zone length and the complete set of tangential and normal displacements for the actual load step can be calculated. The range of ΔCTSD is obtained as the difference of the tangential displacements at the crack tip for maximum load $\mathrm{CTSD_{max}}$ and minimum load $\mathrm{CTSD_{min}}$. Hence, crack advance by Δa can

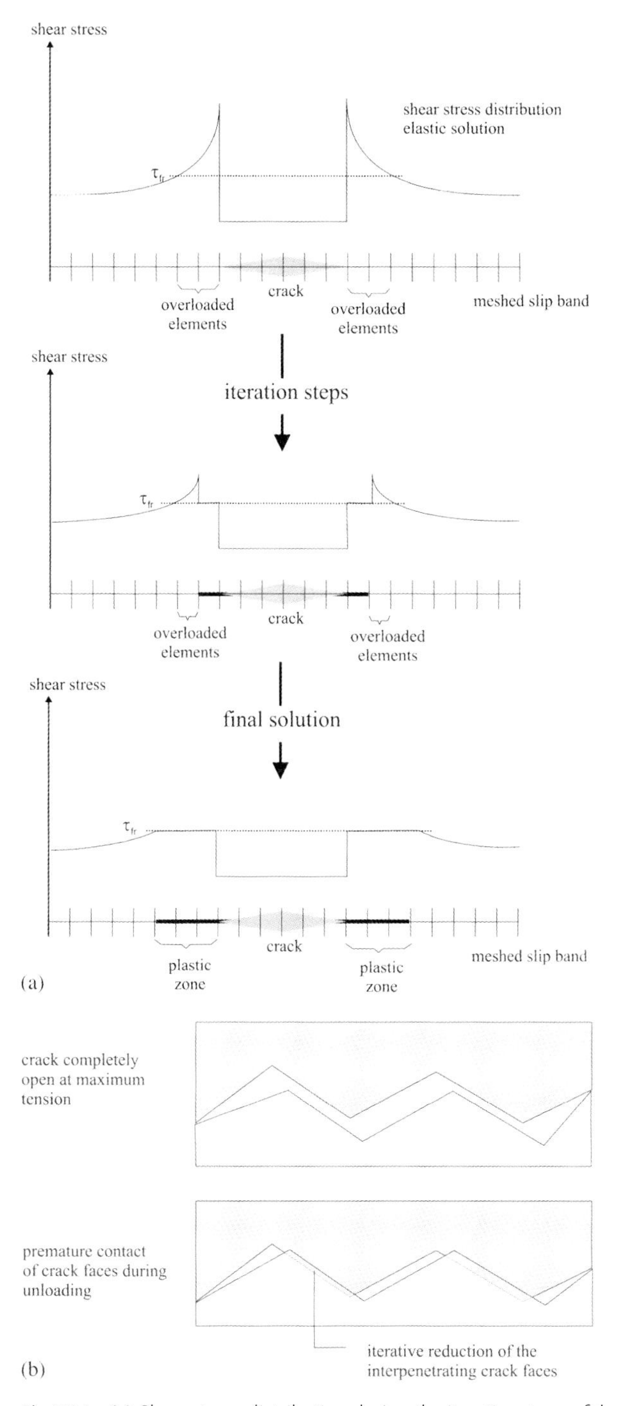

Fig. 7.14 (a) Shear-stress distribution during the iterative steps of the slip-band calculation; (b) schematic representation of the algorithm to account for crack closure (according to Schick [515]).

be computed using Eq. (7.16) in combination with ΔCTSD. Once the accumulated crack-advance intervals exceed a predefined value, a slip-band element is reassigned to a crack element for the subsequent calculation steps. It should be mentioned here that applying Eq. (7.16) can be considered as defining a certain fraction B of the cyclic slip displacement at the crack tip as irreversible (cf. Section 6.2.1 and Fig. 6.8).

The advantage of the numerical approach as compared to the analytical model of Navarro and de los Rios is the fact that it can be applied to two-dimensional microstructures with an arbitrary distribution of grain sizes, phases and crystallographic orientations as they were obtained from experiments. This allows a more realistic simulation of microcrack propagation accounting for microstructural features and roughness-induced crack-closure effects.

7.3.2
Slip Transmission in Polycrystalline Microstructures

Transfer of the plastic zone across grain boundaries is treated in a similar way as in the model of Navarro and de los Rios [324]: When a critical stress at a dislocation source on one of the slip systems in the adjacent grains (grains B and C in Fig. 7.15a) is exceeded, the respective slip band is meshed. The resolved shear stress

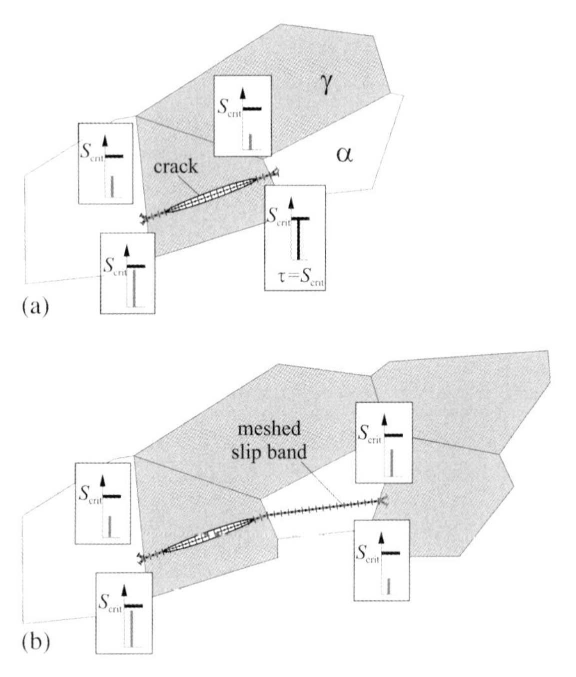

Fig. 7.15 (a) Crack in one grain of a model microstructure with shear stresses acting on the sensor elements of the neighboring grains; (b) meshed slip band representing the highest-stressed sensor element.

acting in the direction of these slip bands is measured by sensor elements, which are placed immediately behind the grain or phase boundary. The sensor elements are single boundary elements representing the possible course of the crack path along the slip bands of the adjacent grains or along the grain boundary (intercrystalline crack propagation). This is shown in Fig. 7.15. The length of the sensor elements reflects the distance r_0 of the dislocation source from the boundary (analogous to the model of Navarro and de los Rios), and is set to $2r_0$, i.e., the dislocation source is assumed to be located in the center of the sensor elements.

The critical shear stress to activate the dislocation source S_{crit} can be estimated as follows. According to the model of Navarro and de los Rios [324], the stress on the dislocation source at $x = d/2 + r_0$ is given by the following expression:

$$S_{crit} - \tau_{fr} = \frac{1}{2}\sqrt{\frac{d}{r_0}}\left(\frac{\sigma_Y}{M_{Sa}} - \tau_{fr}\right) \tag{7.29}$$

The friction stress (here considered as the sum of resistance to dislocation motion on a slip plane) can be assumed to be equivalent to the difference between the cyclic yield strength σ_{Yc} and the fatigue limit σ_{FL}, both referring to deformation along slip bands by means of the Sachs factor M_{Sa} [515]:

$$\tau_{fr} = \frac{\sigma_{Yc} - \sigma_{FL}}{M_{Sa}} \tag{7.30}$$

Equation (7.29) can be rearranged, implementing the Hall–Petch relationship with the Hall–Petch constant k' (cf. Section 4.3) as

$$\frac{\sigma_Y}{M_{Sa}} = \tau_{fr} + \frac{k'}{M_{Sa}\sqrt{d}} \tag{7.31}$$

eventually leading to

$$S_{crit} \approx \frac{1}{2}\frac{1}{\sqrt{r_0}}\frac{k'}{M_{Sa}} + \tau_{fr} \tag{7.32a}$$

With the definition of the fatigue limit σ_{FL} as the stress that is just not sufficient to drive an existing crack across a microstructural barrier, Eq. (7.29) can be written as

$$S_{crit} = \frac{1}{2}\sqrt{\frac{d}{r_0}}\frac{\sigma_{FL}}{M_{Sa}} + \tau_{fr} \tag{7.32b}$$

When furthermore assuming that the distance of the dislocation source corresponds to 1% of the slip band length ($0.01d$), the critical stress should correspond to five times the fatigue limit. Schick [515] applied Eq. (7.29) to estimate the critical stress for the β-titanium alloy LCB in the solution heat-treated condition assuming r_0 to be equal to an estimate value of the slip band spacing, $r_0 = 0.25$ μm. With a grain size of $d = 100$ μm, a cyclic yield strength of $\sigma_{Yc} = 800$ MPa and a fatigue limit of $\sigma_{FL} = 300$ MPa, he obtained a value for $S_{crit} = 1565$ MPa.

Figure 7.16 shows an example of a microstructurally short fatigue crack in the β-titanium alloy LCB for which the crack length vs. the number of cycles was determined experimentally (using optical and scanning electron microscopy) and calculated by means of the numerical model introduced above [561]. For the given crack path and taking geometry-induced crack closure (boundary condition $b_n \geq 0$) into account, simulation and experiment are in excellent agreement. Obviously, the effect of geometry-induced crack closure leads to retardation of microstructurally short-fatigue-crack propagation in a similar way to the case for long fatigue cracks due to plasticity-induced crack closure after Elber. Parameter studies in [515, 561] revealed that geometry-induced crack closure during the short-crack regime depends strongly on the slip-band inclination with respect to the applied-loading axis β^i.

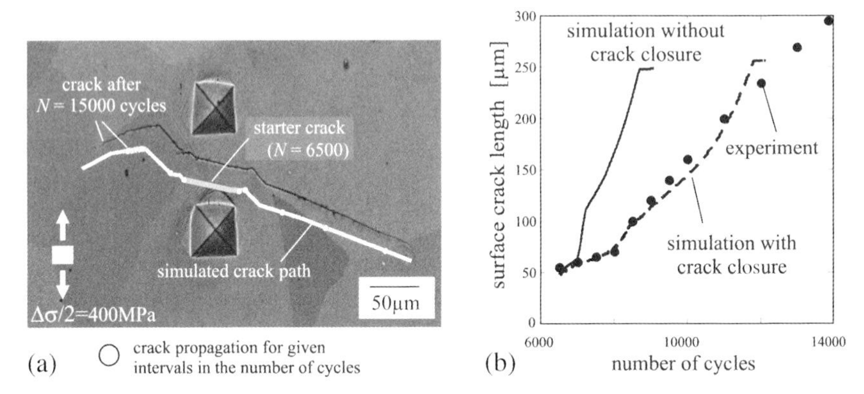

Fig. 7.16 Microcrack in the β-titanium alloy LCB: (a) scanning electron micrograph showing the crack path, (b) comparison of the calculated and measured values of the crack lengths vs. the number of cycles.

7.3.3 Simulation of Microcrack Propagation in Synthetic Polycrystalline Microstructures

While the original model of Navarro and de los Rios is limited to one-dimensional crack propagation with constant grain size d, or in further developments to idealized model microstructures, e.g., by Tanaka et al. [538], Olfe et al. [542], de los Rios et al. [562] or Ahmadi and Zenner [563], the numerical approach using the boundary element method allows the simulation of microcracks in arbitrary two- and three-dimensional microstructures (the latter being subject of ongoing research [564]). Hence, material and loading parameters can be chosen that are close to real service conditions.

By means of automated EBSD, characteristic microstructural features like grain size, grain geometry, contiguous and noncontiguous phase areas or the crystallographic orientation can be quantitatively evaluated for a representative area (cf. Section 3.3.2). As an example, Fig. 7.17 shows the γ-austenite and a-ferrite phase distribution for an AISI F51 duplex steel determined by EBSD (orientation imag-

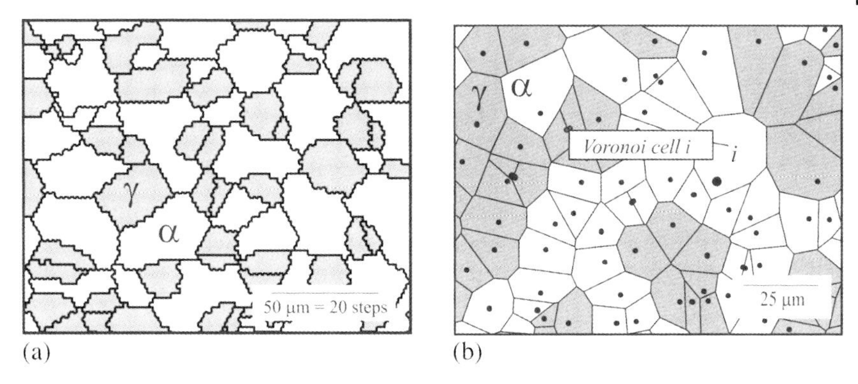

Fig. 7.17 Austenite/ferrite phase distribution in an AISI F51 duplex steel: (a) measured by means of automated EBSD in a scanning electron microscope; (b) corresponding synthetic microstructure, generated by the Voronoi algorithm (according to Künkler [421]).

ing microscopy, OIM™). To implement the measured distribution into the numerical model, Schick [515] and Künkler [421] used the Voronoi algorithm [544] to create appropriate synthetic microstructures. Voronoi cells are generated on the basis of a random two-dimensional distribution of points. These points are enclosed by areas, fulfilling the condition that all possible positions must be closer to the enclosed point than to any other point. According to Weinhandl [565], the Voronoi algorithm represents material microstructures in a realistic way, since (1) each grain is bounded by six boundaries on average and (2) the mean angle between the boundaries is 120°. Figure 7.17b shows a section of a Voronoi diagram to generate the two-phase microstructure of the AISI F51 duplex steel represented in Fig. 7.17a, using a tailor-made algorithm [421] to adjust the grain size and phase distribution according to the stereographic parameters obtained experimentally by Düber [317].

Together with the synthetic Voronoi microstructure being adapted to EBSD measurements of the crystallographic orientation distribution, the numerical model is capable of predicting the propagation behavior of microstructurally short fatigue cracks, based on the dislocation pileup at grain and phase boundaries (depending on the misorientation angle of the slip planes), the contribution of geometry-induced crack-closure effects and the values of the critical stress to activate dislocation sources S_{crit}, and the friction stress τ_{fr}.

As an alternative to the application of Eqs. (7.29) to (7.32), which cannot be used to derive data for multi-phase materials, Düber [317, 344] proposed a procedure to estimate the microstructural parameters S_{crit} and τ_{fr} for two-phase materials: According to a concept by Werner and Stüwe [566], further developed by Fan et al. [567], the classical Hall–Petch relationship between the monotonic yield strength and the mean grain size of polycrystalline metals and alloys can be applied to two-phase materials by using a weighted distribution of the grain and phase boundaries. As illustrated in Fig. 7.18a, the real two-phase microstructure is composed of

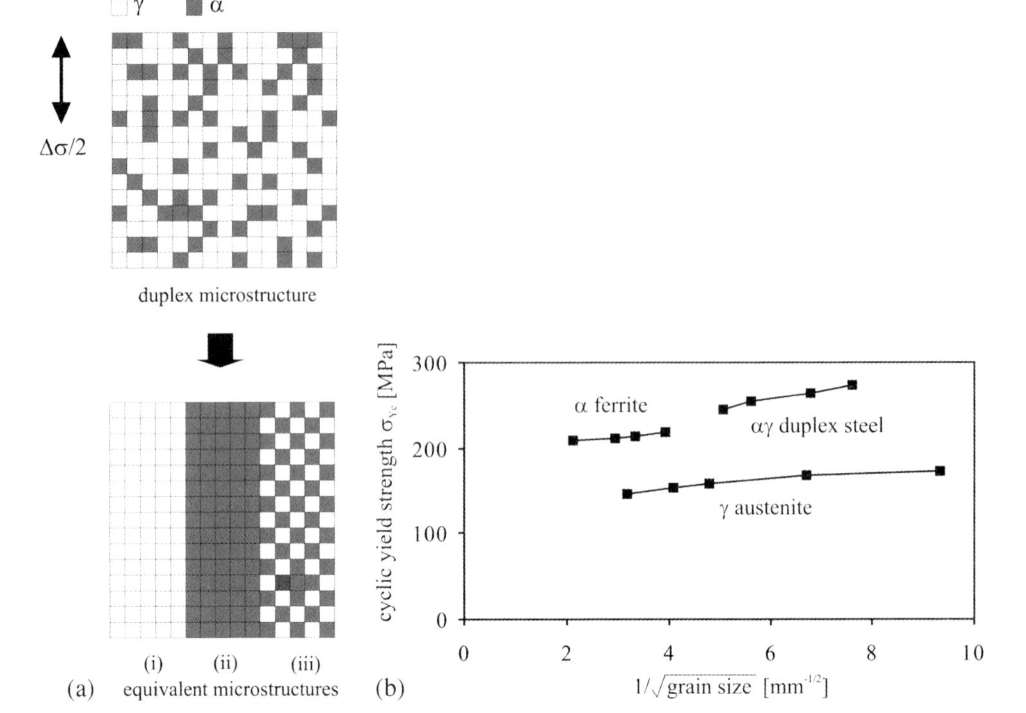

Fig. 7.18 (a) Subdivision of the two-phase γ-austenite–α-ferrite microstructure in three equivalent microstructures with only (i) γγ grain boundaries, (ii) αα grain boundaries, and (iii) αγ phase boundaries; (b) Hall–Petch plot of the cyclic yield strength vs. the reciprocal square root of the grain size of three kinds of stainless steels.

(1) two distinct single-phase areas and (2) an area where exclusively phase boundaries are present. As an example, the concept of Fan et al. [567] is applied to the cyclic yield strength σ_{Yc} and the cyclic Hall–Petch constant k_c of γ-austenitic/α-ferritic AISI F51 duplex steel:

$$\sigma_{Yc} = \sigma_{Yc}^{\alpha} \cdot f^{\alpha\,\mathrm{cont}} + \sigma_{Yc}^{\gamma} \cdot f^{\gamma\,\mathrm{cont}} + \sigma_{Yc}^{\alpha\gamma} \cdot F^{\mathrm{sep}} \tag{7.33}$$

$$k_c = k_c^{\alpha} \cdot f^{\alpha\,\mathrm{cont}} + k_c^{\gamma} \cdot f^{\gamma\,\mathrm{cont}} + k_c^{\alpha\gamma} \cdot F^{\mathrm{sep}} \tag{7.34}$$

with $f^{\alpha\,\mathrm{cont}}$ and $f^{\gamma\,\mathrm{cont}}$ being the relative contiguous volumes of the α and the γ phases. They can be obtained by referring the number of single-phase grain boundaries $N^{\alpha\alpha}$ and $N^{\gamma\gamma}$, respectively, to the total number of boundaries of the respective phase area $N^{\alpha\alpha}+N^{\alpha\gamma}$ or $N^{\gamma\gamma}+N^{\alpha\gamma}$ multiplied with the total volume fraction f^{α} or f^{γ}:

$$f^{\gamma\,\mathrm{cont}} = \frac{N^{\gamma\gamma}}{N^{\gamma\gamma}+N^{\alpha\gamma}} f^{\gamma} \quad \text{or} \quad f^{\alpha\,\mathrm{cont}} = \frac{N^{\alpha\alpha}}{N^{\alpha\alpha}+N^{\alpha\gamma}} f^{\alpha} \tag{7.35}$$

The degree of separation F^{sep} corresponds to the separated volume fraction of the two phases according to

$$F^{sep} = f^{\alpha\,sep} + f^{\gamma\,sep} = (f^{\alpha} - f^{\alpha\,cont}) + (f^{\gamma} - f^{\gamma\,cont}) \tag{7.36}$$

The values of the cyclic yield strength of (1) the AISI F51 duplex steel σ_{Yc}, (2) a single-phase ferritic steel σ_{Yc}^{α} (340LX, X3CrNb17) and (iii) a single-phase austenitic steel (316L, X2CrNiMo17 12 2) σ_{Yc}^{γ} with various grain sizes (adjusted by heat treatment [317]) were determined experimentally by means of incremental step tests (ISTs, see Section 3.1.1). The results are summarized in a Hall–Petch-type representation in Fig. 7.18b, showing the values of the cyclic yield strength vs. the reciprocal square root of the grain size d.

Of course, the Hall–Petch constant $k_c^{\alpha\gamma}$ and the cyclic yield strength $\sigma_{Yc}^{\alpha\gamma}$ for the completely separated fraction, composed exclusively of $\alpha\gamma$ phase boundaries (Fig. 7.18a), cannot be determined experimentally. However, these data can be extracted from Eqs. (7.33) and (7.34). The cyclic critical stress for the completely separated fraction $\sigma_{0c}^{\alpha\gamma}$ is obtained by means of the respective Hall–Petch relationship

$$\sigma_{Yc}^{\alpha\gamma} = \sigma_{0c}^{\alpha\gamma} + \frac{k_c^{\alpha\gamma}}{\sqrt{\bar{d}_{\alpha\gamma}}} \tag{7.37}$$

with $\bar{d}_{\alpha\gamma}$ being the mean cluster size, i.e., the arithmetic average of the mean grain size and the size of the phase patches: $\bar{d}_{\alpha\gamma} = (\bar{d}_{\alpha\alpha} + \bar{d}_{\gamma\gamma} + \bar{d}_{\alpha} + \bar{d}_{\gamma})/4$. The critical cyclic yield stresses and the Hall–Petch constants, obtained by the analysis introduced above, are summarized in Table 7.1. Assuming planar slip to prevail during fatigue, because only then does it makes sense to apply the Hall–Petch relationship to cyclic loading conditions, the cyclic Hall–Petch constant can be considered as a measure of the different barrier strengths of the $\alpha\alpha$ grain boundaries (k_c^{α}), the $\gamma\gamma$ grain boundaries ($k_c^{\gamma\gamma}$) and the $\alpha\gamma$ phase boundaries ($k_c^{\alpha\gamma}$). The critical cyclic yield stresses are set equal to the friction stress within the respective phases. Together with Eq. (7.31) the data can be used to derive the critical stress S_{crit} to activate dislocation sources in the numerical short-crack model.

Figures 7.19 and 7.20 (after [334]) show examples of the application of the numerical short-crack model in combination with the experimentally determined microstructural data (grain size, crystallographic orientation relationship, critical stress, frictions stress). The simulated crack lengths vs. number of cycles are in ex-

Table 7.1 Cyclic critical stresses and cyclic Hall–Petch constants for four kinds of boundaries: (1) $\gamma\gamma$ grain boundaries (316L steel), (2) $\alpha\alpha$ grain boundaries (340LX steel), (3) $\alpha\gamma$ phase boundaries (calculated) and (4) grain and phase boundaries (AISI F51 duplex steel).

	$\gamma\gamma$	$\alpha\alpha$	$\alpha\gamma$	**Duplex**
Cyclic critical stress σ_{Yc} (MPa)	137	198	212	196
Cyclic Hall–Petch constant k_c (MPa mm$^{1/2}$)	4.2	5.0	15.8	10.1

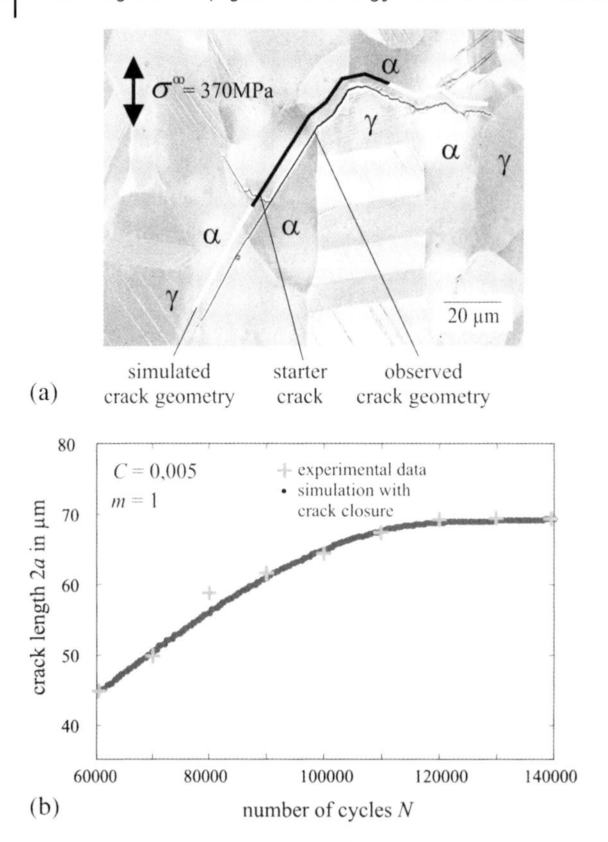

Fig. 7.19 Microcrack in AISI F51 duplex steel: (a) scanning electron micrograph showing the crack path; (b) comparison of the calculated and measured values of the crack lengths vs. the number of cycles (after [334]).

cellent agreement with the experimentally observed ones. In both figures the decrease in the crack-propagation rate when approaching boundaries is clearly represented; the difference in crack-propagation rate of the right- and the left-hand tip in Fig. 7.20 can be attributed to the difference in shear stresses acting on the respective slip planes: While the left-hand crack branch is almost parallel to the 45° inclination of maximum shear with respect to the applied loading axis, the right-hand crack branch is loaded under mode I being reflected by a very small crack-propagation rate.

7.3.4
Transition from Mode II to Mode I Crack Propagation

As mentioned in Sections 6.2.2 and 6.4, the transition in the crack-propagation mode is of substantial significance for the overall crack-propagation rate and the transition from short- to long-crack behavior. Particularly in the case of the two-

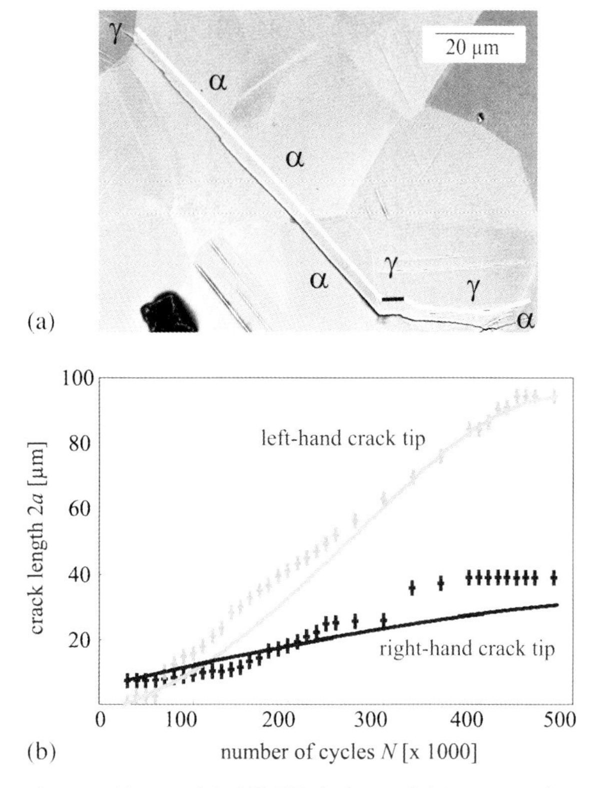

Fig. 7.20 Microcrack in AISI F51 duplex steel: (a) scanning electron micrograph showing the crack path; (b) comparison of the calculated and measured crack-propagation data for the right-hand and the left-hand crack tip vs. the number of cycles (after [334]).

phase AISI F51 duplex steel, early transition in mode I crack propagation by operating alternate slip systems was observed [317, 318]. Different from the long-crack propagation mechanism driven by multiple slip (plastic blunting) and leading to the formation of striations, microcrack propagation in mode I operates slip on two distinct slip planes, which has been implemented in the model by Künkler [421] in a similar way to that described above (cf. also [568]): during crack advance along a slip band the elastic shear stresses in the vicinity of the crack tip increase. When the stress acting on a second slip plane exceeds a critical stress for slip deformation it is meshed in the same way as described in Section 7.3.2 for transition of the plastic zone across a grain boundary. The direction of subsequent crack propagation then follows the vector addition of the displacements along the two slip systems, thus turning perpendicular to the loading direction (mode I). This procedure is illustrated in Fig. 7.21.

In accordance to the observations discussed in Section 6.2.2 this kind of multiple-slip crack propagation has been termed stage Ia, owing to the fact that a transi-

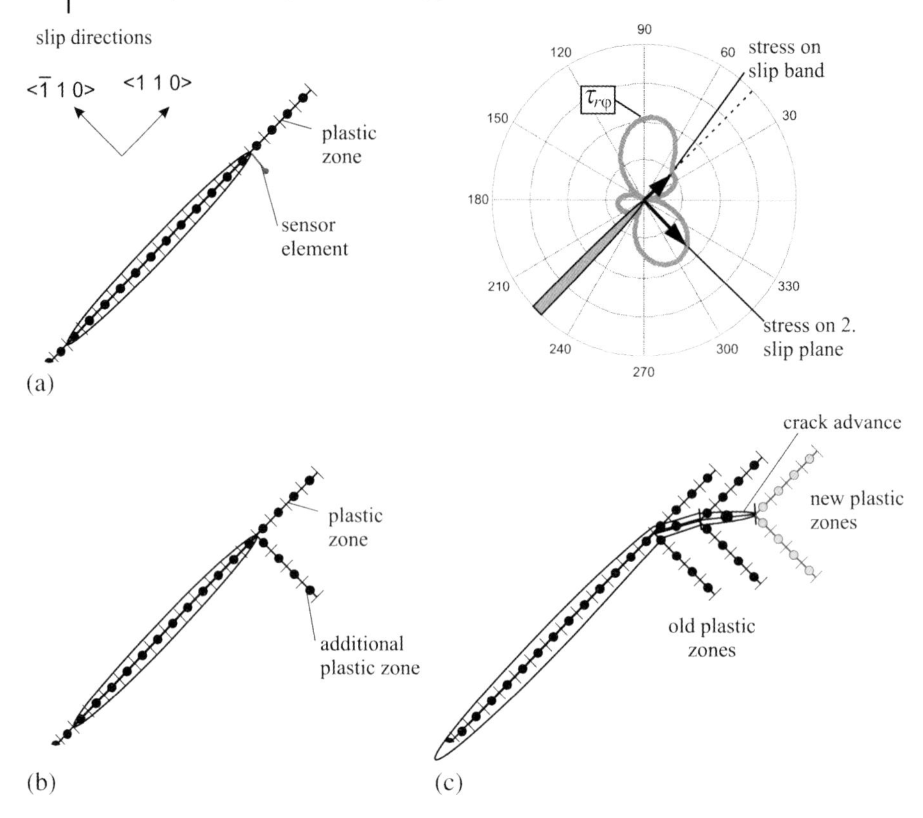

Fig. 7.21 Schematic representation of the implementation of the transition from single- to double-slip crack propagation: (a) stress evaluation at the crack tip by sensor elements, (b) addition of a second plastic zone, (c) crack propagation by alternate activation of two slip planes (according to Künkler [421]).

tion back to crystalline mode II crack propagation is possible. Only when the plastic-zone size ahead of the crack tip exceeds one grain diameter, is a crack considered to be a long fatigue crack. An example for the application of the double-slip crack model is shown in Fig. 7.22.

Figure 7.23a shows the evolution of the calculated propagation of several microcracks through a synthetic microstructure, clearly revealing the rate-decreasing effect of mode I/mode II transition [569]. When taking both the mode II (stage Ia) to mode I (stage Ib/stage II) transition and the development of crack-closure effects into account (cf. Chapter 6) it is possible to simulate the overall fatigue-crack propagation process from early initiation to final fracture. Figure 7.23b shows the respective da/dN vs. ΔK simulation results being qualitatively in excellent agreement with experiments on microcrack and long-crack propagation [421].

Crack-propagation simulations applied to a great number of arbitrary synthetic microstructures revealed, for example, an increase in the fatigue life for decreasing

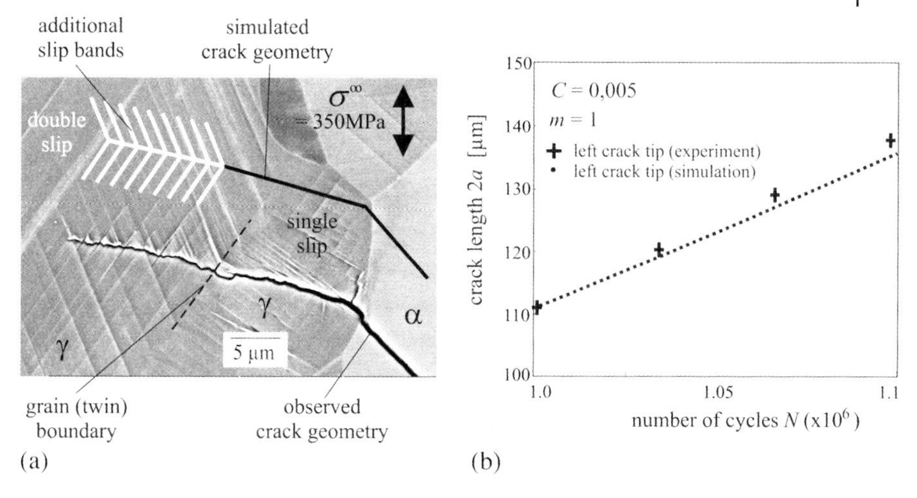

Fig. 7.22 Microcrack in AISI F51 8 duplex steel propagating by alternate operating slip systems: (a) scanning electron micrograph showing the crack path and activated slip bands; (b) comparison of the calculated and measured crack length vs. the number of cycles (after [334]).

grain size and for decreasing the block parameter of the γ-austenite phase in austenitic–ferritic duplex steel, i.e., it is beneficial for the fatigue behavior if the softer phase acts as a matrix [421]. Hence the model cannot only be used for fatigue-life prediction, but also for designing microstructures with a higher resistance to fatigue damage.

7.3.5
Future Aspects of Applying the Boundary Element Method to Short-Fatigue-Crack Propagation

The model originally developed by Schick [515] has been widely modified and extended. The focus of ongoing research is placed on the following aspects:

- Crack-closure effects. It should be mentioned that implementing crack propagation by operating two slip planes also allows one to account for plasticity-induced crack-closure effects. Plastic slip occurring along the two slip bands in tension forces the crack to stay open after complete removal of the remote load. To close the crack additional compressive load is required. This effect vanishes when the crack grows longer, eventually leading to the establishment of a plastic wake that can be considered by means of the model of Newman ([570], Fig. 7.24a), where the crack is assumed to grow into a plastically stretched plastic zone.
- Crack-coalescence effects (cf. Section 6.4.2). Interaction of two mutual approaching crack tips must be considered in an additional variable stress term in Eqs. (7.27) and (7.28). Furthermore, the gradual development of a new three-dimensional equilibrium shape of two coalescing cracks needs to be taken into ac-

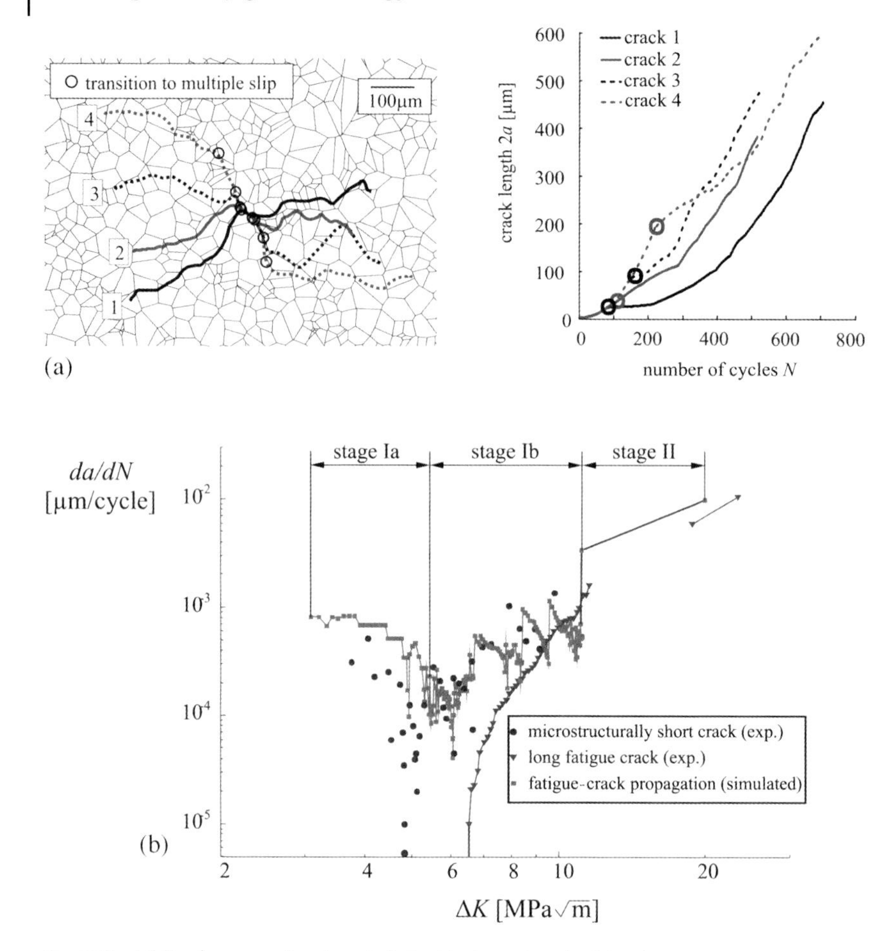

Fig. 7.23 (a) Crack propagation in a statistical body-centered cubic microstructure generated by the Voronoi algorithm: microcrack length vs. number of cycles, for different critical stresses required to activate multiple slip systems (after [569]). (b) Experimentally measured and simulated crack-propagation vs. stress-intensity range data (*da*/*dN* vs. ΔK, after [421]).

count. The crack density depends strongly on the load amplitude and the mechanical behavior of the material and should be considered as an essential element governing crack coalescence.

- Load history effects. Generally, the numerical short-crack model is capable of being applied to random load spectra. According to Schick [515], the calculation is carried out from load peak to load peak (see Fig. 7.24b): if the first loading step yields a displacement vector $\vec{b}_1$, every kth loading or unloading step is superimposed by the displacement interval $\Delta\vec{b}_k$ resulting in the displacement $\vec{b}_k = \vec{b}_{k-1} + \Delta\vec{b}_k$. Calculation of the displacement interval $\Delta\vec{b}_k$ at the time k is done

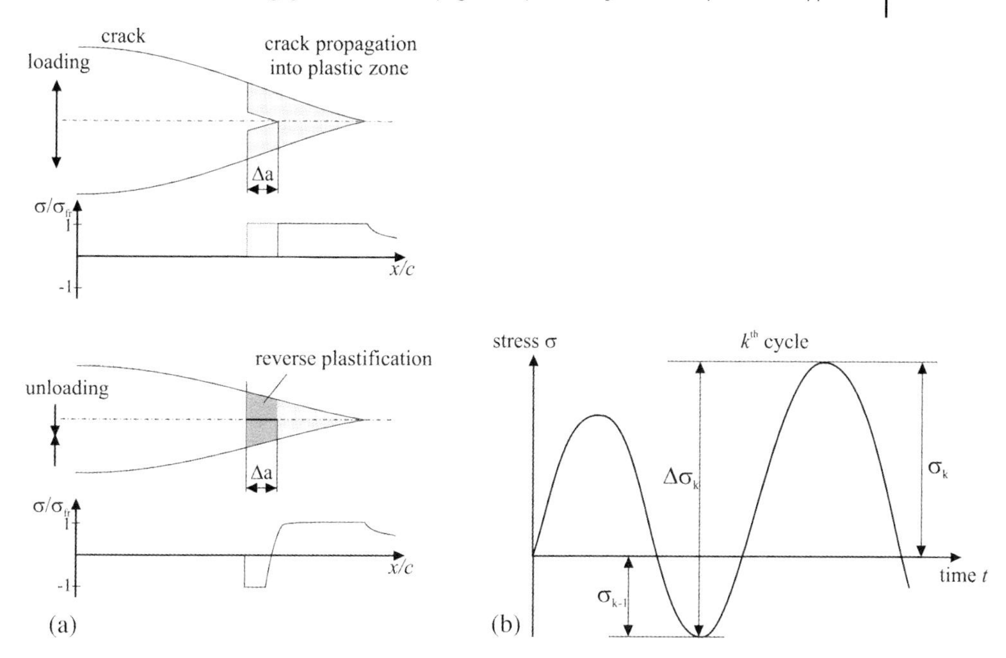

Fig. 7.24 Schematic representation of implementation of (a) plasticity-induced crack closure (according to the model of Newman [570]); (b) load-history effects (according to Schick [515]).

for the change in load $\Delta\sigma_k$. Eventually, the boundary condition that the normal displacement within the crack area cannot be negative, $b_n \geq 0$, must be fulfilled for the resulting total displacement $\vec{b}_k$. Together with the implementation of geometry and plasticity-induced crack-closure concepts the simulation of overload effects, owing to local microstructural features in a way close to reality, is possible.

- As the most striking challenge in modeling the propagation behavior of microstructurally short fatigue cracks the implementation of the third dimension can be considered. The examples in Chapter 6 reveal clearly the pronounced significance of the triaxiality of the relevant stress–strain states and the three-dimensional microstructure. In general, a three-dimensional modification of the boundary element approach by using two-dimensional slip-plane elements is possible (cf. [571]) and together with experimental studies using synchrotron computer tomography is subject of ongoing research [564]. However, it should be emphasized that the two-dimensional simulation results are in surprisingly good agreement with the experimentally determined crack-propagation rates. Obviously, the mechanisms of short-fatigue-crack propagation are mainly governed by the arrangement of the surface grains.

7.4
Modeling Dwell-Time Cracking: A Grain-Boundary Diffusion Approach

The following sections focus on only a few aspects of the wide field of theoretical assessments of fatigue-crack propagation at high temperatures. More detailed overviews are given, for example, by Riedel [80] or Evans [175] (creep damage); some recent modeling approaches to thermomechanical fatigue of metals and alloys are introduced by Teteruk [572].

The main differences between fatigue-crack propagation at elevated temperatures and at room temperature are changes in the cyclic-plasticity mechanisms (increasing dislocation mobility, dislocation climb) and the superposition time-dependent deformation and damage processes, i.e., creep and environmental attack. To consider empirically the time dependence of fatigue damage, Woodford and Coffin [573] suggest a simple relationship between the number of cycles to fracture N_f, the loading frequency f and the plastic strain amplitude $\Delta\varepsilon_{pl}$:

$$N_f = \frac{F(\Delta\varepsilon_{pl})}{f^{k-1}} \tag{7.38}$$

with k being a measure of the influence of the frequency with the limiting cases $k = 0$: strong time-dependent behavior, and $k = 1$: only cycle-dependent behavior. According to Ostergren [574], $F(\Delta\varepsilon_{pl})$ is a damage function related to the detrimental effect of the plastic strain ($\Delta\varepsilon_{pl}$) particularly in the tensile regime (σ_{max}):

$$F(\Delta\varepsilon_{pl}) = L(\sigma_{max}\Delta\varepsilon_{pl})^{-\eta} \tag{7.39}$$

where L and η are material constants.

During fatigue loading of creep-resistant polycrystalline nickel-based superalloys at very low frequencies, or/and superposition of dwell times at maximum tension, a transition from transgranular to intergranular crack propagation can be observed in combination with a substantial increase in the crack-propagation rate. According to the remarks made in Section 6.5, this effect can be attributed to environmentally assisted embrittlement of the grain boundaries. Instead of applying crack-propagation equations based on the Paris–Erdogan approach (Eq. 2.5), as has been done, for example, by Ghonem and Zheng [432] or Osinkolu et al. [440], it seems to be more useful to consider a damage-development model, where crack advance by Δa results from a superimposition of time-dependent crack propagation and cycle-dependent crack propagation, in a form similar to that suggested by Saxena [575]:

$$\Delta a = a_1 - a_0 = \int_{N_0}^{N_1}\left(\frac{da}{dN}\right)dN + \int_{t_0}^{t_1}\left(\frac{da}{dt}\right)dt \tag{7.40}$$

where the contributions of cycle- and time-dependent damage are treated separately for the time interval $t_1 - t_0$. The experimental results discussed in Section 6.5 suggest that, at least at moderate temperatures and high yield strength, the detrimental effect of dwell times and/or low frequencies can be attributed to the dynamic embrittlement mechanism, where a high tensile stress ahead of an intercrystalline crack tip drives oxygen diffusion into the cohesive zone, subsequently leading to decohesion. The main difference between dynamic embrittlement and stress-assisted grain boundary oxidation (SAGBO; cf. Section 6.5.1) is the fact that dynamic embrittlement does not involve the formation of a thermally grown oxide layer ahead of the crack tip. According to the author's opinion, SAGBO may govern dwell-time cracking only at very high temperatures, above approximately 800 °C in the case of nickel-based superalloys.

In any case, interfacial diffusion of an embrittling element into an elastically stretched cohesive zone is the core mechanism of time-dependent intergranular cracking, as was observed, for example, for oxygen in nickel-based alloys, sulfur in steel or tin in Cu–Sn alloys (cf. Section 6.5.2). In the following, a model for dynamic embrittlement is introduced; this has been developed in two steps: first by Bika and McMahon [453, 576] and later by Xu and Bassani [577, 578].

Quantification of crack propagation due to dynamic embrittlement requires knowledge of both the time- and location-dependent concentration c and the stress-dependent decohesive effect of the embrittling species. It is assumed that the supply of the embrittling species to the crack tip is sufficiently high and has no effect on the crack-propagation velocity; i.e., the concentration c_0 at the crack tip is considered to be constant. However, in the case of oxygen-induced intercrystalline crack propagation in nickel-based alloys, this assumption is only a rough estimate, since oxidation of the continuously generated new surfaces behind the crack tip can lead to a considerable decrease in the oxygen potential at the actual crack tip.

The first model for dynamic embrittlement was developed by Bika and McMahon [453, 576] for stress-relief cracking of steels. They used the differential equation for solid-state diffusion along grain boundaries under the influence of stress, which is introduced and discussed later in this section. To solve this differential equation numerically a parabolic stress distribution ahead of the crack tip is assumed:

$$\sigma(x) = \sigma_0 + x(2 - x)(\alpha - \sigma_0) \tag{7.41}$$

where σ_0 is the stress at the cavity tip and α is a constant that can be estimated by considering $\sigma_b = 3\sigma_Y$ as the mean stress over the diffusive length λ. This stress field represents the equilibrium conditions at the tip of a growing cavity, according to Chen and Argon [579]. The stress field has to be considered in combination with an assumed fracture criterion, as derived by Shin and Meshii [580], that relates the fracture stress σ_{fr} to the concentration of the embrittling element:

$$\frac{\sigma_{fr}}{\sigma_Y} = \left(\frac{\sigma_{fr,min}}{\sigma_Y}\right)\left[\left(\frac{\sigma_{fr,max}}{\sigma_{fr,min}} - 1\right)\left(\frac{c^{min}}{c}\right)^p + 1\right] \tag{7.42}$$

with $\sigma_{fr,max}$ being the fracture stress corresponding to a minimum embrittler concentration c_{min}. The fracture stress decreases with increasing embrittler concentration to a minimum value of $\sigma_{fr,min}$. Hence, for the calculated embrittler concentration profiles, as shown in Fig. 7.25a for the example of sulfur in steel, the corresponding critical stress σ_{crit} acting normal to the grain boundary can be calculated and compared with the stress distribution according to Eq. (7.41). The point x_{crit} where the stress gradient ahead of the crack tip intersects with the critical stress for a critical time x_{crit} (according to the calculated embrittler concentration profiles) defines the crack-propagation velocity by $da/dt = x_{crit}/t_{crit}$. This is shown in Fig. 7.25b.

Using appropriate parameters for sulfur-induced stress-relief cracking in steel [453], Bika found reasonable agreement with experimental crack-propagation data (cf. Section 6.5.2). The observed oscillation of the crack-propagation velocity might be attributed to a variation in both the grain-boundary diffusivity and the tensile stress acting normal to the boundaries, depending on the grain-boundary structure and geometry.

Later, Xu and Bassani [577, 578] proposed a model for dynamic embrittlement based on steady-state crack propagation (da/dt = constant [581]) in which the stress gradient within the cohesive zone ahead of the crack tip is treated by means of constitutive equations for creep deformation and decohesion. The main development steps of this model are given in the following.

From a thermodynamic point of view, the gradient in the chemical potential μ is the driving force for solid-state diffusion [577]. This is represented by the following partial differential equation for the concentration c of one diffusing species:

$$\frac{\partial c}{\partial t} = \nabla\left(\frac{cD}{kT}\nabla\mu\right) \tag{7.43}$$

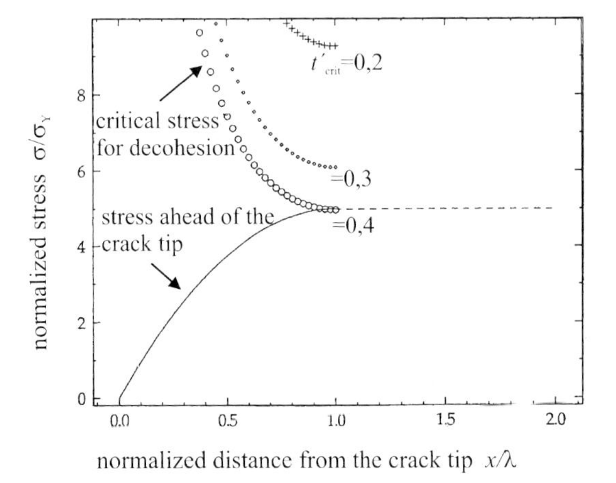

Fig. 7.25 Stress profile in the cohesive zone ahead of a crack tip according to the model of Bika and McMahon [453, 576]. The upper curves represent the fracture criterion depending on the normalized time $t'_{crit} = t \cdot D_{GB}\Omega\sigma_Y/kT\lambda^2$, λ being the diffusion length (after [576]).

To consider the influence of stress in crystalline solids on the chemical potential the governing equation for the chemical potential μ, following the analysis of Larché and Cahn [582, 583], can be expressed as follows:

$$\mu = \mu^0 + kT\ln(\gamma c) - \sigma\Omega \tag{7.44}$$

Here, Ω denotes the atomic volume, γ the chemical activity coefficient and σ the normal stress on the grain boundary. Hence, the term

$$\nabla\mu = -\Omega\nabla\sigma \tag{7.45}$$

can be considered as the driving force for interfacial diffusion without a concentration gradient being present. The respective equation had been introduced earlier by Herring [584] and Hull and Rimmer [469] to describe the diffusion-controlled growth of creep pores; in this case, however, the outward diffusion of metal atoms was considered as rate determining for cavity growth. Under the simplifying assumptions that the grain-boundary diffusion coefficient D_{GB}, as well as the activity coefficient γ, are constant, inserting Eq. (7.44) in Eq. (7.43) yields the following governing differential equation for dynamic embrittlement:

$$\frac{\partial c}{\partial t} = D_{GB}\frac{\partial^2 c}{\partial x^2} - \frac{D_{GB}\Omega}{kT}\frac{\partial}{\partial x}\left(c\frac{\partial\sigma}{\partial x}\right) \tag{7.46}$$

The second term of the right-hand side in Eq. (7.46) represents the contribution of a stress-induced diffusive flux into the cohesive zone ahead of the crack tip, as represented schematically in Fig. 7.26b. The gradient of the stress acting normal to the intercrystalline crack plane can be considered to result from the change in the equilibrium atomic spacing a_0 (Fig. 7.26a) and the corresponding interatomic force.

Formation of an intercrystalline crack (theoretically reversible formation of two new surfaces) requires overcoming the grain-boundary cohesive energy E_c:

$$E_c = 2\gamma_O - \gamma_{GB} \tag{7.47}$$

with the surface energy γ_O and the grain-boundary energy γ_{GB}.

According to McClintock and Bassani [581], a steady-state crack velocity $v = da/dt$ can be assumed to prevail, i.e., the derivative of the concentration with respect to time is proportional to the derivative with respect to location, keeping the origin of the coordinate system at the crack tip (Fig. 7.27):

$$\frac{\partial c}{\partial t} = -v\frac{\partial c}{\partial x} \tag{7.48}$$

With the steady-state condition, the partial differential Eq. (7.46) becomes a second-order ordinary differential equation that can be solved in a closed form, assuming that the stress σ and the grain-boundary diffusivity D_{GB} do not depend on the embrittler concentration c:

$$c(x) = \exp\left(-\frac{v}{D_{GB}}x + \frac{\Omega}{kT}\sigma(x)\right)\left[C_1 + C_2\int_0^x \exp\left(\frac{v}{D_{GB}}\xi - \frac{\Omega}{kT}\sigma(\xi)\right)d\xi\right] \tag{7.49}$$

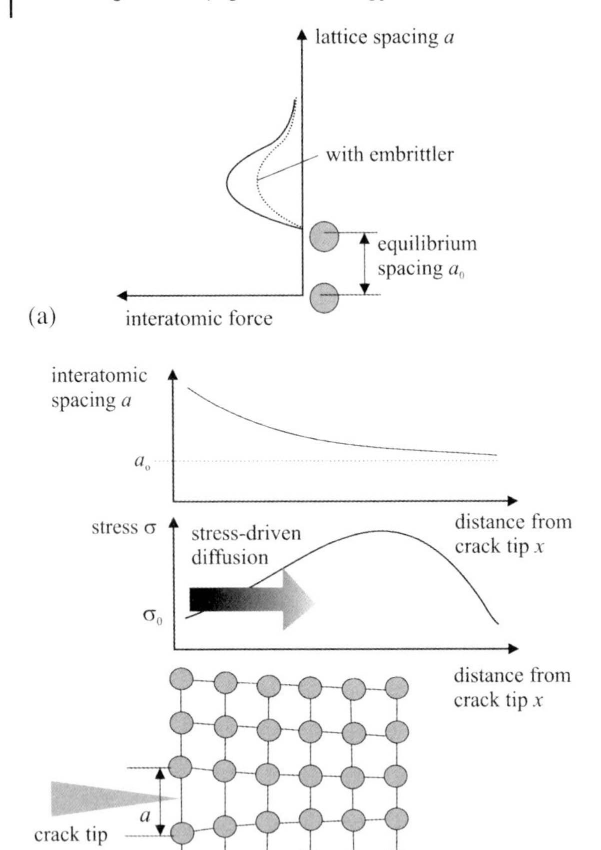

Fig. 7.26 Schematic representation of (a) the interatomic force vs. the lattice spacing *d* and (b) corresponding stress distribution ahead of a sharp crack tip.

With the boundary conditions for the embrittler concentration $c(x = \infty) = 0$ and $c(x = 0) = c_0$, the solution simplifies to

$$c(x) = c_0 \exp\left[-\frac{v}{D_{\mathrm{GB}}}x + \frac{\Omega}{kT}(\sigma(x) - \sigma_0)\right] \tag{7.50}$$

with σ_0 being the normal stress at the crack tip. The difference $\sigma(x) - \sigma_0$ can be roughly estimated to be of the order of the yield strength of the particular material (according to Xu und Bassani [585]).

When applying the approximate solution in Eq. (7.50) to experimental results and to literature data on dynamic embrittlement of the nickel-based superalloy alloy 718, such as the atomic volume $\Omega = 2.3 \times 10^{29}\ \mathrm{m}^3$, the product $kT(650\,°\mathrm{C}) = 1.27 \times 10^{-20}\ \mathrm{J}$, a grain-boundary-diffusion coefficient of $D_{\mathrm{GB}} = (650\,°\mathrm{C}) = 10^{-15}\ \mathrm{m}^2\ \mathrm{s}^{-1}$, a steady-state crack velocity during sustained loading of $da/dt = 10^{-5}$ m s^{-1} (cf. Section 6.5.3) and using the difference $(\sigma(x) - \sigma_0) = 10^3$ MPa, one obtains for the oxygen concentration in the cohesive zone

$$\frac{c}{c_0} = \exp\left(-10^{10}\mathrm{m}^{-1}x + 1.8\right) \tag{7.51}$$

The estimation would yield a theoretical process zone of $x = 0.28$ nm wherein the oxygen concentration drops to approximately one-third of the original value. This would mean that the dynamic embrittlement mechanism is restricted to the atomic length scale. It should be noted that in this case the diffusion calculation would not be strictly applicable. However, the grain-boundary diffusion coefficient D_{GB} used in Eq. (7.51) results from an extrapolation of data derived from studies of Bricknell and Woodford [586] and Caplan et al. [587] of pore formation during high-temperature oxidation of nickel at temperatures between 900 and 1100 °C, and should be considered as a very rough and uncertain approximation. The activation energy for oxygen grain-boundary diffusion was determined as to be $Q = 268$ kJ mol^{-1}, which is close to the activation energy Pfaendtner [473] found for four-point bending tests on alloy 718 specimens at various temperatures, $Q = 290$ kJ mol^{-1}l. On the other hand, it is also close to the activation energy of nickel self-diffusion, $Q = 285.9$ kJ mol^{-1} [588]. Therefore, the crack propagation mechanism might be partially determined by creep of the unbroken ligaments remaining behind the advancing crack tip. Quantification of the contribution of creep and the determination of reliable values for the grain-boundary diffusion coefficient for oxygen in nickel-based alloys are subject of ongoing research. Experiments on bicrystalline and grain-boundary-engineering-type processed polycrystalline bend specimens may reveal the relationship between the grain-boundary structure, the distribution of "special" CSL grain boundaries and the time-dependent interfacial cracking behavior (cf. Section 6.5.4).

Certainly, the assumption that the tolerable stress acting normal to the grain boundary is independent of the embrittler concentration (see Eqs. 7.49 and 7.50) is an oversimplification. Xu and Bassani [577, 578] developed a cohesive-zone model wherein the relationship between the concentration of the embrittling element and the stress state is represented by constitutive equations for crack opening due to creep δ_{cr} and crack opening due to the grain-boundary separation δ_d within the cohesive zone of length L. The total opening of the cohesive zone is then just the sum of creep and crack separation, $\delta = \delta_{cr} + \delta_d$ (cf. Fig. 7.27). Following Norton's creep law, the crack opening rate $\dot{\delta}_{cr}$ due to power-law creep can be expressed as

$$\dot{\delta}_{cr} = B\left(\sigma^*\right)^n \overbrace{\left(\frac{\sigma}{\sigma^*} - 1\right)^n}^{=0 \text{ when } \sigma/\sigma^* < 1} \tag{7.52}$$

where B is a constant, n is the Norton exponent and σ^* is the creep stress, above which creep deformation sets in. The relationship between the applied stress σ and

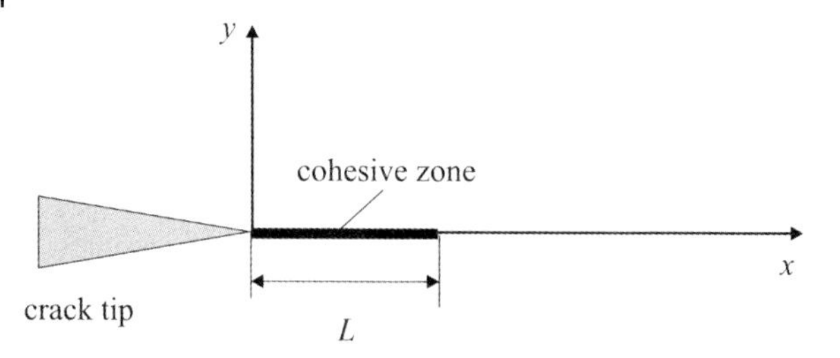

Fig. 7.27 Schematic representation of the steady-state cohesive-zone model of dynamic embrittlement (according to Xu and Bassani [577, 578]).

the opening δ_{d} in the cohesive zone can be given by a phenomenological equation according to Needleman [589]:

$$\sigma=\left(A\,\sigma_{\max}f_{\mathrm{Ds}}\frac{\delta_{\mathrm{d}}}{\delta_{\mathrm{crit}}}+\sigma^{*}\right)\overbrace{\left(1-\frac{\delta_{\mathrm{d}}}{f_{\mathrm{D}d}\delta_{\mathrm{crit}}}\right)^{2}}^{=0\ \text{when}\ \delta_{\mathrm{d}}/f_{\mathrm{D}\delta}\delta_{\mathrm{d}}>1} \tag{7.53}$$

with A being a constant, $\sigma_{\max}$ being the maximum tolerable tensile stress and δ_{crit} being the critical normal opening displacement in the cohesive zone. At the tip of the cohesive zone the opening displacement becomes zero, $\delta_{\mathrm{d}}=0$, and the stress is equal to the creep stress $\sigma=\sigma^{*}$. When δ_{d} reaches a critical opening displacement δ_{crit}, the grain boundary cannot carry any load and intercrystalline decohesion sets in. To address the dependence of grain-boundary embrittlement by a penetrating species on its concentration c, Xu and Bassani [577, 578] used Kachanov-type damage functions $f_{\mathrm{D}\sigma}$ (influence of the embrittler on $\sigma_{\max}$) and $f_{\mathrm{D}\delta}$ (influence of the embrittler on δ_{crit}):

$$f_{\mathrm{D}\sigma/\delta}=\left(1-\frac{c}{c_{\sigma/\delta}}\right)^{k_{\sigma/\delta}} \tag{7.54}$$

The concentrations c_{σ} and c_{δ} as well as the exponents k_{σ} and k_{δ} define the extent by which an embrittling element lowers the grain-boundary cohesion.

To determine the stress σ and the contributions δ_{cr} and δ_{d} of the cohesive-zone-opening displacement, a further equation besides Eqs. (7.52) and (7.53) is required. The stress in the cohesive zone can be considered as a superimposition of the remote tensile stress σ_{a} and the stress gradient due to the dislocation distribution within the cohesive zone, $\mu(x,t)$ [577, 590]. This yields the following integral equation for the dislocation density within the cohesive zone:

$$\mu(x,t)=-\frac{\partial\delta}{\partial x}=\frac{4}{E'\pi}\int_{0}^{L}\frac{\sqrt{L-x}}{\sqrt{L-\xi}}\frac{\sigma(x,t)}{\xi-x}d\xi \tag{7.55}$$

The equations given above form a complete set to solve numerically the governing differential Eq. (7.46) for dynamic embrittelement. This was done by Xu and Bassani [577, 578], who applied the solution to tin-induced dynamic embrittlement

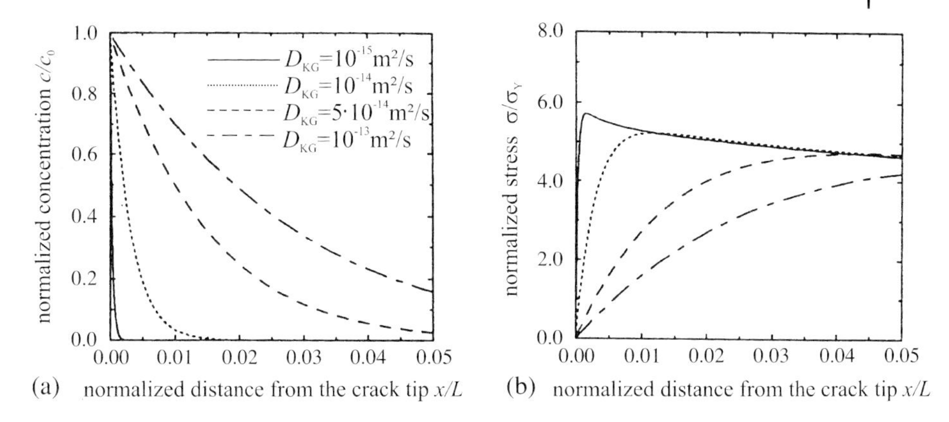

Fig. 7.28 (a) Calculated concentration profiles of the embrittling element tin; (b) corresponding normal-stress distribution in the cohesive zone ahead of an intercrystalline crack tip in a Cu-7Sn model alloy for different grain-boundary diffusion coefficients D_{GB} (after [578]).

of a Cu-7Sn model alloy [591]. Figure 7.28 shows exemplary results of these calculations, representing the normalized concentration c of the embrittling element tin and the normalized tensile stress (related to the yield strength σ_Y) in the cohesive zone acting normal to the crack plane (grain boundary) vs. the normalized distance x from the crack tip (relative to the length of the cohesive zone), here $L = 23.8$–24.4 µm [578], for various grain-boundary diffusion coefficients for tin in Cu-7Sn. It can be seen clearly that both the extension of the dynamic embrittlement process zone and the tolerable stress depend strongly on the grain-boundary diffusion coefficient.

It should be noted that also the grain-boundary diffusion coefficient in the cohesive zone cannot be considered as constant. It should rather exhibit a transition from diffusion in a nondamaged grain boundary D_{GB}^0 towards diffusion in a completely cracked grain boundary, the latter corresponding to surface diffusion D_S. Owing to this transition, Xu and Bassani [578] suggested a simple relationship:

$$D_{GB} = D_{SF} + \left(D_{GB}^0 - D_{SF}\right)\left(1 - \frac{\delta}{\delta_{\text{crack tip}}}\right) \tag{7.56}$$

8
Concluding Remarks

In a slightly simplified manner, fatigue-life-design concepts can be subdivided in the total-life approach and the damage-tolerant approach. While the total-life approach has been used for more than 100 years based on the work of August Wöhler [4] and the *S/N* plot (stress vs. numbers of cycles to fracture, or strain-based modifications according to the Manson–Coffin concept [18, 19]), damage-tolerant fatigue-life assessment became essential for the design of lightweight structures, particularly in the aerospace industry. In many cases, however, a material is considered as a continuum, and the significance of microstructural parameters on the fatigue damage process is accounted for by empirical fit parameters. The large scatter in crack-propagation data and *S/N* data in the high-cycle fatigue regime being a consequence of local microstructural features requires the consideration of safety factors. But one should be aware of the fact that using safety factors is essentially the same as wasting a part of the material's strength potential.

This book gives evidence for the fact that the scatter in fatigue life is directly linked to (i) local deviations in stress and strain from the remote loading conditions due to elastic and plastic anisotropy of the material, which is important for the crack-initiation period, (ii) interactions between the crack tip and the crystallographic orientation and shape of the adjacent grains, which is important for the early stage I of fatigue-crack propagation and (iii) the transient stage of crack-closure effects, which is important for the transition from microstructurally short to short and eventually to long-fatigue-crack propagation.

It was shown in several studies that loading materials exhibiting a nonisotropic Young's modulus (elastic anisotropy) leads to stress concentrations along grain boundaries with unfavorable crystallographic orientation relationships. Such stress concentrations, which certainly can also arise from second-phase precipitates or pores, cause local plastic deformation to an extent that depends on the local slip geometry with respect to the local direction of maximum shear stresses. During cyclic loading dislocation motion along the slip bands is partly irreversible. Dislocation pairing and pileup at obstacles lead to crack initiation and propagation. When the crack tip approaches a grain or phase boundary dislocation motion on the ligament is increasingly hindered and the back-stress acting at the crack tip increases. The crack-propagation rate decreases. At the same time the stress in the vicinity of the respective grain boundary increases. As soon as plastic flow is re-initiated at a slip band in the adjacent grain, the stress on the ligament between crack

Fatigue Crack Propagation in Metals and Alloys: Microstructural Aspects and Modelling Concepts. Ulrich Krupp

ISBN: 978-3-527-31537-6

tip and grain boundary is relieved and the crack-propagation rate increases again. Hence, the barrier strength of a phase or a grain boundary is directly related to the crystallographic twist and tilt misorientation between the slip systems of the neighboring grains. Direct slip transmission by dislocation motion across grain boundaries was reported for small-angle grain boundaries only.

These kinds of oscillating propagation rates have been observed by several authors to be characteristic for microstructurally short fatigue cracks, which can grow well below the threshold ΔK_{th} for long fatigue cracks.

Particularly for high-strength, planar-slip alloys, microstructurally short fatigue cracks grow in a crystalline manner (stage Ia). This results in a mutual displacement of the crack faces, lading to premature crack closure before the specimen/component is completely unloaded. Following Elber's concept of plasticity-induced crack closure [30], this phenomenon has been termed roughness-induced crack closure. Plasticity-induced crack closure according to Elber [30] requires the formation of a plastic wake, being formed during preceding cycles and causing premature closure by acting as a wedge. By definition, short cracks should not have a pronounced wake and hence plasticity-induced crack closure is negligible. On the other hand, plastic deformation ahead of the crack tip must be balanced by some additional amount of compression, when crack propagation is no more crystallographic but driven by multiple slip on alternate slip systems (stage Ib). Under such conditions, the crack might stay open for a substantially larger interval than would be predicted according to Elber [30]. Only when the fatigue crack is long enough (stage II) can the contribution of crack closure be considered as constant and easily be implemented in Paris-like crack-propagation laws [26].

There is no doubt that it is impossible to describe the complex interactions between microstructure, load history and crack propagation rate by simple phenomenological equations. Based on the idea of Navarro and de los Rios [324] and Taira et al. [535] that the plastic zone (yield strip) ahead of the crack tip is blocked by a grain or a phase boundary resulting in an increase of both the back-stress acting on the crack tip and the stress acting on a dislocation source in the adjacent grain, a numerical model has been developed at the University of Siegen (cf. Schick [515], Künkler [421] and Chapter 7), where the crack and the plastic zone is represented by boundary elements. This model allows the prediction of two-dimensional crack propagation accounting for real microstructure data and the development of crack-closure effects. Crack advance is considered as a function of the range of crack-tip-slide displacement on a single slip band or on a combination of alternating operated slip systems (multiple slip). It was shown that the simulations are in excellent agreement with experimental results and that they are capable of representing the typical oscillations in the crack propagation rate below the threshold ΔK_{th} for long fatigue cracks. Since the model can be applied to synthetic microstructures it may serve not only for mechanism-based fatigue-life prediction but also for optimization of tailored microstructures.

At elevated temperatures or in aggressive environments, fatigue crack propagation becomes increasingly time-dependent. Particularly operation conditions, where ingress of a gaseous atmosphere is promoted by high tensile stresses, e.g., hold times at maximum tension or out-of-phase thermomechanical fatigue leading to tensile mean stresses, the crack-propagation rate may increase dramatically by a transition from transgranular to quasi-brittle intergranular crack propagation. This effect was attributed to fast interfacial diffusion of an embrittling species lowering the grain-boundary cohesion giving rise to fast time-dependent intergranular cracking, i.e., dynamic embrittlement. By means of experiments on bicrystalline and grain-boundary-engineering-type processed specimens it became evident that environmentally assisted intergranular cracking is strongly dependent on the grain-boundary structure. Small-angle and special CSL grain boundaries are highly resistant to this kind of quasi-brittle crack propagation.

Modern techniques of material characterization, like the automated electron backscatted diffusion technique for crystallographic orientation mapping in the scanning electron microscope or *in situ* observation of fatigue damage using electron microscopy or synchrotron computer tomography, in combination with sophisticated computer modeling concepts, define the road map for efficient materials design and optimization in the 21st century.

References

1 Lund, J.R.; Byrne, J.P.: Leonardo da Vinci's Tensile Strength Tests: Implications for the Discovery of Engineering Mechanics, *Civil Engineering and Environmental Systems*, **18** (2001) 243

2 Smith, R.A.: The Versailles Railway Accident of 1842 and the First Research into Metal Fatigue, *Proc. Fatigue '90* (1990) 2033

3 Braithwaite, F.: On the Fatigue and Consequent Fracture of Metals, *Minutes of Proc. ICE* (1854) 463

4 Wöhler, A.: Über die Festigkeits-Versuche mit Eisen und Stahl, *Zeitschrift für Bauwesen*, **20** (1870) 73

5 Schütz, W.: Zur Geschichte der Schwingfestigkeit, *Materialwissenschaft und Werkstofftechnik*, **24** (1993) 203

6 Bauschinger, J.: Über die Veränderung der Elastizitätsgrenze und der Festigkeit des Eisens und Stahls durch Strecken und Quetschen, durch Erwärmen und Abkühlen und durch oftmals wiederholte Beanspruchung, *Mitteilungen des mechanisch-technischen Laboratoriums der Königlich Technischen Hochschule München*, **13** (1886) 1

7 Ewing, J.A.; Humfrey, J.C.W.: The Fracture of Metals under Repeated Alternations of Stress, *Philosophical Transactions, Royal Society London*, **CC** (1903) 241

8 Basquin, O.H.: The Exponential Law of Endurance Tests, *Proc. Annual Meeting ASTM*, **10** (1919) 625

9 Palmgren, A.: Die Lebensdauer von Kugellagern, *VDI-Zeitschrift*, **68** (1924) 339

10 Miner, M.A.: Cumulative Damage in Fatigue, *Transactions ASME – Journal of Applied Mechanics*, **12** (1954) A159

11 Inglis, C.E.: Stresses in a Plate due to the Presence of Cracks and Sharp Corners, *Transactions of the Institute of Naval Architects*, **55** (1913) 219

12 Griffith, A.A.: The Phenomena of Rupture and Flow in Solids, *Philosophical Transactions A*, **221** (1920) 163

13 Neuber, H.: *Kerbspannungslehre*, 3rd edition, Springer, Berlin/Heidelberg (1985)

14 Neuber, H.: Theory of Stress Concentration for Shear-Strained Prismatical Bodies with Arbitrary Nonlinear Stress Strain Law, *Journal of Applied Mechanics*, **28** (1961) 544

15 Polanyi: Über eine Art Gitterstörung, die einen Kristall plastisch machen könnte, *Zeitschrift für Physik*, **89** (1934) 660

16 Orowan, E.: Theory of the Fatigue of Metals, *Proc. Royal Society A*, **171** (1939) 79

17 Hirsch, P.B.; Horne, R.W.; Whelan, M.J.: Direct Observation of the Arrangement and Motion of Dislocations in Aluminum, *Philosophical Magazine*, **1** (1956) 677

18 Manson, S.S.: Behaviour of Materials under Conditions of Thermal Stress, *Proc. Heat Transfer Symposium*, University of Michigan (1953) 9

19 Coffin, L.E.: The Problem of Thermal Stress Fatigue in Austenitic Steels at Elevated Temperatures, *Proc. Symposium on Effects of Cyclic Heating and Stressing on Metals at Elevated Temperatures*, ASTM STP 165 (1954) 31

20 Thompson, N.; Wadsworth, N.; Louat, N.: XI. The Origin of Fatigue Fracture in Copper, *Philosophical Magazine*, **1** (1956) 113

Fatigue Crack Propagation in Metals and Alloys: Microstructural Aspects and Modelling Concepts. Ulrich Krupp

ISBN: 978-3-527-31537-6

21 Zappfe, C.A.; Worden, C.O.: Fractographic Registrations of Fatigue, *Transactions ASM*, **43** (1951) 958
22 Forsyth, P.J.E.; Ryder, D.A.: Fatigue Fracture, *Aircraft Engineering*, **32** (1960) 96
23 Laird, C.: The Influence of Metallurgical Structure on the Mechanism of Fatigue Crack Propagation, *Proc. Fatigue Crack Propagation*, ASTM STP 415, (1967) 131
24 Neumann, P.: Coarse Slip Model of Fatigue, *Acta Metallurgica*, **17** (1969) 1219
25 Irwin, G.R.: Fracture, in: *Handbuch der Physik*, Bd. IV, Springer-Verlag, Berlin/Heidelberg (1958) 551
26 Paris, P.C.; Gomez, M.P.; Anderson, W.E.: A Rational Analytic Theory in Engineering, *The Trend in Engineering*, **13** (1960) 9
27 Paris, P.C., Erdogan, F.: A Critical Analysis of Crack Propagation Laws, *Journal of Basic Engineering*, **85** (1963) 528
28 Jung, A.: Charakterisierung des Ausbreitungsverhaltens von Ermüdungsrissen, in: *Ermüdungsverhalten metallischer Werkstoffe*, H.-J. Christ (Ed.), Werkstoff-Informationsgesellschaft, Frankfurt (1998) 137
29 Elber, W.: Fatigue Crack Propagation: Some Effects of Crack Closure on the Mechanism of Fatigue Crack Propagation under Cyclic Tension Loading, PhD Thesis, University of New South Wales (1968)
30 Elber, W.: The Significance of Fatigue Crack Closure, Proc. *Damage Tolerance in Aircraft Structures*, ASTM STP 486 (1971) 230
31 Pearson, S.: Initiation of Fatigue Cracks in Commercial Aluminum Alloys and the Subsequent Propagation of Very Short Cracks, *Engineering Fracture Mechanics*, **7** (1975) 235
32 Shiozawa, K.; Matsushita, H.: Crack Initiation and Small Fatigue Crack Growth Behaviour of Beta Ti-15V-3Cr-3Al-3Sn Alloy, *Proc. Fatigue '96*, G. Lütjering, H. Nowack (Eds.), Berlin (1996) 301
33 Tokaji,K.; Takafiji, S.; Ohya.K.; Kato, Y.; Mori,K.: Fatigue Behavior of Beta Ti-22V-4Al Alloy Subjected to Surface-Microstructural Modification, *Journal of Materials Science*, **38** (2003) 1153
34 Mann, J.Y.: *Bibliography on the Fatigue of Materials Components and Structures*, Vols. 1–4, Pergamon Press, Oxford (1970–1990)
35 Richard, H.A.; Sander, M.; Fulland, M.; Kullmer, G.: Numerical and Experimental Investigations on Fatigue Crack Growth in a Wheel of the German High-Speed Train ICE, *Proc. Int. Conference on Fracture ICF11*, Turin 2005, on CD-ROM
36 Suresh, S.: *Fatigue of Materials*, 2nd edition, Cambridge University Press, Cambridge (1998)
37 Ramberg, W.; Osgood, W.R.: Description of Stress-Strain Curves by Three Parameters, Technical Report No 902 NACA (1943)
38 Masing, G.: Eigenspannungen und Verfestigung beim Messing, Proc. *2nd Int. Congress of Applied Mechanics*, Zürich (1926) 332
39 Radaj, D.: *Ermüdungsfestigkeit, Grundlagen für Leichtbau, Maschinen- und Stahlbau*, Springer-Verlag, Berlin/Heidelberg (1995)
40 Haibach, E.: *Betriebsfestigkeit, Verfahren und Daten zur Bauteilberechnung*, 2nd edition, Springer-Verlag, Berlin/Heidelberg (2002)
41 Naubereit, H.: *Einführung in die Ermüdungsfestigkeit*, Hanser-Verlag, Munich (1999)
42 Dengel, D.: Die arcsin√P-Transformation – ein einfaches Verfahren zur grapischen und rechnerischen Auswertung geplanter Wöhler-Versuche, *Zeitschrift für Werkstofftechnik*, **6** (1975) 253
43 Deubelbeiss, E.: Dauerfestigkeitsversuche mit einem modifizierten Treppenstufenverfahren, *Materialprüfung*, **16** (1974) 240
44 Morrow, J.D.: Cyclic Plastic Strain Energy and Fatigue of Metals, ASTM STP 378 (1965) 45
45 Smith, K.N.; Watson, P.; Topper, T.H.: A Stress–Strain Function for the Fatigue of Materials, *Journal of Materials*, **4** (1970) 767
46 Haibach, E.; Lehrke, H.P.: Das Verfahren der Amplitudentransformation zur Lebensdauerberechnung bei Schwingbeanspruchung, *Archiv für das Eisenhüttenwesen*, **47** (1976) 623

47 Schöler, K.: Einfluß der Mikrostruktur auf das Wechselverformungsverhalten teilchengehärteter Legierungen bei hohen Temperaturen nach einer Vorverformung, Fortschritt Berichte VDI, Nr. 574, VDI-Verlag, Düsseldorf (1999)

48 Schöler, K.; Christ, H.-J. Influence of Prestraining on Cyclic Deformation Behaviour and Microstructure of a Single-Phase Ni-Base Superalloy, *International Journal of Fatigue*, **23** (2001) 767

49 Matsuishi, M.; Endo, T.: Fatigue of Metals Subjected to Varying Stress, *Proc. Kyushu Branch of Japan Society of Mechanical Engineering* (1968) 37

50 Dowling, S.D.; Socie, D.F.: Simple Rainflow Counting Algorithms, *International Journal of Fatigue*, **4** (1982) 31

51 Anthes, R.J.: Modified Rainflow Counting Keeping the Load Sequence, *International Journal of Fatigue* **7** (1997) 529

52 Foreman, R.G.; Keary, V.E.; Eagle, R.M.: Numerical Analysis of Crack Propagation in Cyclic-Loaded Structures, *Journal of Basic Engineering*, **89** (1967) 459

53 Weertman, J.: Rate of Growth of Fatigue Cracks Calculated from the Theory of Infinitesimal Dislocations Distributed on a Plane, *International Journal of Fracture*, **2** (1966) 460

54 Klesnil, M.; Lukas, P.: Influence of Strength and Stress History on Growth and Stabilisation of Fatigue Cracks, *Engineering Fracture Mechanics*, **4** (1972) 209

55 McEvily, A.J.: On Closure in Fatigue Crack Growth, ASTM STP 982 (1988) 35

56 Anderson, T.L.: *Fracture Mechanics: Fundamentals and Applications*, 2nd edition, CRC Press, Boca Raton, FL (1995)

57 Broek, D.: *Elementary Engineering Fracture Mechanics*, 4th revised edition, Leyden, Noordhof (1985)

58 Gross, D.: *Bruchmechanik*, 2nd edition, Springer-Verlag, Berlin/Heidelberg (1996)

59 Schwalbe, K.-H.: *Bruchmechanik metallischer Werkstoffe*, Hanser-Verlag (1980)

60 Gumbsch, P.: Modelling Brittle and Semi-Brittle Fracture Processes, *Materials Science and Engineering A*, **319–321** (2001) 1

61 Elsäser, C.; Gumbsch, P.: Risse und Grenzflächen im Computer, *Physik-Journal*, **4** (2005) 23

62 Orowan, E.: Fracture and Strength of Solids, *Reports on Progress in Physics*, **12** (1948) 185

63 Irwin, G.R.: Onset of Fast Crack Propagation in High Strength Steel and Aluminium Alloys, *Proc. Sagamore Research Conf.*, **2** (1956) 289

64 Westergaard, H.M.: Bearing Pressures and Cracks, *Journal of Applied Mechanics*, **6** (1939) 49

65 Tada, H.; Paris, P.C.; Irwin, G.R.: *The Stress Analysis Handbook*, 3rd edition, ASME Press, New York (2000)

66 Murakami, Y: *The Stress Intensity Factors Handbook*, Pergamon Press, New York (1987)

67 Sommer, E.: *Bruchmechanische Bewertung von Oberflächenrissen*, Springer-Verlag. Berlin/Heidelberg (1984)

68 Irwin, G.R.: Plastic Zone Near a Crack and Fracture Toughness, *Proc. Sagamore Research Conf.*, **4** (1961)

69 Dugdale, D.S.: Yielding of Steel Sheets Containing Slits, *Journal of the Mechanics and Physics of Solids*, **8** (1960) 100

70 Barenblatt, G.I.: The Mathematical Theory of Equilibrium Cracks in Brittle Fracture, *Proc. Advances in Applied Mechanics*, Vol. VII, Academic Press, New York/London (1962) 55

71 Bilby, B.A.; Cottrell, A.H.; Swinden, K.H.: The Spread of Plasticity Yield from a Notch, *Proceedings of the Royal Society London*, **A272** (1963) 304

72 Burdekin, F.M.; Stone, D.E.W.: The Crack Opening Displacement Approach to Fracture Mechanics in Yielding Materials, *Journal of Stress Analysis*, **1** (1966) 145

73 Rice, J.R.: A Path Independent Integral and the Approximate Analysis of Strain Concentration by Notches and Cracks, *Journal of Applied Mechanics*, **35** (1968) 379

74 Hutchinson, J.W.: Singular Behavior at the End of a Tensile Crack Tip in a Hardening Material, *Journal of the Mechanics and Physics of Solids*, **16** (1968) 13

75 Rice, J.R.; Rosengren, G.F.: Plane Strain Deformation Near a Crack Tip in a Power Law Hardening Material, *Journal* *of the Mechanics and Physics of Solids,* **16** (1968) 1

76 Dowling, N.E.; Begley, J.A.: Fatigue Crack Growth during Gross Plasticity and the J-Integral, ASTM STP 590, Philadelphia (1976) 82

77 Wüthrich, C.: The Extension of the J-Integral Concept to Fatigue Cracks, *International Journal of Fracture,* **20** (1982) R35

78 H.H. Heitmann, H. Vehoff, P. Neumann: Life Prediction for Random Load Fatigue Based on the Growth Behavior of Microcracks, *Proc. 6th Int. Conf. on Fracture (ICF 6),* S.R. Valluri, D.M.R. Taplin, P. Rama Rao, J.F. Knott, R. Dubey (Eds.), Pergamon Press, Oxford (1984) 3599

79 Merkle, J.G.; Corten, H.T.: A J-Integral Analysis of the Compact Specimen Considering Axial Force as Well as Bending Effects, *Journal of Pressure Vessel Technology,* **96** (1974) 286

80 Riedel, H.: *Fracture at High Temperatures,* Springer-Verlag, Berlin (1987)

81 Miller, M.P.; McDowell, D.L.; Oehmke, R.L.T.; Antolovich, S.D.: A Life Prediction Model for Thermomechanical Fatigue Based on Microcrack Propagation, *Proc. Thermomechanical Fatigue Behavior of Metals,* ASTM STP 1186, H. Sehitoglu (Ed.), ASTM (1993) 35

82 Jung, A.: Einfluss einer Partikelverstärkung auf das Hochtemperaturermüdungsverhalten einer dispersionsgehärteten Aluminiumlegierung, Fortschritt-Berichte VDI, Reihe 5 Nr. 600, VDI-Verlag, Düsseldorf (2000)

83 Schöler, K., Christ, H.-J.: Servohydraulische Prüfsysteme vergleichen – Untersuchungen zum Regelverhalten, *Zeitschrift Materialprüfung* **38** (1996) 488

84 Christ, H.-J.; Mughrabi, H.: Cyclic Stress–Strain Response and Microstructure under Variable Amplitude Loading, *Fatigue and Fracture of Engineering Materials and Structures,* **19** (1996) 335

85 Christ, H.-J.: *Wechselverformung von Metallen,* Springer-Verlag, Berlin/Heidelberg (1991)

86 Polak, J.; Hajek, M.: Cyclic Stress–Strain Curve Evaluation Using Incremental Step Test Procedure, *International Journal of Fatigue,* **13** (1991) 216

87 Christ, H.-J.; Mughrabi, H.; Kraft, S.; Petry, F.; Zauter, R.; Eckert, K.: The Use of Plastic Strain Control in Thermomechanical Fatigue Testing, *Proc. Fatigue under Thermal and Mechanical Loading,* J. Bressers, L. Remy (Eds.) (1996) 119

88 Yu, X.; Abel, A.: Mixed-Mode Crack Surface Interference under Cyclic Shear Loads, *Fatigue and Fracture of Engineering Materials and Structures,* **23** (2000) 1515

89 Sharpe Jr., W.N.; Grandt Jr., A.F.: A Preliminary Study of Fatigue Crack Retardation Using Laser Interferometry to Measure Crack Surface Displacements, *Proc. Mechanics of Crack Growth,* ASTM STP590 (1976) 302

90 Floer, W.: Untersuchungen zur mechanismenorientierten Lebensdauervorhersage an einer β-Titanlegierung, VDI-Verlag Reihe 5, Düsseldorf (2003)

91 Heinz, A.; Neumann, P.: Crack Initiation During High Cycle Fatigue of an Austenitic Steel, *Acta Metallurgica et Materialia,* **38** (1990) 1933

92 Ebi, G: Ausbreitung von Mikrorissen in duktilen Stählen, Dissertation, RWTH Aachen (1987)

93 Im, S.-W.: Untersuchung von Mikrorissen bei Wechselbeanspruchung durch Laserinterferometrie, Fortschritt-Berichte VDI, Nr. 201, VDI-Verlag Düsseldorf (1990)

94 Lenczowski, B.: *Das Verhalten von Mikrorissen bei Ermüdung von Werkstoffen,* Shaker-Verlag, Aachen (1992)

95 Ashbaugh, N.E.; Dattaguru, B.; Khobaib, M.; Nicholas, T.; Prakash, R.V.; Ramamurthy, T.S.; Seshardi, B.P.; Sunder, R.: Experimental and Analytical Estimates of Fatigue Crack Closure in an Aluminium Copper Alloy: I. Laser Interferometry and Electron Fractography, *Fatigue and Fracture of Engineering Materials and Structures,* **20** (1997) 951

96 Akiniwa, Y.; Harada, S.; Fukushima, Y.: Dynamic Measurement of Crack Closure Behaviour of Small Fatigue Cracks by an Interferometric Strain/Displacement Gauge wit a Laser Diode, *Fatigue and Fracture of Engineering Materials and Structures,* **14** (1991) 317

97 Ravichandran, K.S.: Three-Dimensional Crack Shape Effects During the Growth of Small Surface Cracks in a Titanium Base Alloy, *Fatigue and Fracture of Engineering Materials and Structures*, **20** (1997) 1423

98 Larsen, J.M.; Jira, J.R.; Weerasooriya, T.: Crack Opening Displacement Measurements on Small Cracks in Fatigue, Proc. *18th Symposium on Fracture Mechanics*, D.T. Read, P.P. Reed (Eds.), ASTM STP 945 (1988) 896

99 Zupan, M.; Hemker, K.J.: Application of Fourier Analysis to the Laser Based Interferometric Strian/Displacement Gage, *Experimental Mechanics*, **42** (2002) 214

100 Knobbe, H.: Modifizierung eines ISDDG Systems zur hochaufgelösten Dehnungsmessung, diploma thesis, Universität Siegen (2005)

101 Fax, J.C.; Edwards, R.L.; Sharpe, W.N.: Thin-Film Gage Markers for Laser-Based Strain Measurement on MEMS Materials, *Experimental Techniques*, **5/6** (1999) 28

102 von Heimendahl, M.: *Einführung in die Transmissionselektronenmikroskopie*, Vieweg-Verlag, Braunschweig (1970); English version: *Electron Microscopy of Materials*, Academic Press, New York (1980)

103 Eisenmeier, G.: Ermüdungsverhalten der Magnesiumlegierung AZ91: Experimentelle Untersuchungen und Lebensdauervorhersage, Dissertation, Universität Erlangen-Nürnberg (2001)

104 Bauschke, H.-M.; Schwalbe, K.-H.: Measurement of the Depth of Surface Cracks Using the Direct Current Potential Drop Method, *Zeitschrift für Werkstofftechnik*, **16** (1985) 156

105 Taylor, D.: *Fatigue Thresholds*, Butterworths, London (1989)

106 Zeller, R.: Risstiefenmessung nach dem Potentialsondenverfahren mit Wechselstrom, *Materialprüfung*, **23** (1981) 85

107 Ono, K.: Current Understanding of Mechanisms of Acoustic Emission, *Journal of Strain Analysis for Engineering Design*, **40** (2005) 1

108 Saxena, A.; Hudak, S.J.; Donald, J.K.; Schmid, D.W.: Computer-Controlled Decreasing Stress Intensity Techniques for Low Rate Fatigue Crack Growth Testing, *Journal of Testing and Evaluation*, **6** (1978) 167

109 Bressers, J.; Timm, J.; Williams, S.J.; Bennett, A.; Affeldt, E.E.: Effects of Cycle Type and Coating on the TMF Lives of a Nickel-Based Gas Turbine Blade Alloy, *Proc. Thermomechanical Fatigue Behavior of Materials*, Vol. 2, M.J. Verrilli, M.G. Castelli (Eds.), ASTM STP 1263 (1996) 56

110 Bowles, C.Q.; Schijve, J.: Crack Tip Geometry for Fatigue Cracks Grown in Air and Vacuum, *Proc. Advances in Quantitative Measurement of Physical Damage*, ASTM STP 811 (1983) 400

111 Maire, E.: Computed X-Ray Tomography, in: *Handbook of Cellular Metals*, H.-P. Degischer, B. Kriszt (Eds.), Wiley-VCH, Weinheim (2002)

112 Ludwig, W.; Buffière, J.-Y.; Savelli, S.; Cloetens, P.: Study of a Short Fatigue Crack with Grain Boundaries in a Cast Al Alloy Using X-Ray Microtomography, *Acta Materialia*, **51** (2003) 585

113 Marrow, T.J; Buffiere, J.-Y.; Withers, P.J.; Johnson, G.; Engelberg, D.: High Resolution X-ray Tomography of Short Fatigue Crack Nucleation in Austempered Ductile Cast Iron, *International Journal of Fatigue*, **26** (2004) 717

114 Toda, H.; Sinclair, I.; Buffiére, J.-Y.; Maire, E.; Khor, K.H.; Gregson, P.; Kobayashi, T.: A 3D Measurement Procedure for Internal Local Crack Driving Forces via Synchrotron X-Ray Microtomography, *Acta Materialia*, **52** (2004) 1305

115 Ludwig, W.; Bellet, D.: Penetration of Liquid Gallium into the Grain Boundaries of Aluminium: A Synchroton Radiation Microtomographic Investigation, *Materials Science and Engineering*, **A281** (2000) 198

116 Nielsen, S.F.; Ludwig, W.; Bellet, D.; Lauridsen, E.M.; Poulsen, H.F.; Juul Jenseen, D.: Three Dimensional Mapping of Grain Boundaries, in: *Proc. 21st Int. RISO Symposium on Materials Science*, N. Hansen et al. (Eds.), Roskilde, Denmark (2000) 473

117 Margulies, L.; Poulsen, H.F.: Three-Dimensional X-Ray Diffraction (3DXRD) Analysis, in: *Proc. 25th Int. RISO*

Symposium on Materials Science, C. Gundlach et al. (Eds.), Roskilde, Denmark (2004) 61

118 Wetzig, K.; Schulze, D. (Eds.): *In Situ Scanning Electron Microscopy in Materials Research*, Akademie-Verlag, Berlin (1995)

119 Weidner, A.; Tirschler, W.; Blcohwitz, C.: Overstraining Effects on the Crack Opening Displacment of Microstructurally Short Cracks, *Materials Science and Engineering A*, **390** (2005) 414

120 Blochwitz, C.; Tirschler, W.: In-Situ Scanning Electron Microscope Observations of the Deformation Behaviour of Short Cracks, *Materials Science and Engineering*, **A276** (2000) 273

121 Nakajima, K.; Terao, K.; Miyata, T.: The Effect of Microstructure on Fatigue Crack Propagation of α+β Titanium Alloys, *Materials Science and Engineering*, **A243** (1998) 176

122 Li, W.-F.; Zhang, X.-P.: Investigation of Initiation and Growth Behavior of Short Fatigue Cracks Emanating from a Single-Edge Notch Specimen Using in-situ SEM, *Materials Science and Engineering*, **A318** (2001) 129

123 Richter, R.; Tirschler, W.; Blochwitz, C.: In-Situ Scanning Electron Microscopy of Fatigue Crack Behaviour in Ductile Materials, *Materials Science and Engineering*, **A213** (2001) 237

124 Zhang, X.P.; Wang, C.H.; Chen, W.; Ye, L.; Mai, Y.-W.: Investigation of Short Fatigue Cracks in Nickel-Based Single Crystal Superalloy SC16 by In-Situ SEM Fatigue Testing, *Scripta Materialia*, **44** (2001) 2443

125 Sickert, M.: Konstruktion und Inbetriebnahme einer Ermüdungsprüfmaschine für das Rasterelektronenmikroskop, Diplomarbeit, Universität Siegen (2002)

126 Dewald, D.K.; Lee, T.C.; Robertson, I.M.; Birnbaum, H.K.: Dislocation Structure of Advancing Cracks, *Metallurgical Transactions A*, **21A** (1990) 2411

127 Gemperlová, J.; Polcraová, M.; Gemperle, A.; Zárubová, N.: Slip Transfer Across Grain Boundaries in Fe-Si Bicrystals, *Journal of Alloys and Compounds*, **378** (2004) 97

128 Ohmura, T.; Minor, A.M.; Stach, E.A.; Morris Jr., J.W.: Dislocation–Grain Boundary Interactions in Martensitic Steel Observed through In Situ Nanoindentation in a Transmission Electron Microscope, *Journal of Materials Research*, **19** (2004) 3926

129 Lee, T.C.; Robertson, I.M.; Birnbaum, H.K.: An In Situ Transmission Electron Microscope Deformation Study of the Slip Transfer Mechanisms in Metals, *Metallurgical Transactions*, **21A** (1990) 2437

130 Goldstein, J.I.; Newbury, D.E.; Echlin, P.; Joy, D.C.; Romig Jr., A.D.; Lyman, C.E.; Fiori, C.; Lifshin, E.: *Scanning Electron Microscopy and X-Ray Microanalysis*, Plenum Press, New York/London (1992)

131 Hull, D.: *Fractography*, Cambridge University Press (1999)

132 Flügge, J. (Ed.): *The Appearance of Cracks and Fracture in Metallic Materials* [bilingual in German and English], Verlag Stahleisen (1996)

133 Bichler, C.; Pippan, R.: Direct Observation of the Residual Plastic Deformation Caused by a Single Tensile Overload, Proc. *2nd Symp. on Fatigue Crack Closure, Measurement and Analysis*, ASTM STP 1343 (1999) 270

134 Richter, R., personal communication, Technische Universität Dresden (2000)

135 Schmidt, P.F. (Ed.): *Praxis der Rasterelektronenmikroskopie und Mikrobereichsanalyse*, Expert-Verlag, Munich (1994)

136 Schwartz, A.J.; Kumar, M.; Adams, B.L. (Eds.): *Electron Backscatter Diffraction in Materials Science*, Kluwer, New York (2000)

137 Randle, V.: *The Measurement of Grain Boundary Geometry*, Institute of Physics, Bristol (1993)

138 Wright, S.I.; Field, D.P.: Recent Studies of Local Texture and its Influence on Failure, *Materials Science and Engineering*, **A257** (1998) 165

139 Dingley, D.J.: *Atlas of Backscattering Kikuchi Diffraction Patterns*, Institute of Physics, Bristol (1995)

140 Katrakova, D.; Mücklich, F.: Specimen Preparation and Electron Backscatter Diffraction, *Proc. European Metallographic Conference 2000: Fortschritte in der Metallographie*, G. Petzow (Ed.) (2001) 355

141 Wilkinson, A.J.: Measurement of Elastic Strains and Small Lattice Rotations

Using Electron Back Scatter Diffraction, *Ultramicroscopy*, **62** (1996) 237

142 Ullrich, H.-J.; Bauch, J.; Brechbühl, J.; Bretschneider, I.; Lin, H.: Grundlagen und neue Anwendungen der Kossel-Technik in der Materialforschung, *Zeitschrift für Metallkunde*, **89** (1998) 106

143 Wright, S.: Fundamentals of Automated EBSD, in: *Electron Backscatter Diffraction in Materials Science*, A.J. Schwartz, M. Kumar, B.L. Adams (Eds.), Kluwer, New York (2000) 51

144 Engler, O.; Gottstein, G.: A New Approach in Texture Research: Local Orientation Determination with EBSP, *Steel Research*, **63** (1992) 413

145 van der Wal, D.; Dingley, D.J.: An Introduction to EBSP, *Philips Electron Optics Bulletin*, **134** (1996) 19

146 Dingley, D.J.; Randle, V.: Review: Microtexture Determination by Electron Back-Scatter Diffraction, *Journal of Materials Science*, **27** (1992) 4545

147 Wiliams, D.B.; Carter, C.B.: *Transmission Electron Microscopy: A Textbook for Materials Science*, Plenum Press, New York (1996)

148 Cullity, B. D.: *Elements of X-ray Diffraction*, 2nd edition, Addison-Wesley, Reading, MA (1978)

149 Randle, V.: A Methodology for Grain Boundary Plane Assessment by Single Section Trace Analysis, *Scripta Materialia*, **44** (2001) 2789

150 Zhang, C.; Suzuki, A.; Ishimaru, T.; Enomoto, M.: Characterization of Three-Dimensional Grain Structure in Polycrystalline Iron by Serial Sectioning, *Metallurgical and Materials Transactions A*, **35A** (2004) 1927

151 D.M. Saylor, B.S. El-Dasher, B.L. Adams, G.S. Rohrer: Measuring the Five Parameter Grain Boundary Distribution From Observations of Planar Sections, *Metallurgical and Materials Transactions A*, **35A** (2004) 1981

152 Randle, V.: *Role of the Coincidence Site Lattice in Grain Boundary Engineering*, Institute of Materials, London (1996)

153 Palumbo, G.; Lehockey, E.M.; Lin, P.: Applications for Grain Boundary Engineered Materials, *Journal of Materials*, **2** (1998) 40

154 Brandon, D.G.: The Structure of High-Angle Grain Boundaries, *Acta Metallurgica*, **14** (1966) 1479

155 Sarid, D.: *Scanning Force Microscopy*, Oxford University Press, New York (1991)

156 Bennewitz, K.; Haaks, M.; Staab, T.; Eisenberg, S.; Lampe, T.; Maier, K.: Positron Annihilation Spectroscopy: A Non-Destructive Method for Lifetime Prediction in the Field of Dynamical Material Testing, *Zeitschrift für Metallkunde*, **93** (2002) 8

157 Jiang, L.; Liaw, P.K.; Brooks, C.R.; Somieski, B.; Klarstrom, D.L.: Nondestructive Evaluation of Fatigue Damage in UDIMET Superalloy, *Materials Science and Engineering A*, **313** (2001) 153

158 Hartley, J.H.; Howell, R.H.; Asoka-Kumar, P.; Sterne, P.A.; Akers, D.; Denison, A.: Positron Annihilation Studies of Fatigue in 304 Stainless Steel, *Applied Surface Science*, **149** (1999) 204

159 Moorthy, V.; Choudhary, B.K.; Vaidyanathan, S.; Jayakumar, T.; Bhanu Sankara Rao, B.; Raj B.: An Assessment of Low-Cycle Fatigue Damage Using Magnetic Barkhausen Emission in 9Cr-1Mo Ferritic Steel, *International Journal of Fatigue*, **21** (1999) 263

160 Li, Y.; Wang, J.; Wang, M.; Shalakov, P.; Witney, A.; de Luccia, J.; Laird, C.: A Demonstration of the Capability of the Electrochemical Fatigue Sensor, *Proc. Fatigue in the Very High Cycle Regime*, Stanzl-Tschegg, S.; Mayer, H. (Eds.), Vienna (2001) 215

161 Guo, Y.; Guo, W.: Mechanical and Electrostatic Properties of Carbon Nanotubes under Tensile Loading and Electrical Field, *Journal of Physics*, **36** (2003) 805

162 Gottstein, G.: *Physikalische Grundlagen der Materialkunde*, 2nd edition, Springer-Verlag, Berlin/Heidelberg (2001)

163 Haasen, P.: *Physical Metallurgy*, 3rd edition, Cambridge University Press, Cambridge, UK (1996)

164 Haasen, P.; Cahn, R. W. (Eds.): *Physical Metallurgy*, 5th revised edition, North Holland (1996)

165 Hull, D.; Bacon, D.J.: *Introduction to Dislocations*, 3rd edition, Pergamon Press, Oxford (1984)

166 Nabarro, F.R.N.: *Theory of Crystal Dislocations*, Oxford University Press, Oxford (1967); Dover Publications (1987)

167 Sommer, C.; Christ, H.-J.; Mughrabi, H.: Non-Linear Elastic Behaviour of the Roller Bearing Steel SAE 52100 during Cyclic Loading, *Acta Metallurgica et Materialia*, **39** (1991) 1177

168 Hertzberg, R.W.: *Deformation and Fracture Mechanics of Engineering Materials*, Wiley, New York (1996)

169 Frenkel, J.: Zur Theorie der Elastizitätsgrenze und der Festigkeit kristallinischer Körper, *Zeitschrift für Physik*, **37** (1926) 572

170 Peierls, R.: The Size of Dislocations, *Proceedings of the Physical Society*, **52** (1940) 34

171 Nabarro, F.R.N.: Dislocations in a Simple Cubic Lattice, *Proceedings of the Physical Society*, **59** (1947) 256

172 Read, W.T.: *Dislocations in Crystals*, McGraw-Hill, New York (1953)

173 Hong, S.I.; Laird, C.: Mechanisms of Slip Mode Modification in F.C.C. Solid Solutions *Acta Metallurgica et Materialia*, **38** (1990) 1581

174 Polmear, I.J.: *Light Alloys, Metallurgy of the Light Metals*, Edward Arnold, London/New York (1988)

175 Evans, H.E.: *Mechanisms of Creep Fracture*, Elsevier, London (1984)

176 Frost, H.J.; Ashby, M.F.: *Deformation-Mechanism Maps: The Plasticity and Creep of Metals and Ceramics*, Pergamon, Oxford (1982)

177 Askeland, D.: *Materialwissenschaften*, Spektrum-Verlag, Heidelberg/Berlin/Oxford (1996)

178 Schmid, E., Boas, W.: *Plasticity of Crystals*, Hughes Press, London (1950)

179 Gemperlová, J.; Polcarová, M.; Gemperle, A.; Zárubová, N.: Slip Transfer Across Grain Boundaries in Fe-Si Bicrystals, *Journal of Metals and Alloys*, **378** (2004) 97

180 Hall, E. O.: The Deformation and Aging of Mild Steel: II. Characteristics of the Lüders Deformation, *Proceedings of the Royal Society*, **64B** (1951) 747

181 Petch, N.J.: The Cleavage Strength of Polycrystals, *Journal of the Iron and Steel Institute*, **5** (1953) 25

182 Schijve, J.: *Fatigue of Structures and Materials*, Kluwer, Dordrecht (2001)

183 Bílý, M. (Ed.): Cyclic Deformation and Fatigue of Materials, *Materials Science Monographs 78*, Elsevier, Amsterdam (1993)

184 Ellyin, F.: *Fatigue Damage, Crack Growth and Life Prediction*, Chapman and Hall, London (1997)

185 Feltner, C.E.; Laird, C.: Factors Influencing the Dislocation Structures in Fatigued Metals, *TMS-AIME* **242** (1968) 1253

186 Lukas, P.; Klesnil, M.: Fatigue Damage and Dislocation Substructure, *Proc. 2nd Int. Conf. on Corrosion Fatigue*, O.J. Devereux et al. (Eds.), NACE, Houston, TX (1972) 118

187 Mughrabi, H.: Microscopic Mechanisms of Metal Fatigue, in: *The Strength of Metals and Alloys*, Vol. 3, P. Haasen, V. Gerold, G. Kostorz (Eds.), Pergamon, Oxford (1980) 1615

188 Wilkens, M.; Herz, K.; Mughrabi, H.: An X-Ray Diffraction Study of Cyclically and Unidirectionally Deformed Copper Single Crystals, *Zeitschrift für Metallkunde*, **71** (1980) 376

189 Winter, A.T.: A Model for the Fatigue of Copper at Low Plastic Strain Amplitudes, *Philosophical Magazine*, **30** (1974) 719

190 Blochwitz, C.; Veit, U.: Plateau Behaviour of Fatigued fcc Single Crystals, *Crystal Research and Technology*, **17** (1982) 219

191 Basinski, Z.S.; Korbel, A.S.; Basinski, S.J.: The Temperature Dependence of the Saturation Stress and Dislocation Structure in Fatigued Copper Single Crystals, *Acta Metallurgica*, **28** (1980) 191

192 Holzwart, U.; Essmann, U.: The Evolution of Persistent Slip Bands in Copper Single Crystals, *Applied Physics A*, **57** (1993) 131

193 Wang, Z.G.; Gong, B.; Wang, Z.: Cyclic Deformation Behavior and Dislocation Structures of [001] Copper Single Crystals: II. Characteristics of Dislocation Structures, *Acta Metallurgica*, **45** (1997) 1379 (see also pp. 1365ff.)

194 Essmann, U.; Mughrabi, H.: Annihilation of Dislocations During Tensile and Cyclic Deformation and Limits of Dislocation Densities, *Philosophical Magazine A*, **40** (1979) 731
195 Plumtree, A.; Pawlus, L.D.: Substructural Developments during Strain Cycling of Wavy Slip Mode Metals, *Proc. Basic Questions in Fatigue*, J.T. Fong, R.J. Fields (Eds.), ASTM STP 924 (1988) 81
196 Sestak, B.; Seeger, A.: Gleitung und Verfestigung in kubisch raumzentrierten Metallen und Legierungen (I–III), *Zeitschrift für Metallkunde*, **69** (1978) 195 (I), 355 (II), 425 (III)
197 Mughrabi, H.; Ackermann, F.; Herz, K.: Persistent Slip Bands in Fatigued Face-Centered and Body-Centered Metals, *Proc. Fatigue Mechanisms*, ASTM STP 675 (1979) 69
198 Pohl, K.; Mayr, P.; Macherauch, E.: Persistent Slip Bands in the Interior of a Fatigued Low-Carbon Steel, *Scripta Metallurgica*, **14** (1980) 1167
199 Lukas, P.: Fatigue Crack Nucleation and Microstructure, in: *ASM Handbook: Fatigue and Fracture*, Vol. 19, ASM (1996) 96
200 Lindley, T.C.; Nix, K.J.: Metallurgical Aspects of Fatigue Crack Growth, *Proc. Fatigue Crack Growth*, R.A. Smith (Ed.), Pergamon (1986) 53
201 Newman Jr., J.C.; Phillips, E.P.; Swain, M.H.: Fatigue-Life Prediction Methodology Using Small Crack Theory, *International Journal of Fatigue*, **21** (1999) 109
202 Lankford, J.: The Growth of Small Fatigue Cracks in 7075-T6 Aluminium, *Fatigue of Engineering Materials and Structures*, **5** (1982) 233
203 Polak, J.: Cyclic Strain Localisation, Crack Nucleation and Short Crack Growth, *Proc. Low Cycle Fatigue and Elasto Plastic Behaviour of Materials LCF4*, K.T. Rie, P.D. Portella (Eds.) (1998) 493
204 Vallellano, C.; Navarro, A.; Domínguez, J.: Fatigue Crack Growth Threshold Conditions at Notches: I. Theory , *Fatigue and Fracture of Engineering Materials and Structures*, **23** (2000) 113
205 Vallellano, C.; Navarro, A.; Domínguez, J.: Fatigue Crack Growth Threshold Conditions at Notches: II. Generalization and Application to Experimental Results, *Fatigue and Fracture of Engineering Materials and Structures*, **23** (2000) 123
206 Andrews, S.; Sehitoglu, H.: A Computer Model for Fatigue Crack Growth from Rough Surfaces, *International Journal of Fatigue*, **22** (2000) 619
207 Dörr, T.; Wagner, L.: Mechanische Oberflächenbehandlung von Titanlegierungen: Grundsätzliche Mechanismen, *Forum der Forschung*, **5.2** (1997) 138
208 Almer, J.D.; Cohen, J.B.; Moran, B.: The Effects of Residual Macrostresses on Fatigue Crack Initiation, *Materials Science and Engineering*, **A284** (2000) 268
209 Kocan, M.; Rack, H.J.; Wagner, L.: Fatigue Performance of Metastable Beta Titanium Alloys: Effect of Microstructure and Surface Finish, *Journal of Materials Engineering and Performance*, **14** (2005) 765
210 Peters, J.O.; Roder, O.; Boyce, B.L.; Thompson, A.W.; Ritchie, R.O.: Role of Foreign-Object Damage on Thresholds for High-Cycle Fatigue in Ti-6Al-4V, *Metallurgical and Materials Transactions*, **31A** (2000) 1571
211 Peters, J.O.; Ritchie, R.O.: Influence of Foreign-Object Damage on Crack Initiation and Early Growth during High-Cycle Fatigue of Ti-6Al-4V, *Engineering Fracture Mechanics*, **67** (2000) 193
212 Stanzl-Tschegg, S.; Mayer, H. (Eds.): *Proc. Fatigue in the Very High Cycle Regime*, Vienna (2001)
213 Argon, A.S.: Effect of Surfaces on Fatigue Crack Initiation, *Proc. Corrosion Fatigue*, O.F. Devereux, A.J. McEvily, R.W. Staehle (Eds.), NACE (1972) 176
214 Morris, W.L.: The Noncontinuum Crack Tip Deformation Behavior of Surface Microcracks, *Metallurgical Transactions A*, **11** (1980) 1117
215 Morris, W.L.: Microcrack Closure Phenomena for Al2219-T851, *Metallurgical Transactions A*, **10** (1979) 5
216 Differt, K.; Essmann, U.; Mughrabi, H.: A Model of Extrusions and Intrusions in Fatigued Metals: II. Surface Roughening by Random Irreversible Slip, *Philosophical Magazine A*, **54** (1986) 237
217 Essmann, U.; Gösele, U.; Mughrabi, H.: A Model of Extrusions and Intrusions in Fatigued Metals: I. Point-Defect Produc-

tion and the Growth of Extrusions, *Philosophical Magazine A*, **44** (1981) 405

218 Ma, B.-T.; Laird, C.: Overview of Fatigue Behavior in Copper Single Crystals: II. Population, Size, Distribution and Growth Kinetics of Stage I Cracks for Tests at Constant Strain Amplitude, *Acta Metallurgica*, **37** (1989) 349

219 Ma, B.-T.: Strain Localization and Short Crack Growth Kinetics under Variable Loads: The Physical Basis of Damage Summation in Cyclic Deformation, Dissertation, University of Pennsylvania (1987)

220 Basinski, Z.S.; Pascual, R.; Basinski, S.J.: Low Amplitude Fatigue of Copper Single Crystals: I. The Role of the Surface in Fatigue Failure, *Acta Metallurgica*, **31** (1983) 191

212 Witmer, D.E.; Farrington, G.C.; Laird, C.: Changes in Strain Localization Behavior Induced by Fatigue in Inert Environments, *Acta Metallurgica*, **35** (1987) 1895

222 Figuera, J.C.; Laird, C.: Crack Initiation Mechanisms in Copper Polycrystals Cycled under Constant Strain Amplitudes and in Step Tests, *Materials Science and Engineering*, **60** (1983) 45

223 Kim, W.H.; Laird, C.: Crack Nucleation and Stage I Propagation in High-Strain Fatigue: I. Microscopic and Interferometric Observations, *Acta Metallurgica*, **26** (1978) 777

224 Sauzay, M.; Gilormini, P.: Influence of Surface Effects on Fatigue of Microcracks Nucleation, *Theoretical and Applied Fracture Mechanics*, **38** (2002) 53

225 Kim, W.H.; Laird, C.: Crack Nucleation and Stage I Propagation in High-Strain Fatigue: II. Mechanism, *Acta Metallurgica*, **26** (1978) 789

226 Huang, H.L.; Ho, N.J.: The Study of Fatigue in Polycrystalline Copper under Various Strain Amplitude at Stage I: Crack Initiation and Propagation, *Materials Science and Engineering*, **A293** (2000) 7

227 Krupp, U., Floer, W.; Lei, J.; Hu, Y.M.; Christ, H.-J.; Schick, A.; Fritzen, C.-P.: Mechanisms of Short-Fatigue-Crack Initiation and Propagation in a β-Ti Alloy, *Philosophical Magazine A*, **82** (2002) 3321

228 Bürgel, R.: *Handbuch Hochtemperaturwerkstofftechnik*, Vieweg-Verlag, Braunschweig Wiesbaden (2006)

229 Sims, C.T.; Stoloff, N.S.; Hagel, W.C.: *Superalloys II*, John Wiley, New York (1987)

230 Simmons, G.; Wang, H.: *Single-Crystal Elastic Constants and Calculated Aggregate Properties: A Handbook*, 2nd edition, MIT Press, Cambridge, MA/London (1971)

231 Trubitz, P.; Rehmer, B.; Pusch, G.: Die Ermittlung elastischer Konstanten von Gusseisenwerkstoffen, *Proc. Werkstoffprüfung 2004*, M. Pohl (Ed.), DGM Werkstoffinformationsgesellschaft, Frankfurt (2004)

232 Schick, A.; Fritzen, C.-P.; Floer, W.; Hu, Y.-M.; Krupp, U.; Christ, H.-J.: Stress Concentrations at Grain Boundaries due to Anisotropic Elastic Materials Behavior, *Proc. Damage and Fracture Mechanics VI 2000*, A.P.S. Selvadurai, C.A. Brebbia (Eds.), WIT Press, Southampton (2000) 393

233 Voigt, W.: *Lehrbuch der Kristallphysik*, Teubner-Verlag, Leipzig (1928)

234 Reuss, A.: Berechnung der Fließgrenze von Mischkristallen auf Grund der Plastizitätsbedingungen für Einkristalle, *Zeitschrift für angewandte Mathematik und Mechanik*, **9** (1929) 49

235 Kuhn, H.-A.: Anwendung von Grenzwertkonzepten und Phasenmischungsregeln auf die elastischen Eigenschaften von Superlegierungen zwischen Raumtemperatur und 1200°C, Dissertation, Universität Erlangen-Nürnberg (1987)

236 Paufler, P., Schulze, G.E.R.: *Physikalische Grundlagen mechanischer Festkörpereigenschaften*, Vieweg-Verlag, Braunschweig (1978)

237 Leyens, C.; Peters, M. (Eds.): *Titanium and Titanium Alloys: Fundamentals and Applications*, Wiley-VCH, Weinheim (2003)

238 Kocan, M.; Yazgan-Kokuoz, B.; Rack, H.J.; Wagner, L.: High-Cycle Fatigue Performance of Lightweight Titanium Automotive Suspensions, *Journal of Materials*, **57** (2005) 66

239 Quazi, J.I.; Rack, H.J.: Metastable Titanium Alloys for Orthopedic Applications, *Advanced Engineering Materials*, **7** (2005) 993

240 Gregory, J.: The Influence of Cold Work on Mechanical Properties of High Strength TIMETAL LCB, *Proc. Titanium '95: Science and Technology* (1995) 1288

241 Weiss, I.; Srinivasan, R.; Saqib, M.; Stefansson, N.; Jackson, A.G.; LeClair, S.R.: Bulk Deformation of Ti-6.8Mo-4.5Fe-1.5Al (Timetal LCB) Alloy, *Journal of Materials Engineering and Performance*, **5** (1996) 335

242 Weiss, I.; Srinivasan, R.; Saqib, M.; Stefansson, N.; Jackson, A.; LeClair, S.R.: Cold and Warm Working of LCB Titanium Alloy, *Proc. Advances in the Science and Technology of Titanium Alloy Processing*, I. Weiss, R. Srinivasan, P.J. Bania, D. Eylon, S.L. Semiatin (Eds.), TMS, Warrendale, PA (1997) 241

243 Allen, P.G.; Bania, P.J.; Combres, Y.: TIMETAL®LCB: A Low Cost Beta Alloy for Automotive and Other Industrial Applications, *Proc. 8th World Conference on Titanium*, P.A. Blenkinsop, W.J. Evans, H.M. Flower (Eds.), TMS, Warrendale, PA (1995) 1680

244 Schick, A.: unpublished research, University Siegen, Germany (2001)

245 Ahlberg, L.A.; Buck, O.; Paton, N.E.: Effects of Anisotropic Elastic Properties of BCC Ti Alloys, *Scripta Metallurgica*, **12** (1978) 1051

246 Ankem, S.; Margolin, H.: The Role of Elastic Interaction Stresses on the Onset of Plastic Flow of Oriented Two Ductile Phase Structures, *Metallurgical Transactions A*, **11** (1980) 963

247 Vehoff, H.; Nykyforchyn, A.; Metz, R.: Fatigue Crack Nucleation at Interfaces, *Materials Science and Engineering*, **A387–389** (2004) 546

248 Anagnostou, E.; Brahme, A.; Cornwell, C.; El-Dasher, B.S.; Fridy, J.; Horstemeyer, M.F.; Ingraffea, A.R.; Lee, S.-B.; Noack, R.; Papazian, J.; Rollett, A.D.; Saylor, D.; Weiland, H.: Simulation of Fatigue Crack Initiation and Propagation in Aluminum Alloys Using Realistic Microstructures, *Proc. Int. Conference on Fracture ICF11*, Turin 2005, on CD-ROM

249 Peralta, P.; Llanes, L.; Bassani, J.; Laird, C.: Deformation from Twin-Boundary Stresses and the Role of Texture: Application to Fatigue, *Philosophical Magazine A*, **70** (1994) 219

250 Neumann, P.; Tönnessen, A.: Crack Initiation at Grain Boundaries in F.C.C. Materials, *Proc. Int. Conference on the Strength of Metals and Alloys (ICSMA)*, Tampere, Finland (1987) 743

251 Neumann, P.; Tönnessen, A.: Cyclic Deformation and Crack Initiation, *Proc. Fatigue '87*, Vol. 1, Charlottesville, USA (1987) 3

252 Llanes, L.; Laird, C.: The Role of Twin Boundaries in the Cyclic Deformation of fcc Materials, *Materials Science and Engineering*, **A157** (1992) 21

253 Gemperlova; J.; Paidar, V.; Kroupa, F.: Compatibility Stresses in Deformed Bicrystals, *Czechoslovakian Journal of Physics B*, **39** (1989) 427

254 Hu, Y.M.; Floer,.W.; Krupp, U.; Christ, H.-J.: Microstructurally Short Fatigue Crack Initiation and Growth in Ti-6.8Mo-4.5Fe-1.5Al, *Materials Science and Engineering A*, **278** (2000) 180

255 Köster, P.: Kristallplastische Simulation des lokalen Verformungsverhaltens eines Duplex-Stahls mit Hilfe der Finite-Elemente Methode, diploma thesis, University, Siegen Germany (2006)

256 Hu, Y.-M.: unpublished research, University Siegen, Germany (2000)

257 Blochwitz, C.; Tirschler, W.: Influence on Twin Boundary Cracks in Fatigued Austenitic Steels, *Materials Science and Engineering A*, **339** (2003) 318

258 Boettner, R.C.; McEvily, A.J.; Liu, Y.C.: On the Formation of Fatigue Cracks at Twin Boundaries, *Philosophical Magazine*, **10** (1964) 95

259 Muskelishivili, N.-J.: *Some Basic Problems of Mathematical Theory of Elasticity*, Noordhoff, Leyden (1975)

260 Blochwitz, C.; Richter, R.; Tirschler, W.; Obrtlik, K.: The Effect of Local Textures on Microcrack Propagation in Fatigued fcc Metals, *Materials Science and Engineering*, **A234–236** (1997) 563

261 Peralta, P.; Laird, C.: Compatibility Stresses in Fatigued Bicrystals: Dependence on Misorientation and Small Plastic Deformations, *Acta Materialia*, **45** (1997) 5129

262 Zhang, Z.F.; Wang, Z.G.: Comparison of Fatigue Cracking Possibility Along

Large- and Low- Angle Grain Boundaries, *Materials Science and Engineering*, **A284** (2000) 285

263 Zhang, Z.F.; Wang, Z.G.: Fatigue-Cracking Characteristics of a Copper Bicrystal when Slip Bands Transfer through the Grain Boundary, *Materials Science and Engineering*, **A343** (2003) 308

264 Zhang, Z.F.; Wang, Z.F.; Eckert J.: What Types of Grain Boundaries can be Passed through by Persistent Slip Bands, *Journal of Materials Research*, **18** (2003) 1031

265 Hook, R.E.; Hirth, J.P.: On the Deformation Behavior of Non-Isoaxial Bicrystals of Fe-3%Si, *Acta Metallurgica*, **15** (1967) 1099

266 Zhai, T.; Wilkinson, A.J.; Martin, J.W.: A Crystallographic Mechanism for Fatigue Crack Propagation Through Grain Boundaries, *Acta Materialia*, **48** (2000) 4917

267 Zhai, T.; Jiang, X.P.; Li, J.X.; Garratt, M.D.; Bray, G.H.: The Grain Boundary Geometry for Optimum Resistance to Growth of Short Fatigue Cracks in High Strength Al Alloys, *International Journal of Fatigue*, **27** (2005) 1202

268 Argon, A.; Qiao, Y.: Cleavage Cracking Resistance of Large-Angle Grain Boundaries in Fe-3wt%Si Alloy, *Philosophical Magazine A*, **82** (2002) 3333

269 Gysler, A.; Terlinde, G.; Lütjering, G.: Influence of Grain Size on the Ductility of Age-Hardened Titanium Alloys, *Proc. Titanium and Titanium Alloys: Scientific and Technological Aspects*, Vol. 3, J.C. Williams, A.F. Belov (Eds.), Plenum, New York (1982) 1919

270 Bowen, A.W.: Omega Phase Formation in Metastable β-Titanium Alloys, in: *Beta Titanium Alloys in the 80s*, Metallurgical Society of AIME, Warrendale, PA (1984) 85

271 Krupp, U.; Orosz, R.; Christ, H.-J., Buschmann, U.; Wiechert, W.: Internal Nitridation during Creep Loading of Polycrystalline Ni-Base Superalloys, *Materials Science Forum*, **461–464** (2004) 37

272 Krupp, U.; Chang, S.Y.; Christ, H.-J.: Modelling Internal Corrosion Processes as a Consequence of Oxide Scale Failure, *Proc. EFC-Workshop Life Time Modelling of High-Temperature Corrosion Processes*, M. Schütze, W.J. Quadakkers, J.R. Nichols (Eds.), ECF Publications No. 34 (2001) 148

273 Kofstad, P.: *High Temperature Corrosion*, Elsevier, London (1988)

274 Fisher, J.C.: Calculation of Diffusion Penetration Curves for Surface and Grain Boundary Diffusion, *Journal of Applied Physics*, **22** (1951) 74

275 Mishin, Y.; Herzig, C.: Grain Boundary Diffusion: Recent Progress and Future Research, *Materials Science and Engineering*, **A260** (1999) 55

276 Min, B.K.; Raj, R.: A Mechanism for Intergranular Fracture during High Temperature Fatigue, ASTM STP 675 (1979) 569

277 Fujita, F.E.: Dislocation Theory of Fracture of Crystals, *Acta Metallurgica*, **6** (1958) 543

278 Varvani, A.; Topper, T.H.: Crack Growth and Closure Mechanisms of Shear Cracks under Constant Amplitude Biaxial Straining and Periodic Compressive Overstraining in 1945 Steel, *International Journal of Fatigue*, **7** (1997) 589

279 Mott, N.F.: A Theory of the Origin of Fatigue Cracks, *Acta Metallurgica*, **6** (1958) 195

280 Olson, G.B.; Cohen, M.: Kinetics of Strain-Induced Martensite Nucleation, *Metallurgical Transactions A*, **6** (1975) 791.

281 Spencer, K.; Embury, J.D.; Conlon, K.T.; Véron, M.; Bréchet, Y.: Strengthening Via the Formation of Strain-Induced Martensite in Stainless Steels, *Materials Science and Engineering A*, **387–389** (2004) 873

282 Long, M.; Crooks, R.; Rack, H.J.: High Cycle Fatigue Performance of Solution-Treated Metastable-β Titanium Alloys, *Acta Materialia*, **47** (1999) 661

283 Miller, K.J.; Mohamed, H.J.; de los Rios, E.R.: Fatigue Damage Accumulation Above and Below the Fatigue Limit, *Proc. The Behaviour of Short Fatigue Cracks*, K.J. Miller, E.R. de los Rios (Eds.), ESIS, London (1986) 491

284 Radhakrishnan, V.M.; Mutoh, Y.: On Fatigue Crack Growth in Stage I, *Proc. The Behaviour of Short Fatigue Cracks*,

K.J. Miller, E.R. de los Rios (Eds.), ESIS, London (1986) 87

285 Bathias, C.: Designing Components Against Gigacycle Fatigue, *Proc. Fatigue in the Very High Cycle Regime*, Stanzl-Tschegg, S.; Mayer, H. (Eds.), Vienna (2001) 97

286 Mughrabi, H.: Fatigue Mechanisms in the Ultrahigh Cycle Regime, *Proc. 9th International Fatigue Congress*, Atlanta, GA, 2006, CD-ROM

287 Miller, K.J.; O'Donnell, W.J.: The Fatigue Limit and its Elimination, *Fatigue and Fracture of Engineering Materials and Structures*, **22** (1999) 545

288 Miller, K.J.: The Two Thresholds of Fatigue Behaviour, *Fatigue and Fracture of Engineering Materials and Structures*, **16** (1993) 931

289 Bathias, C.: There Is No Infinite Fatigue Life in Metallic Materials, *Fatigue and Fracture of Engineering Materials and Structures*, **22** (1999) 559

290 Laird, C.: Strain Localization in Cyclic Deformation and How Environments Affect it, *Proc. Fatigue 2002*, A.F. Blohm (Ed.), Stockholm (2002) 1

291 Degischer, H.-P.; Kriszt, B.(Eds.): *Handbook of Cellular Metals*, Wiley-VCH, Weinheim (2002)

292 Wagner, I.; Hintz, C.; Sahm, P.R.: Precision Casting Near Net Shape Components Based on Cellular Metal Material, *Proc. Euromat 99, Vol. 5: Metal Matrix and Metallic Foams*, T.W. Clyne, F. Simancík (Eds.), Wiley-VCH, Weinheim (2000) 40

293 Stanzick, H.; Duarte, I.; Banhart, J.: Der Schäumprozeß von Aluminium, *Materialwissenschaft und Werkstofftechnik*, **31** (2000) 409

294 Asholt, P.: Manufacturing of Aluminium Foams from PMMC Melts: Material Characteristics and Typical Properties, *Proc. Metallschäume*, J. Banhart (Ed.), Verlag MIT, Bremen (1997) 27

295 Demirai, S.; Hohe, J.; Becker, W.: Strain-Energy Based Homogenization of 2-Dimensional and 3-Dimensional Hyperelastic Solid Foams, *Journal of Materials Science*, **40** (2005) 5839

296 Ohrndorf, A.; Krupp, U.; Christ, H.-J.: Fatigue Behaviour of Open and Closed Cell Al Foams, *Proc. MetFoam 2001*, J. Bahnhart, M.F. Ashby, N.A. Fleck (Eds.), MIT-Verlag, Bremen (2001) 311

297 Krupp, U.; Ohrndorf, A.; Guillén, T.; Christ, H.-J.; Demiray, S.; Hohe, J.; Becker, W.: Isothermal and Thermo-mechanical Fatigue of Open-Cell Metal Sponges: Experimental Analysis and Homogenization Modeling, *Advanced Engineering Materials* **8** (2006) 821

298 Onck, P.R.; van Merkerk, R.; Raaijmakers, A.; de Hosson, J.Th.M.: Fracture of Open and Closed-Cell Metal Foams, *Journal of Materials Science*, **40** (2005) 5821

299 Ohrndorf, A.; Krupp, U.; Christ, H.-J.: Charakterisierung der mechanischen Eigenschaften geschlossenporiger Aluminiumschäume, *Materialprüfung*, **3** (2002) 78

300 Stolarz, J.: Multicracking in Low Cycle Fatigue: A Surface Phenomenon?, *Materials Science and Engineering*, **A234–236** (1997) 861

301 Suresh, S.; Ritchie, R.O.: Propagation of Short Fatigue Cracks, *International Metals Reviews*, **29** (1984) 445

302 Hussain, K.: Short Fatigue Crack Behaviour and Analytical Models. A Review, *Engineering Fracture Mechanics*, **58** (1997) 327

303 Tokaji, K.; Ogawa, T.: The Growth Behaviour of Microstructurally Small Fatigue Cracks in Metals, *Proc. Short Fatigue Cracks*, ESIS 13, K.J. Miller, E.R. de los Rios (Eds.), Mechanical Engineering Publications, London (1992) 85

304 Taylor, D.; Knott, J.F.: Fatigue Crack Propagation Behaviour of Short Cracks: The Effect of Microstructure, *Fatigue and Fracture of Engineering Materials and Structures*, **4** (1981) 147

305 Rodopoulus, C.A.; de los Rios, E.R.: Theoretical Analysis on the Behaviour of Short Fatigue Cracks, *International Journal of Fatigue*, **24** (2002) 719

306 Okazaki, M.; Yamada, H.; Nohmi, S.: Temperature Dependence of the Intrinsic Small Fatigue Crack Growth Behaviour in Ni-Base Superalloys Based on the Measurement of Crack Closure, *Proc. Fatigue under Thermal and Mechanical Loading*, J. Bressers, L. Remy (Eds.) (1996) 119

307 Newman Jr., J.C.: FASTRAN II: A Fatigue Crack Growth Structural Analysis Program, NASA TM 104159 (1992) 1

308 Miller, K.J.; de los Rios E.R. (Eds.): *Proc. The Behaviour of Short Fatigue Cracks*, Mechanical Engineering Publications, London (1986)

309 Miller, K.J. (Eds.): *Proc. Short Fatigue Cracks*, in: *Fatigue and Fracture of Engineering Materials and Structures* (special issue) **14** (1991) 143

310 Miller, K.J.; de los Rios E.R. (Eds.): *Proc. Short Fatigue Cracks*, ESIS Publication 13, London (1992)

311 Ravichandran, K.S.; Ritchie, R.O.; Murakami, Y. (Eds.): *Proc. Small Fatigue Cracks: Mechanics, Mechanisms and Applications*, Elsevier (1999)

312 Hudak, S.J.: Small Crack Behavior and the Prediction of Fatigue Life, *Journal of Engineering Materials Technologies*, **103** (1981) 26

313 Ritchie, R.O.; Peters, J.O.: Small Fatigue Cracks: Mechanics, Mechanisms and Engineering Applications, *Materials Transactions*, **42** (2001) 58

314 Miller, K.J.: The Short Crack Problem, *Fatigue of Engineering Materials*, **5** (1982) 223

315 McEvily, A.: The Growth of Short Fatigue Cracks: A Review, *Materials Science Research International*, **4** (1998) 3

316 Dubey, S.; Soboyejo, A.B.O.; Soboyejo, W.O.: An Investigation of Stress Ratio and Crack Closure on the Mechanism of Fatigue Crack Growth in Ti-6Al-4V, *Acta Materialia*, **45** (1997) 2777

317 Düber, O.: Ausbreitungsverhalten kurzer Ermüdungsrisse in einem austenitisch-ferritischen Duplexstahl, Dissertation, Universität Siegen (2006)

318 Düber, O.; Künkler, B.; Krupp, U.; Christ, H.-J.; Fritzen, C.-P.: Short Crack Propagation in Duplex Stainless Steel, *Proc. Fatigue 2006*, Atlanta, GA (on CD ROM)

319 Ravichandran, K.S.; Li, X.D.: Fracture Mechanical Character of Small Cracks in Polycrystalline Materials: Concept and Numerical K Calculations, *Acta Materialia*, **48** (2000) 525

320 Newman Jr., J.C.; Raju, I.S.: Stress Intensity Factor for Cracks in Three-Dimensional Finite Bodies, *Proc. Fracture Mechanics, 14th Symp.*, Vol. 1, J.C. Lewis, G. Sines (Eds.), ASTM STP791 (1983) 238

321 Cottrell, A.H.: *Theory of Crystal Dislocations*, Gordon and Breach, New York/London (1964)

322 Liang, F.L.; Laird, C.: Control of Intergranular Fatigue Cracking by Slip Homogeneity in Copper: I. Effect of Grain Size, *Materials Science and Engineering*, **A117** (1989) 95

323 Blom, A.F.; Hedlund, A.; Zhao, W.; Fathulla, A.; Weiss, B.; Stickler, R.: Short Fatigue Crack Growth in Al2024 and Al 7475, *Proc. The Behaviour of Short Fatigue Cracks*, Miller, K.J.; de los Rios, E.R. (Eds.), Mechanical Engineering Publications, London (1986) 37

324 Navarro, A.; de Los Rios, E.R.: Short and Long Fatigue Crack Growth: A Unified Model, *Philosophical Magazine A*, **57** (1988) 15

325 Zaefferer, S.; Kuo, J.-C.; Zhao, Z.; Winning, M.; Raabe, D.: On the Influence of the Grain Boundary Misorientation on the Plastic Deformation of Aluminum Bicrystals, *Acta Materialia*, **51** (2003) 4719

326 Ruppen, J.; Bhowal, P.; Eylon, D.; McEvily, A.J.: On the Process of Subsurface Fatigue Crack Initiation in Ti-6Al-4V, *Proc. Fatigue Mechanisms*, ASTM STP 675 (1979) 47

327 Blochwitz, C.; Richter, R.: Plastic Strain Amplitude Dependent Surface Path of Microstructurally Short Fatigue Cracks in Face-Centred Cubic Metals, *Materials Science and Engineering*, **A267** (1999) 120

328 Turnbull, A.; de los Rios, E.R.: The Effect of Grain Size on Fatigue Crack Growth in an Aluminium Magnesium Alloy, *Fatigue and Fracture of Engineering Materials and Structures*, **18** (1995) 1355

329 Zhang, Y.H.; Edwards, L.: On the Blocking Effect of Grain Boundaries on Small Crystallographic Fatigue Crack Growth, *Materials Science and Engineering*, **A188** (1994) 121

330 Wilkinson, A.J.; Roberts, S.G.; Hirsch, P.B.: Modelling of the Threshold Conditions for Propagation of Stage I Fatigue Cracks, *Acta Materialia*, **46** (1998) 379

331 Weertman, J.: Crack Tip Stress Intensity Factor of the Double Slip Plane Crack Model: Short Cracks and Short Short Cracks, *International Journal of Fracture*, **26** (1984) 31
332 Caracostas, C.A.; Shodja, H.M.; Weertman, J.: The Double Slip Plane Model for the Study of Short Cracks, *Mechanics of Materials*, **20** (1995) 195
333 Laird, C.: The Influence of Metallurgical Structures on Fatigue Crack Propagation – A Review, ASTM STP415 (1967) 131
334 Düber, O.; Künkler, B.; Krupp, U.; Christ, H.-J.; Fritzen, C.-P.: Experimental Characterization and Two-Dimensional Simulation of Short-Crack Propagation in an Austenitic-Ferritic Duplex Steel, *International Journal of Fatigue* **28** (2006) 983
335 Forsyth, P.J.E: A Two Stage Process of Fatigue Crack Growth, in: *Proc. Crack Propagation, Cranfield-Symposium*, Her Majesty's Stationery Office, London (1962) 76
336 McEvily, A.J.; Cai, H.: On the Formation of Striations during Fatigue Crack Growth, in: *Proc. International Conference on Advanced Technology in Experimental Mechanics 2003*, Nagoya, Japan (2003) (on CD-ROM)
337 Cai, H.; McEvily, A.J.: On Striations and Fatigue Crack Growth in 1018 Steel, *Materials Science and Engineering A*, **314** (2001) 86
338 Laird, C.; Smith, G.C.: Crack Propagation in High Stress Fatigue, *Philosophical Magazine*, **77** (1962) 847
339 Neumann, P.: New Experiments Concerning the Slip Processes at Propagating Fatigue Cracks: I, *Acta Metallurgica*, **22** (1974) 1155
340 Pippan, R.: Threshold and Effective Threshold of Fatigue Crack Propagation in ARMCO Iron: I. The Influence of Grain Size and Cold Working, *Materials Science and Engineering*, **138** (1991) 1
341 Briggs, R.D.; Taggart, R.; Polonis, D.H.: Crack Growth in the Ti-15-3 Beta Titanum Alloy, *Proc. Effect of Microstructure on Fracture Toughness and Fatigue Crack Growth in Titanium Alloys*, A.K. Chakrabarti, J.S. Chesnutt (Eds.) Denver (1987) 65
342 Pippan, R.; Flechsig, K.; Riemelmoser, F.O.: Fatigue Crack Propagation Behavior in the Vicinity of an Interface Between Materials with Different Yield Stresses, *Materials Science and Engineering*, **A283** (2000) 225
343 Pippan, R.; Riemelmoser, F.O.: Fatigue of Bimaterials. Investigation of the Plastic Mismatch in Case of Cracks Perpendicular to the Interface, *Computational Materials Science*, **13** (1998) 108
344 Krupp, U.; Düber, O.; Christ, H.-J.; Künkler, B.; Schick, A.; Fritzen, C.-P.: Application of the EBSD Technique to Describe the Initiation and Growth Behaviour of Microstructurally Short Fatigue Cracks in a Duplex Steel, *Journal of Microscopy*, **213** (2004) 313
345 Stolarz, J.; Focht, J.: Specific Features of Two Phase Alloy Response to Cyclic Deformation, *Materials Science and Engineering A*, **319–321** (2001) 501
346 Stolarz, J.; Baffle, N.: Microstructural Barriers to the Short Crack Propagation in Two Phase Materials and Low Cycle Fatigue Behaviour of Duplex Steels, *Proc. Fatigue 2002*, A.F. Blohm (Ed.), Stockholm (2002) 2831
347 Johansson, J.; Odén, M.: Load Sharing between Austenite and Ferrite in a Duplex Stainless Steel during Cyclic Loading, *Metallurgical and Materials Transactions*, **31A** (1999) 1557
348 Stolarz, J.: Influence of Microstructure on Low-Cycle Fatigue in Some Single-Phase and Biphasic Stainless Steels, *Proc. Intl. Conf. on Low Cycle Fatigue LCF5*, P.D. Portella, H. Sehitoglu, K. Hatanaka (Eds.), Berlin (2004) 3
349 Marrow, T.J.; King, J.E.: Fatigue Crack Propagation Mechanisms in Thermally Aged Duplex Stainless Steels, *Materials Science and Engineering A*, **183** (1994) 91
350 Llanes, L.; Mateo, A.; Violan, P.; Méndez, J.; Anglada, M.: On the High-Cycle Fatigue Behavior of Duplex Stainless Steels: Influence of Thermal Aging, *Materials Science and Engineering*, **A234–236** (1997) 850
351 Fissolo, A.; LeRoux, J.C.; Mottot, M.; Maillot, V.; Jayet-Gendrot, S.; Degallaix, S.; Francois, D.: Fatigue Behaviour of Aged Austeno-Ferritic Steels, *Proc. 13th European Conference on Fracture*, Elsevier, Amsterdam (2000) (on CD-ROM)

352 Hornbogen, E.; Zum Gahr, K.-H.: Microstructure and Fatigue Crack Growth in a γ-Fe-Ni-Al Alloy, *Acta Metallurgica*, **24** (1976) 581

353 Elber, W.: Fatigue Crack Closure under Cyclic Tension, *Engineering Fracture Mechanics*, **2** (1979) 37

354 Newman, J.C.; Elber, W. (Eds.): Mechanics of Fatigue Crack Closure, AST STP 982 (1988)

355 Kaynak, C.; Ankara, A.; Baker, T.J.: A Comparison of Short and Long Fatigue Crack Growth in Steel, *International Journal of Fatigue*, **18** (1996) 17

356 Ritchie, R.O.; Suresh, S.: Mechanics and Physics of the Growth of Small Cracks, *Proc. AGARD 55th SMP Meeting*, Toronto (1982) 1.1

357 Schijve, J.: Fatigue Crack Closure: Observations and Technical Significance, *Proc. Mechanics of Fatigue Crack Closure*, J.C. Newman Jr., W. Elber (Eds.), ASTM STP 982 (1988) 5

358 Liaw, P.K.: Overview of Crack Closure at near-Threshold Fatigue Crack Growth Levels, *Proc. Mechanics of Fatigue Crack Closure*, J.C. Newman Jr., W. Elber (Eds.), ASTM STP 982 (1988) 62

359 Fleck, N.A.: Fatigue Crack Growth: The Complications, *Proc. Fatigue Crack Growth*, R.A. Smith (Ed.), Pergamon (1986) 75

360 Shin, C.S.; Cai, C.Q.: A Model for Evaluating the Effect of Fatigue Crack Repair by the Infiltration Method, *Fatigue and Fracture of Engineering Materials and Structures*, **23** (2000) 835

361 Vasudevan, A.K.; Sadananda, K.; Louat, N.: A Review of Crack Closure, Fatigue Crack Threshold and Related Phenomena, *Materials Science and Engineering*, **A188** (1994) 1

362 Pippan, R.; Riemelmoser, F.O.: Visualization of the Plasticity-Induced Crack Closure under Plane Strain Conditions, *Engineering Fracture Mechanics*, **3** (1998) 315

363 Pippan, R.; Riemelmoser, F.O.; Weinhandl, H.; Kreuzer, H.: Plasticity Induced Crack Closure under Plane Strain Condition in the Near Threshold Regime, *Philosophical Magazine* **82** (2002) 3299

364 Pippan, R.; Riemelmoser, F.O.; Motz, Ch: Plasticity-Induced Crack Closure Under Plane Strain Condition: A Simple Explanation, *Proc. Fatigue '99*, Vol. 1, X.-R. Wu and Z.-G. Wang (Eds.), Peking (1999) 521

365 Riemelmoser, F.O.; Pippan, R.: Investigation of a Growing Fatigue Crack by Means of a Discrete Dislocation Model, *Materials Science and Engineering*, **A234–236** (1997) 135

366 Pokluda, J.; Sandera, P.; Pippan, R.: Analysis of Crack Closure Level in Terms of Crack Wake Plasticity, *Proc. 9th International Fatigue Congress*, Atlanta, GA (2006) CD ROM

367 Gall, K.; Sehitoglu, H.; Kadioglu, Y.: FEM Study of Fatigue Crack Closure under Double Slip, *Acta Materialia*, **44** (1996) 3955

368 Pippan, R.; Kolednik, O.; Lang, M.: A Mechanism for Plasticity-Induced Crack Closure under Plane Strain Conditions, *Fatigue and Fracture of Engineering Materials and Structures*, **17** (1994) 721

369 Nicholas, T.; Palazotto, A.N.; Bednarz, E.: An Analytical Investigation of Plasticity Induced Closure Involving Short Cracks, *Proc. Mechanics of Fatigue Crack Closure*, J.C. Newman Jr., W. Elber (Eds.), ASTM STP 982 (1988) 361

370 Newman Jr., J.C.: A Finite Element Analysis of Fatigue Crack Closure, in: *Proc. Mechanics of Crack Growth*, ASTM STP 590 (1976) 280

371 Newman, J.C.; Wu, X.R.; Swain, M.H.; Zhao, W.; Phillips, E.P.; Ding, C.F.: Small Crack Growth and Fatigue Life Predictions for High-Strength Aluminium Alloys: II. Crack Closure and Fatigue Analyses, *Fatigue and Fracture of Engineering Materials and Structures*, **23** (2000) 59

372 Daniewicz, S.R.; Bloom, J.M.: An Assessment of Geometry Effects on Plane Stress Fatigue Crack Closure Using a Modified Strip Yield Model, *International Journal of Fatigue*, **18** (1996) 483

373 Xu, Y.; Gregson, P.J.; Sinclair, I.: Systematic Assessment and Validation of Compliance-Based Crack Closure Measurements in Fatigue, *Materials Science and Engineering*, **A284** (2000) 114

374 Ritchie, R.O.; Zaiken, E.; Blom, A.F.: Is the Concept of Fatigue Threshold Meaningful in the Presence of Com-

pression Cycles?, *Proc. Workshop on Fundamental Questions and Critical Experiments on Fatigue*, J. Fong, R.J. Fields (Eds.), ASTM STP 924 (1988) 337

375 Liaw, P.K.; Leax, T.R.; Donald, J.K.: Fatigue Crack Growth Behavior of 4340 Steels, *Acta Metallurgica*, **35** (1987) 1415

376 Lang, M.; Marci, G.: The Influence of Single and Multiple Overloads on Fatigue Crack Propagation, *Fatigue and Fracture of Engineering Materials and Structures*, **22** (1999) 257

377 Ellyin, F.; Wu, J.: A Numerical Investigation on the Effect of an Overload on Fatigue Crack Opening and Closure Behaviour, *Fatigue and Fracture of Engineering Materials and Structures*, **22** (1999) 835

378 Lang, M.: A Model for Fatigue Crack Growth, Part I: Phenomenology, *Fatigue and Fracture of Engineering Materials and Structures*, **23** (2000) 587

379 Lang, M.: A Model for Fatigue Crack Growth, Part II: Modelling, *Fatigue and Fracture of Engineering Materials and Structures*, **23** (2000) 603

380 Petit, J.; Zeghloul, A.: On the Effect of Environment on Short Crack Growth Behaviour and Threshold, *Proc. The Behaviour of Short Fatigue Cracks*, K.J. Miller, E.R. de los Rios (Eds.), ESIS, London (1986) 163

381 Wang, S.-H.; Müller, C.: Fracture Surface Roughness and Roughness-Induced Fatigue Crack Closure in Ti-2.5 wt% Cu, *Materials Science and Engineering*, **A255** (1998) 7

382 Haberz, K.; Pippan, R.; Stüwe, H.P.: The Threshold of Stress Intensity Range of Iron, *Proc. Fatigue '93*, J.-P. Bailon, Dickson, J.I..Dickson (Eds.), Montreal (1993) 525

383 Jung, H.Y.; Antolovich, S.D.: Fatigue Crack Closure as a Function of Crack Length in Al-Li Alloys, *Engineering Fracture Mechanics*, **3** (1996) 307

384 Suresh, S.: Fatigue Crack Deflection and Fracture Surface Contact: Micromechanical Models, *Metallurgical Transactions A*, **16A** (1985) 249

385 Suresh, S.; Ritchie, R.O.: A Geometrical Model for Fatigue Crack Closure Induced by Fracture Surface Roughness, *Metallurgical Transactions A*, **13A** (1982) 1627

386 Pokluda, J.; Sandera, P.; Hornikova, J.: Statistical Model of Roughness-Induced Crack Closure, *Proc. 13th European Conference on Fracture*, San Sebastian, Spain (2000) CD-ROM

387 Petit, J; Mendez, J.; Berata, W.; Legendre, L.; Müller, C.: Influence of Environment on the Propagation of Short Fatigue Cracks in a Titanium Alloy, *Proc. Short Fatigue Cracks*, ESIS 13, K.J. Miller, E.R. de los Rios (Eds.), Mechanical Engineering Publications, London (1992) 235

388 Ma, L.; Chang, K.-M.; Mannan, S.K.: Oxide-Induced Crack Closure: An Explanation for Abnormal Time-Dependent Fatigue Crack Propagation Behavior in INCONEL Alloy 783, *Scripta Materialia*, **48** (2003) 583

389 Liaw, P.K.; Leax, T.R.; Williams, R.S.; Peck, M.G.: Influence of Oxide-Induced Closure on Near Threshold Fatigue Crack Growth, *Acta Metallurgica*, **30** (1982) 2071

390 Bleck, W.; Papaefthymiou, S.; Frehn, A.: Microstructure and Tensile Pproperties in Dual phase and TRIP steels, *Steel Research*, **75** (2004) 704

391 Hornbogen, E.: Martensitic Transformation at a Propagating Crack, *Acta Metallurgica*, **26** (1978) 147

392 Mayer, H.R.; Stanzl-Tschegg, S.E.; Sawaki, Y.; Huhner, M.; Hornbogen, E.: Influence of Transformation-Induced Crack Closure on Slow Fatigue Crack Growth Under Variable Amplitude Loading, *Fatigue and Fracture of Engineering Materials and Structures*, **18** (1995) 935

393 Mei, Z.; Morris Jr., J.W.: Analysis of Transformation-Induced Crack Closure, *Engineering Fracture Mechanics*, **39** (1991) 569

394 Vasudevan, A.K.; Sadananda, K.: Application of Unified Fatigue Damage Approach to Compression-Tension Region, *International Journal of Fatigue*, **21** (1999) 263

395 Sadananda, K.; Vasudevan, A.K.; Holtz, R.L.; Lee, E.U.: Analysis of Overload Effects and Related Phenomena, *International Journal of Fatigue*, **21** (1999) 233

396 Krenn, C.R.; Morris Jr., J.W.: The Compatibility of Crack Closure and K_{max} Dependent Models of Fatigue Crack

Growth, *International Journal of Fatigue*, **21** (1999) 147

397 Anthes, R.J.: Ein neuartiges Kurzriß-fortschrittsmodell zur Anrißlebens-dauervorhersage bei wiederholter Beanspruchung, Veröffentlichung des Instituts für Stahlbau und Werkstoff-mechanik der TH Darmstadt, Heft 57, Darmstadt (1997)

398 Larsen, J.M.; Williams, J.C.; Thompson, A.W.: Crack-Closure Effects on the Growth of Small Surface Cracks in Titanium-Aluminum Alloys, *Proc. Mechanics of Fatigue Crack Closure*, J.C. Newman Jr., W. Elber (Eds.), ASTM STP 982 (1988) 149

399 James, M.N.; Sharpe Jr., W.N.: Closure Development and Crack Opening in the Short Crack Regime for Fine and Coarse Grained A533B Steel, *Fatigue and Fracture of Engineering Materials and Structures*, **12** (1989) 347

400 James, M.N.; Graz, R.E.: Relating Closure Development in Long Cracks to the Short Crack Regime, *International Journal of Fatigue*, **13** (1991) 169

401 Jira, J.R.; Nicholas, T.; Larsen, J.M.: Crack Closure Development and its Relation to the Small-Crack Effect in Titanium Alloys, *Proc. 4th Int. Conf. on Fatigue and Fatigue Thresholds, Fatigue '90*, Vol. 2, Honolulu (1990) 1295

402 Ritchie, R.O.; Suresh, S.: Some Considerations on Fatigue Crack Closure at Near Threshold Stress Intensities due to Fracture Surface Morphology, *Metallurgical Transactions A*, **13A** (1982) 937

403 Dowson, A.L.; Halliday, M.D.; Beevers, C.J.: In-Situ SEM Studies of Short Crack Growth and Crack Closure in a Near-Alpha Ti Alloy, *Materials and Design*, **14** (1993) 57

404 Morris, W.L.; Buck, O.: Crack Closure Load Measurements for Microcracks Developed During the Fatigue of Al2219-T851, *Metallurgical Transactions A*, **8** (1977) 597

405 Blochwitz, C., Tirschler, W.; Weidner, A.: Crack Opening Displacement and Propagation of Microstructurally Short Cracks, *Materials Science and Engineering A*, **357** (2003) 264

406 McDowell, D.L.: An Engineering Model for Propagation of Small Cracks in Fatigue, *Engineering Fracture Mechanics*, **56** (1997) 357

407 Nisitani, H.; Takao, K.-I.: Significance of Initiation, Propagation and Closure of Microcracks in High Cycle Fatigue of Ductile Metals, *Engineering Fracture Mechanics*, **3–4** (1981) 445

408 Tanaka, K.; Nakai, Y.: Propagation and Non-Propagation of Short Cracks at a Sharp Notch, *Fatigue and Fracture of Engineering Materials and Structures*, **6** (1983) 315

409 Lee, S.-Y.; Song, J.-H.: Crack Closure and Growth of Physically Short Fatigue Crack under Random Loading, *Engineering Fracture Mechanics*, **66** (2000) 321

410 Cheng, A.; Laird, C.: The Transition from Stage I to Stage II Fatigue Crack Propagation in Copper Single Crystals Cycled at Constant Strain Amplitudes, *Materials Science and Engineering*, **60** (1983) 177

411 Yoder, C.A.; Cooley, L.A.; Crooker, T.W.: On Microstructural Control of Near Threshold Fatigue Crack Growth in 7000-Series Aluminium Alloys, *Scripta Materialia*, **16** (1982) 1021

412 Carlson, R.L.: An Examination of Scatter in Small Fatigue Crack Growth, *International Journal of Fracture*, **115** (2002) 55

413 Stolarz, J.; Kurzydlowski, K.J.: Stereological Analysis of Fatigue Short Crack Propagation in Zircaloy-4, Proc. *Low Cycle Fatigue and Elasto Plastic Behaviour of Materials LCF4*, K.T. Rie, P.D. Portella (Eds.), Garmisch-Partenkirchen (1998)

414 Ravichandran, K.S.; Larsen, J.M.: An Approach to Measure the Shapes of Three-Dimensional Surface Cracks During Fatigue Crack Growth, *Fatigue and Fracture of Engineering Materials and Structures*, **16** (1993) 909

415 Fett, T.: The Crack Opening Displacement Field of Semi-Elliptical Surface Cracks in Tension for Weight Functions Applications, *International Journal of Fracture*, **36** (1988) 55

416 Ravichandran, K.S.; Larsen, J.M.; Li, X.-D.: Significance of Crack Shape or Aspect Ratio to the Behavior of Small Fatigue Cracks in Titanium Alloys, in: *Proc. Small Fatigue Cracks: Mechanics, Mechanisms and Applications*, K.S. Ravichandran, R.O. Ritchie, Y. Murakami (Eds.), Elsevier (1999) 95

417 Ravichandran, K.S.; Li X.-D.: Fracture Mechanics Character of Small Cracks in Polycrystalline Materials: Concept and Numerical *K* Calculations, *Acta Materialia*, **48** (2000) 525

418 Bayley, C.J.; Bell, R.: Parametric Investigation into the Coalescence of Coplanar Fatigue Cracks, *International Journal of Fatigue*, **21** (1999) 355

419 Meyer, S.; Diegele, E.; Brückner-Foit, A.; Möslang, A.: Crack Interaction Modelling, *Fatigue and Fracture of Engineering Materials and Structures*, **23** (2000) 315

420 Ochi, Y.; Ishii, A.; Sasaki, S.: An Experimental and Statistical Investigation of Surface Fatigue Crack Initiation and Growth, *Fatigue and Fracture of Engineering Materials and Structures*, **8** (1985) 327

421 Künkler, B.: Mechanismenorientierte Modellierung der Initiierung und Ausbreitung kurzer Ermüdungsrisse, Dissertation, Universität Siegen (2006)

422 Sims, C.T.; Stoloff, N.S.; Hagel, W.C. (Eds.): *Superalloys II*, Wiley, New York (1987)

423 Birks, N.; Meier, G.H.; Pettit, F.S.: *Introduction to the High-Temperature Oxidation of Metals*, Cambridge University Press (2006)

424 Krupp, U.; Wagenhuber, Ph. E.-G.; Kane, W. M.; McMahon Jr., C.J.: Improving the Resistance to Dynamic Embrittlement and Intergranular Oxidation of Ni-Based Superalloys by Grain-Boundary Engineering-Type Processing, *Material Science and Technology*, **21** (2005) 1247

425 Chang, K.-M.; Henry, M.F.; Benz, M.G.: Metallurgical Control of Fatigue Crack Propagation in Superalloys, *Journal of Materials*, **12** (1990) 29

426 Floreen, S.; Kane, R.H.: Effects of Environment on High Temperature Fatigue Crack Growth in a Superalloy, *Metallurgical Transactions A*, **10** (1979) 1745

427 Andersson, H.; Persson, C.; Hansson, T.: Crack Growth in IN718 at High Temperature, *International Journal of Fatigue*, **23** (2001) 817

428 Wanhill, R.J.H.: Significance of Dwell Cracking for IN718 Turbine Discs, *International Journal of Fatigue*, **24** (2002) 545

429 Goswami, T.: Low Cycle fatigue: Dwell Effects and Damage Mechanisms, *International Journal of Fatigue*, **21** (1999) 55

430 Molins, R.; Hochstetter, G.; Chassaigne, J.C.; Andrieu, E.: Oxidation Effects on the Fatigue Crack Growth Behaviour of Alloy 718 at High Temperatures, *Acta Materialia*, **45** (1997) 663

431 Sadananda, K.; Shahinian, P.: High-Temperature Time-Dependent Crack Growth, *Proc. Micro and Macro Mechanics of Crack Growth*, K. Sadananda et al. (Eds.), Metallurgical Society of AIME, Warrendale, PA (1981) 119

432 Ghonem, H.; Zheng, D.: Depth of Intergranular Oxygen Diffusion during Environment-Dependent Fatigue Crack Growth in Alloy 718, *Materials Science and Engineering*, **A150** (1992) 151

433 Zheng, D.; Ghonem, H.: Oxidation-Assisted Fatigue Crack Growth Behavior in Alloy 718: II. Applications, *Fatigue and Fracture of Engineering Materials and Structures*, **14** (1991) 761

434 Lynch, S.P.; Radtke, T.C.; Wicks, B.J.; Byrnes, R.T.: Fatigue Crack Growth in Nickel-Based Superalloys at 500–700 °C. II: Direct-Aged Alloy 718, *Fatigue and Fracture of Engineering Materials and Structures*, **17** (1994) 313

435 Lynch, S.P.; Radtke, T.C.; Wicks, B.J.; Byrnes, R.T.: Fatigue Crack Growth in Nickel-Based Superalloys at 500–700°C. I: Waspaloy, *Fatigue and Fracture of Engineering Materials and Structures*, **17** (1994) 297

436 Ghonem, H.; Nicholas, T.; Pineau, A.: Elevated Temperature Fatigue Crack Growth in Alloy 718: I. Effects of Mechanical Variables, *Fatigue and Fracture of Engineering Materials and Structures*, **16** (1993) 565

437 Ghonem, H.; Nicholas, T.; Pineau, A.: Elevated Temperature Fatigue Crack Growth in Alloy 718: II. Effects of Environmental and Material Variables, *Fatigue and Fracture of Engineering Materials and Structures*, **16** (1993) 577

438 Carpenter, W.; Kang, B. S.-J.; Chang K. M.: SAGBO Mechanism on High Temperature Cracking Behavior of Ni-base Superalloys, *Proc. Superalloys 718, 625, 706 and Various Derivatives*, Loria E.A. (Ed.), TMS, Warrendale, PA (1997) 679

439 Gourgues, A.F.; Andrieu, E.: High-Temperature, Oxidation Assisted Intergranular Cracking of a Solid-Solution Strengthened Nickel-Base Alloy, *Materials Science and Engineering,* **A351** (2003) 39

440 Osinkolu, G.A.; Onofrio, G.; Marchionni, M.: Fatigue Crack Growth in Polycrystalline IN718 Superalloy, *Materials Science and Engineering,* **A356** (2003) 425

441 Browning, M.F.; Henry, M.F.: Oxidation Mechanisms in Relation to High-Temperature Crack Propagation Properties in H_2/H_2O/Inert Environments, *Proc. Superalloys 718, 625 and Various Derivates,* E.A. Loria (Ed.), TMS, Warrendale, PA (1997) 665

442 Hayes, R.W.; Smith, D.F.; Wanner, E.A.; Earthmann, J.C.: Effect of Environment on the Rupture Behavior of Alloys 909 and 718, *Materials Science and Engineering,* **A177** (1994) 43

443 Rösler, J.; Müller, S.: Protection of Ni-Base Superalloys Against Stress Accelerated Grain Boundary Oxidation (SAGBO) by Grain Boundary Chemistry Modification, *Scripta Materialia,* **40** (1998) 257

444 Lyons, J.S.; Reynolds, A.P.; Clawson, J.D.: Effect of Aluminide Particle Distribution on the High-Temperature Crack Growth Characteristics of a Co-Ni-Fe Superalloy, *Scripta Materialia,* **37** (1997) 1059

445 Wei, R.P.; Huang, Z.-F.; Miller, C.F.; Simmons, G.W.: Mechanistic Considerations of Oxygen Enhanced Crack Growth in Inconel 718, *Proc. Superalloys 718, 625 and Various Derivates,* E.A. Loria (Eds), TMS, Warrendale, PA (2001) 691

446 Miller, C.F.; Simmons, G.W.; Wei, R.P.: High Temperature Oxidation of Nb, NbC, Ni_3Nb and Oxygen Enhanced Crack Growth, *Scripta Materialia,* **42** (2000) 227

447 Miller, C.F.; Simmons, G.W.; Wei, R.P.: Mechanism for Oxygen Enhanced Crack Growth in Inconel 718, *Scripta Materialia,* **44** (2001) 2405

448 Chen, S.-F.; Wei, R.P.: Environmentally Assisted Crack Growth in a Ni-18Cr-18Fe Ternary Alloy at Elevated Temperatures, *Materials Science and Engineering,* **A256** (1998) 197

449 Huang, Z.-F.; Iwashita, C.; Chou, I.; Wei, R.P.: Environmentally Assisted, Sustained-Load Crack Growth in Powder Metallurgy Nickel-Based Superalloys, *Metallurgical and Materials Transactions A,* **33** (2002) 1681

450 Andrieu, E.; Pieraggi, B.; Gourgues, A.F.: Role of Metal-Oxide Interfacial Reactions on the Interactions between Oxidation and Deformation, *Scripta Materialia,* **39** (1998) 597

451 Krupp, U.; Kane, W. M.; Liu, X.; Pfaendtner, J.A.; Laird, C.; McMahon Jr., C.J.: Oxygen-Induced Intergranular Fracture of the Nickel-Base Superalloy IN718 during Mechanical Loading at High Temperatures, *Materials Research,* **7** (2004) 35

452 Connolley, T.; Reed, P.A.S.; Starink, M.J.: Short Crack Initiation and Growth at 600°C in Notched Specimens of Inconel 718, *Materials Science and Engineering A,* **340** (2003) 130

453 Bika, D.: Dynamic Embrittlement in Metallic Alloys, Dissertation, University of Pennsylvania (1992)

454 Pfaendtner, J.A.; McMahon Jr., C.J.: Oxygen-Induced Intergranular Cracking of a Ni-Base Alloy at Elevated Temperatures: An Example of Dynamic Embrittlement, *Acta Materialia,* **49** (2001) 3369

455 Krupp, U.: Dynamic Embrittlement: Time-Dependent Brittle Intergranular Fracture at High Temperatures, *International Materials Reviews,* **50** (2005) 83

456 Liu, C.T.; White, C.L.: Dynamic Embrittlement of Boron-Doped Ni_3Al Alloys at 600°C, *Acta Metallurgica,* **35** (1987) 643

457 Birnbaum, H.K.; Sofronis, P.: Hydrogen-Enhanced Localized Plasticity – A Mechanism for Hydrogen-Related Fracture, *Materials Science and Engineering,* **A176** (1994) 191

458 McMahon Jr., C.J.: Hydrogen-Induced Intergranular Fracture of Steels, *Engineering Fracture Mechanics,* **86** (2000) 773

459 Lange, G.: Schäden durch Wasserstoff, in: *Systematische Beurteilung technischer Schadensfälle,* 3rd edition, DGM Informationsgesellschaft, Oberursel (1992)

460 Kameda, J.; Jokl, M.L.: Dynamic Model of Hydrogen-Induced Intergranular Fracture, *Scripta Metallurgica*, **16** (1982) 325

461 Kirchheim, R.: Solid Solutions of Hydrogen in Complex Materials, in: *Solid State Physics*, H. Ehrenreich, F. Spaepen (Eds.), Amsterdam (2004)

462 Gordon, P.; An, H.H.: The Mechanisms of Crack Initiation and Crack Propagation in Metal-Induced Embrittlement of Metals, *Metallurgical Transactions A*, **13A** (1982) 157

463 Rostocker, W.; McCaughey, J.M.; Markus, H.: *Embrittlement by Liquid Metals*, Reinhold, New York (1960)

464 Wolski, K.; Marie, N.; Laporte, V.; Berger, P.; Biscondi, M.: Evidence for a Diffusion-Based Mechanism of Liquid Metal Intergranular Penetration: Case Study of a Ni-Bi Model System, *Defect and Diffusion Forum*, **237–240** (2005) 677

465 Lynch, S.P.: Mechanisms of Intergranular Fracture, *Materials Science Forum*, **46** (1989) 1

466 Lynch, S.P.: Environmentally Assisted Cracking: Overview of Evidence for an Adsorption-Induced Localized Slip Process, *Acta Metallurgica*, **36** (1988) 3295

467 Woodfine, B.C.: Temper-Brittleness: A Critical Review of the Literature, *Journal of the Iron and Steel Institute*, **3** (1953) 229

468 Shin, J.; McMahon, Jr., C.J.: Mechanism of Stress-Relief Cracking in a Ferritic Steel, *Acta Metallurgica*, **32** (1984) 1535

469 Hull, D.; Rimmer, D.E.: The Growth of Grain Boundary Voids under Stress, *Philosophical Magazine*, **4** (1959) 673

470 Bika, D.; Pfaendtner, J.A.; Menyhard, M.; McMahon Jr., C.J.: Sulfur-Induced Dynamic Embrittlement in a Low Alloy Steel, *Acta Metallurgica et Materiulia*, **43** (1995) 1895

471 Misra, R.D.; Prasad, V.S.: On the Dynamic Embrittlement of Copper Chromium Alloys by Sulphur, *Journal of Materials Science*, **35** (2000) 3321

472 Misra, R.D.; McMahon Jr., C.J.; Guha, A.: Brittle Behavior of a Dilute Copper Berylium Alloy at 200°C in Air, *Scripta Metallurgica et Materialia*, **31** (1994) 1471

473 Pfaendtner, J.A.: Oxygen-Induced Diffusion-Controlled Cracking of Structural Alloys, Dissertation, University of Pennsylvania (1998)

474 Krupp, U.; Kane, W.M.; Liu, X.; Laird, C.; McMahon Jr., C.: The Effect of Grain-Boundary Engineering-Type Processing on Oxygen-Induced Cracking of IN718, *Materials Science and Engineering*, **A349** (2003) 213

475 Hippsley, C.A.; Rauh, H.; Bullogh, R.: Stress-Driven Solute Enrichment of Crack-Tips during Low-Ductility Intergranular Fracture of Low-Alloy Steel, *Acta Metallurgica*, **32** (1984) 1381

476 Rauh, H.; Hippsley, C.A.; Bullough, R.: The Effect of Mixed-Mode Loading on Stress-Driven Solute Segregation during High-Temperature Brittle Intergranular Fracture, *Acta Metallurgica*, **37** (1989) 269

477 Chen, I.-W.: Quasi-Static Intergranular Brittle Fracture at 0.5 T_m: A Non-Equilibrium Segregation Mechanism of Sulphur Embrittlement in Stress-Relief Cracking of Low Alloy Steels, *Acta Metallurgica*, **34** (1986) 1335

478 White, C.L.; Schneibel, J.H.; Padgett, R.A.: High Temperature Embrittlement of Ni and Ni-Cr Alloys by Trace Elements, *Metallurgical Transactions A*, **14** (1983) 595

479 Henry, M.F.: personal communication, General Electric Research Laboratory Schenectady (2002)

480 Watanabe, T.: An Approach to Grain Boundary Design for Strong and Ductile Polycrystals, *Res Mechanica*, **11** (1984) 47

481 Gottstein, G.: Tailored Microstructures, Can a Dream Come True?, *Advanced Engineering Materials*, **6** (2004) 617

482 Thaveeprungsriporn, V.; Was G. S.: The Role of Coincidence-Site-Lattice Boundaries in Creep of Ni-16Cr-9Fe at 360 °C, *Metallurgical and Materials Transactions A*, **28A** (1997) 2101

483 Was, G.S.; Thaveeprungsriporn, V.; Crawford, D.C.: Grain Boundary Misorientation Effects on Creep and Cracking in Ni-Based Alloys, *Journal of Materials*, **2** (1998) 44

484 Lehockey, E.M.; Palumbo, G.: On the Creep Behaviour of Grain Boundary Engineered Nickel, *Materials Science and Engineering*, **A237** (1997) 168

485 Lim, L.C.; Raj, R.: Effect of Boundary Structure on Slip-Induced Cavitation in Polycrystalline Nickel, *Acta Metallurgica*, **32** (1984) 1183

486 Lehockey, E.M.; Palumbo, G.; Lin, P.: Improving the Weldability and Service Performance of Nickel- and Iron-Based Superalloys by Grain Boundary Engineering, *Metallurgical and Materials Transactions A*, **29A** (1998) 3069

487 Alexandreanu, B.; Spencer, B.H.; Thaveeprungsriporn, V.; Was, G.S.: The Effect of Grain Boundary Character Distribution on High Temperature Deformation Behavior of Ni-16Cr-9Fe Alloys, *Acta Materialia*, **51** (2003) 3831

488 Palumbo, G.; Aust, K.T.: Structure Dependence of Intergranular Corrosion in High Purity Nickel, *Acta Metallurgica*, **38** (1990) 2343

489 Lin, P.; Palumbo, G.; Erb, U.; Aust, K.T.: Influence of Grain Boundary Character Distribution on Sensitization and Intergranular Corrosion of Alloy 600, *Scripta Metallurgica et Materialia*, **33** (1995) 1387

490 Lehocky, E.M.; Palumbo, G.; Brennenstuhl, A.M.: On the Relationship Between Grain Boundary Character and Intergranular Corrosion, *Scripta Materialia*, **36** (1997) 1211

491 Shimada, M.; Kokawa, H.; Wang, Z.J.; Sato, Y.S.; Karibe, I.: Optimization of Grain Boundary Character Distribution for Intergranular Corrosion Resistant 304 Stainless Steel by Twin-Induced Grain Boundary Engineering, *Acta Materialia*, **50** (2002) 2331

492 Yamaura, S.; Tsurekawa, S.; Watanabe, T.: The Control of Oxidation-Induced Intergranular Embrittlement by Grain Boundary Engineering in Rapidly Solidified Ni-Fe Ribbons, *Materials Transactions*, **44** (2003) 1494

493 Yamaura, S.; Igarashi, Y.; Tsurekawa, S.; Watanabe, T.: Structure of Intergranular Oxidation in Ni-Fe Polycrystalline Alloy, *Acta Materialia*, **47** (1999) 1163

494 Yang, S.; Krupp; U.; Trindade, V.B.; Christ, H.-J.: The Relationship between Grain Boundary Character and the Intergranular Oxide Distribution in IN718 Superalloy, *Advanced Engineering Materials*, **8** (2005) 7

495 Kaur, I.; Gust, W.: *Fundamentals of Grain and Interphase Boundary Diffusion*, Ziegler-Press, Stuttgart (1989)

496 Kane, W.M.: Dynamic Embrittlement of Nickel-Based Alloys, Dissertation, University of Pennsylvania (2005)

497 Lim, L.C.; Watanabe, T.: Fracture Toughness and Brittle-Ductile Transition Controlled by Grain Boundary Character Distribution (GBCD) in Polycrystals, *Acta Metallurgica et Materialia*, **38** (1990) 2507

498 Lim, L.C.; Watanabe, T.: Grain Boundary Character Distribution Controlled Toughness of Polycrystals: A Two-Dimensional Model, *Scripta Metallurgica*, **23** (1989) 489

499 Watanabe, T.: The Impact of Grain Boundary Character Distribution on Fracture in Polycrystals, *Materials Science and Engineering A*, **176** (1994) 39

500 Fortier, P.; Miller, W.A.; Aust, K.T.: Triple Junctions and Grain Boundary Character Distributions in Metallic Materials, *Acta Materialia*, **45** (1997) 3459

501 Aust, K.T.; Erb, U.; Palumbo, G.: Interface Control for Resistance to Intergranular Cracking, *Materials Science and Engineering A*, **176** (1994) 329

502 Schuh, C.; Kumar, M.; King, W.E.: Analysis of Grain Boundary Networks and their Evolution during Grain Boundary Engineering, *Acta Materialia*, **51** (2003) 687

503 Minich, R.W.; Schuh, C.A.; Kumar, M.: The Role of Topological Constraints on the Statistical Properties of Grain Boundary Networks, *Physical Review B*, **66** (2002) 052101

504 Kumar, M.; Schwartz, A.J.; King, W.E.: Microstuctural Evolution during Grain Boundary Engineering of Low to Medium Stacking Fault Energy fcc Materials, *Acta Materialia*, **50** (2002) 2599

505 Palumbo, G.; King, P.J.; Aust, K.T.; Erb, U.; Lichtenberger, P.C.: Grain Boundary Design and Control for Intergranular Stress Corrosion Resistance, *Scripta Metallurgica et Materialia*, **25** (1991) 1775

506 Kumar, M.; King, W.E.; Schwartz, A.J.: Modifications to the Microstructural Topology in F.C.C. Materials Through Thermomechanical Processing, *Acta Materialia*, **48** (2000) 2081

507 Randle, V.: Refined Approaches to the Use of the Coincidence Site Lattice, *Journal of Materials*, **2** (1998) 56

508 Palumbo, G.; Aust, K.T.: Structure Dependence of Intergranular Corrosion in High Purity Nickel, *Acta Metallurgica et Materialia*, **11** (1990) 2343

509 Muthiah, R.C.; Pfaendtner, J.A.; Ishikawa, S.; McMahon Jr., C.J.: Tin-Induced Dynamic Embrittlement of Cu-7%Sn Bicrystals with Σ5 Boundary, *Acta Materialia*, **47** (1999) 2797

510 Kitagawa, H.; Takahashi, S.: Applicability of Fracture Mechanics to Very Small Cracks or the Cracks in the Early Stage, *Proc. 2nd Int. Conf. Mechanical Behavior of Materials*, Boston (1976) 627

511 Brown, M.W.: Interfaces Between Short, Long, and Non-Propagating Cracks, *Proc. The Behaviour of Short Fatigue Cracks*, K.J. Miller, E.R. de los Rios (Eds.), ESIS, London (1986) 423

512 Taylor, D.: Fatigue of Short Cracks: The Limitations of Fracture Mechanics, *Proc. The Behaviour of Short Fatigue Cracks*, K.J. Miller, E.R. de los Rios (Eds.), ESIS, London (1986) 479

513 el Haddad, M.H.; Dowling, N.E.; Topper, T.H.; Smith, K.N.: J Integral Applications for Short Fatigue Cracks at Notches, *International Journal of Fatigue*, **16** (1980) 15

514 el Haddad, Smith, K.N.; Topper, T.H.: Fatigue Crack Propagation of Short Cracks, *Journal of Engineering Materials Technology, Transactions ASME*, **101** (1979) 42

515 Schick, A.: Ein neues Modell zur mechanismenorientierten Simulation der mikrostrukturbestimmten Kurzrissausbreitung, Fortschritt-Berichte VDI, Nr. 292, VDI-Verlag, Düsseldorf (2004)

516 Vehoff, H.; Neumann, P.: Life Prediction based on the Propagation of Short Cracks, *Steel Research*, **63** (1992) 372

517 Mc Evily, A.J.; Eifler, D.; Macherauch, E.: An Analysis of the Growth of Short Fatigue Cracks, *Engineering Fracture Mechanics*, **40** (1991) 571

518 Vormwald, M.: Anrißlebensdauer auf der Basis der Schwingbruchmechanik für kurze Risse, Institut für Stahlbau und Werkstoffmechanik, Heft 47, Technical University Darmstadt (1989)

519 DuQuesnay, D.L.; Topper, T.H.; Yu, M.T.: The Effect of Notch Radius on the Fatigue Notch Factor and the Propagation of Short Cracks, in: *Proc. The Behavior of Short Fatigue Cracks*, K.J. Miller, E.R. de los Rios (Eds.), ESIS, London (1986) 323

520 Grabowski, L.; King, J.E.: Modelling Short Crack Growth Behaviour in Nickel-Base Superalloys, *Fatigue and Fracture of Engineering Materials and Structures*, **15** (1992) 595

521 Bomas, H.; Hünecke, J.; Laue, S.; Schöne, D.: Anrisslebensdauervorhersage am Beispiel glatter und gekerbter Proben des Werkstoffes Cm15, in Proc. *DFG-Kolloquium Mechanismenorientierte Lebensdauervorhersage für zyklisch beanspruchte metallische Werkstoffe*, DVM-Bericht 683, Bremen (2000) 65

522 Bomas, H.; Hünecke, J.; Laue, S.; Mayr, P.; Schöne, D.: Anrisslebensdauervorhersage am Beispiel glatter und gekerbter Proben des Werkstoffs Cm15, *Materialwissenschaft und Werkstofftechnik*, **33** (2002) 230

523 Hobson, P.D.; Brown, M.W.; de los Rios, E.R.: Two Phases of Short Crack Growth in a Medium Carbon Steel, *Proc. The Behaviour of Short Fatigue Cracks*, K.J. Miller, E.R. de los Rios (Eds.), ESIS, London (1986) 441

524 Murtaza, G.; Akid, R.: Modelling Short Fatigue Crack Growth in a Heat-Treated Low-Alloy Steel, *International Journal of Fatigue*, **17** (1995) 207

525 Chan, K.S.; Lankford, J.: A Crack Tip Strain Model for the Growth of Small Fatigue Cracks, *Scripta Metallurgica*, **17** (1983) 529

526 de los Rios, E.R.; Mohamed, H.J.; Miller, K.J.: A Micromechanics Analysis for Short Fatigue Crack Growth, *Fatigue and Fracture of Engineering Materials and Structures*, **8** (1985) 49

527 Doquet, V.: Micromechanical Simulations of Microstructure-Sensitive Stage I Fatigue Crack Growth, *Fatigue and Fracture of Engineering Materials and Structures*, **22** (1999) 215

528 Tanaka, K.; Mura, T.: A Dislocation Model for Fatigue Crack Initiation, *Journal of Applied Mechanics*, **48** (1981) 97

529 Chan, K.S.: A Microstructure-Based Fatigue Crack-Initiation Model, *Metallurgical and Materials Transactions A*, **34** (2003) 43

530 Pippan, R.: The Condition for the Cyclic Plastic Deformation at the Crack Tip: The Influence of Dislocation Obstacles, *International Journal of Fatigue*, **58** (1992) 305

531 Tanaka, K.; Akiniwa, Y.; Nakai, Y.; Wei, R.P.: Modelling of Small Fatigue Crack Growth Interacting With Grain Boundary, *Engineering Fracture Mechanics*, **24** (1986) 803

532 Wilkinson, A.J.; Roberts, S.G.: A Dislocation Model for the Two Critical Stress Intensities Required for Threshold Fatigue Crack Propagation, *Scripta Materialia*, **35** (1996) 1365

533 Tong, Z.-X.; Lin, S.; Hsiao, C.-M.: The Mechanism of Fatigue Crack Propagation in Pure Aluminum Single Crystals, *Scripta Metallurgica*, **20** (1986) 977

534 Lin, I.H.; Thomson, R.: Cleavage, Dislocation Emission, and Shielding for Cracks under General Loading, *Acta Metallurgica*, **34** (1986) 187

535 Taira, S.; Tanaka, K.; Nakai, Y.: A Model of Crack-Tip Slip Band Blocked by Grain Boundary, *Mechanics Research Communications*, **5**(6) (1978) 375

536 Tanaka, K.; Akiniwa, Y.: Mechanics of Small Fatigue Crack Propagation, in: *Proc. Small Fatigue Cracks: Mechanics, Mechanisms and Applications*, K.S. Ravichandran, R.O. Ritchie, Y. Murakami (Eds.), Elsevier (1999) 59

537 Akiniwa, Y.; Tanaka, K.; Kimura, H.: Microstructural Effects on Crack Closure and Propagation Thresholds of Small Fatigue Cracks, *Fatigue and Fracture of Engineering Materials and Structures*, **24** (2001) 817

538 Tanaka, K.; Kinefuchi, M.; Yokomaku, T.: Modelling of Statistical Characteristics of the Propagation of Small Fatigue Cracks, in: *Proc. Short Fatigue Cracks*, K.J. Miller, E.R. de los Rios (Eds.), ESIS, London (1992) 351

539 Tanaka, K.; Akiniwa, Y.: A Model for the Propagation of Small Fatigue Cracks Interacting with Grain Boundary, *Journal of the Society of Materials Science Japan*, **34** (1985) 1310

540 Otsuka, A.; Mori, K.; Tohgo, K.: Mode II Fatigue Cracks in Aluminum Alloys, *Proc. Current Research on Fatigue Cracks*, T. Tanaka, M. Jono, K. Komai (Eds.), Society of Materials Science, Japan (1985) 127

541 Hoshide, T.; Ogaki, T.; Inoue, T.: A Modelling of Fatigue Crack Growth in Notched Specimen Subjected to Complex Stresses and its Application to Life Assessment, *Proc. 3rd Int. Conf. on Low Cycle Fatigue and Elasto-Plastic Behaviour of Materials LCF3*, K.-T. Rie (Ed.), Elsevier, London (1992) 356

542 Olfe, J.; Zimmermann, A.; Rie, K.-T.: Simulation der Mikrorissentwicklung bei zyklischer Belastung unter besonderer Berücksichtigung mikrostruktureller Vorgänge, *Proc. DFG-Kolloquium Mechanismenorientierte Lebensdauervorhersage für zyklisch beanspruchte metallische Werkstoffe*, DVM-Bericht 683, Bremen (2000) 55

543 Johnson, W.A.; Mehl, R.F.: Reaction Kinetics in Processes of Nucleation and Growth, *Metals Technology* (1939) 416

544 Voronoi, G.F.: Nouvelles Applications des Paramètres Continus à la Théorie des Formes Quadratiques, *Journal für reine angewandte Mathematik*, **134** (1908) 198

545 Meyer, S.; Brückner-Foit, A.; Möslang, A.; Diegele, E.: Stochastische Simulation der Schädigungsentwicklung in einem martensitischen Stahl, *Proc. DFG-Kolloquium Mechanismenorientierte Lebensdauervorhersage für zyklisch beanspruchte metallische Werkstoffe*, DVM-Bericht 684, Bremen (2002) 7

546 Brückner-Foit, A.; Huang, X.: Rissinitiierung in einem martensitischen Stahl unter Ermüdungsbelastung, *Proc. DFG-Kolloquium Mechanismenorientierte Lebensdauervorhersage für zyklisch beanspruchte metallische Werkstoffe*, DVM-Bericht 685, Bremen (2004) 7

547 Bertolino, G.; Doquet, V.; Sauzay, M.: Modelling of the Scatter in Short Fatigue Cracks Groth Kinetics in Relation with the Polycrystalline Microstructure, *International Journal of Fatigue*, **27** (2005) 471

548 Navarro, A.; de los Rios, E.R.: An Alternative Model of the Blocking of

Dislocations at Grain Boundaries, *Philosophical Magazine A*, **57** (1988) 37

549 Xin, X.J.; de los Rios, E.R.; Navarro, A.: Modelling Strain Hardening at Short Fatigue Cracks, *Proc. Short Fatigue Cracks*, ESIS 13, K.J. Miller, E.R. de los Rios (Eds.), Mechanical Engineering Publications, London (1992) 369

550 James, M.N.; de Los Rios, E.R.: Variable Amplitude Loading of Small Fatigue Cracks in 6261-T6 Aluminium Alloy, *Fatigue and Fracture of Engineering Materials and Structures*, **19** (1996) 413

551 de Los Rios, E.R.; Rodopoulos, C.A.; Yates, J.R.: Prediction of FCG Behaviour Under Variable Amplitude Loading in MMCs, *Fatigue and Fracture of Engineering Materials and Structures*, **19** (1996) 349

552 de los Rios, E.R.; Xin, X.J.; Navarro, A.: Modelling Microstructurally Sensitive Fatigue Crack Growth, *Proceedings of the Royal Society*, **447** (1994) 111

553 Wilkinson, A.J.: Modelling the Effects of Texture on the Statistics of Stage I Fatigue Crack Growth, *Philosophical Magazine A*, **81** (2001) 841

554 Schick, A.: Modellierung des Kurz- und Langrisswachstums in metallischen Werkstoffen unter Berücksichtigung der Mikrostruktur zur Vorhersage der Ermüdungslebensdauer, diploma thesis, Universität Siegen (1998)

555 de Los Rios, E.R.; Navarro, A.: Considerations of Grain Orientation and Work Hardening on Short Fatigue Crack Modelling, *Philosophical Magazine*, **61** (1990) 435

556 Navarro, A.; de Los Rios, E.R.: A Model for Short Fatigue Crack Propagation with an Interpretation of the Short-Long Crack Transition, *Fatigue and Fracture of Engineering Materials and Structures*, **10** (1987) 169

557 Navarro, A.; de Los Rios, E.R.: A Microstructurally-Short Fatigue Crack Growth Equation, *Fatigue and Fracture of Engineering Materials and Structures*, **11** (1988) 383

558 Riemelmoser, F.O.; Gumbsch, P.; Pippan, R.: Plastic Deformation of Short Edge Cracks under Fatigue Loading, *Engineering Fracture Mechanics*, **66** (2000) 357

559 Blomerus, P.M.; Hills, D.A.: Modelling Plasticity in Finite Bodies Containing Stress Concentrations by a Distributed Dislocation Method, *Journal of Strain Analysis*, **33** (1998) 315

560 Künkler, B.: Simulation eines Dualphasenwerkstoffs: Gefügeerstellung und mikrostrukturelle Kurzrissausbreitung, diploma thesis, Universität Siegen (2002)

561 Schick, A.; Fritzen, C.-P.; Floer, W.; Krupp, U.; Christ, H.-J.: Microstructural Short Fatigue Crack Growth: Effect of Mixed Mode Conditions and Crack Closure, *Proc. European Conference on Fracture: ECF14*, Cracow, Poland (2002) CD-ROM

562 de los Rios, E.R.; Hussain, K.; Miller, K.J.: A Micro-Mechanics Analysis for Short Fatigue Crack Growth, *Fatigue and Fracture of Engineering Materials and Structures*, **8** (1985) 49

563 Ahamadi, A; Zenner, H.: Simulation des Mikrorisswachstums bei zyklischer Beanspruchung, *MP Materialprüfung*, **45** (2003) 189

564 Krupp, U. et al.: Experimental and theoretical quantification of three-dimensional propagation of micro-structurally short fatigue cracks by means of synchrotron computer tomograpy and the three-dimensional boundary-element method, research project in preparation, University Siegen and University of Applied Sciences, Osnabrück, etc.

565 Weinhandl, H.: Über die Computer-Simulation von Nachbarschaftsverhältnissen in Zweiphasengefügen, Dissertation, Montan-Universität Leoben (1996)

566 Werner, E.; Stüwe, H.P.: Phase Boundaries as Obstacles to Dislocation Motion, *Materials Science and Engineering*, **68** (1984–85) 175

567 Fan, Z.; Tsakiropoulos, P.; Smith, P.A.; Miodownik, A.P.: Extension of the Hall-Petch Relation to Two-Ductile-Phase Alloys, *Philosophical Magazine A*, **67** (1993) 515

568 Bjerken, C.; Melin, S.: A Tool to Model Short Crack Fatigue Growth Using a Discrete Dislocation Formulation, *International Journal of Fatigue*, **25** (2003) 559

569 Krupp, U.; Düber, O.; Christ, H.-J.; Köster, P.; Künkler, B.; Fritzen, C.-P.: Propagation Mechanisms of Microstructurally Short Cracks – Factors Governing the Transition from Short-to Long-Crack Behavior, *Materials Science and Engineering* (in press)

570 Newman, J.C.: A Crack Closure model for Predicting Fatigue Crack Growth under Random Loading, ASTM STP 748 (1981) 53

571 Hills, D.A.; Kelly, P.A.; Dai, D.N.; Korsunsky, A.M.: *Solution of Crack Problems: The Distributed Dislocation Technique*, Kluwer, Dordrecht/Boston/London (1995)

572 Teteruk, R.: Modellierung der Lebensdauer bei thermomechanischer Ermüdungsbeanspruchung unter Berücksichtigung der relevanten Schädigungsmechanismen, Fortschritt-Berichte VDI, Nr. 653, VDI-Verlag Düsseldorf (2002)

573 Woodford, D.A.; Coffin Jr., L.F.: The Role of Grain Boundaries in High-Temperature Fatigue, *Proc. Grain Boundaries in Engineering Materials, 4th Bolton Landing Conf.*, J.H. Walter et al. (Eds.), Claitor's Publishing (1975) 421

574 Ostergreen, W.J.: A Damage Function and Associated Failure Equations for Predicting Hold Time and Frequency Effects in Elevated Temperature Low Cycle Fatigue, *Journal of Testing and Evaluation*, **4** (1976) 327

575 Saxena, A.: A Model for Predicting the Environmental Enhanced Fatigue Crack Growth Behavior at High Temperature, *Proc. Thermal and Environmental Effects in Fatigue Research–Design Interface*, Vol. 71, C.E. Jaske, S.J. Hudak, M.E. Mayfield (Eds.), New York (1983) 171

576 Bika, D.; McMahon Jr., C.J.: A Model for Dynamic Embrittlement, *Acta Metallurgica et Materialia*, **43** (1995) 1909

577 Xu, Y.: Crack Propagation Associated with Stress-Assisted Diffusion of Impurities under Creep Conditions, Dissertation, University of Pennsylvania (1999)

578 Xu, Y.: Bassani, J.L., A Steady-State Model for Diffusion-Controlled Fracture, *Materials Science and Engineering*, **A260** (1999) 48

579 Chen, I.-W.; Argon, A.S.: Diffusive Growth of Grain-Boundary Cavities, *Acta Metallurgica*, **29** (1981) 1759

580 Shin, K.S.; Meshii, M.: Effect of Sulfur Segregation and Hydrogen Charging on Intergranular Fracture of Iron, *Acta Metallurgica*, **31** (1983) 1559

581 McClintock, F.A.; Bassani, J.L.: Problems in Environmentally-Affected Creep Crack Growth, in: *Proc. Three-Dimensional Constitutive Relations and Ductile Fracture*, J. Zarka, S. Nemat-Nasser (Eds.), North Holland (1981) 123

582 Lárche, F.; Cahn, J.W.: The Interactions of Composition and Stress in Crystalline Solids, *Acta Metallurgica*, **33** (1985) 331

583 Lárche, F.; Cahn, J.W.: A Linear Theory of Thermochemical Equilibrium of Solids under Stress, *Acta Metallurgica*, **21** (1973) 1051

584 Herring, C.: Diffusional Viscosity of a Polycrystalline Solid, *Journal of Applied Physics*, **21** (1950) 437

585 Xu, Y, Bassani, J.L.: unpublished research, University of Pennsylvania (1998)

586 Bricknell, R.H.; Woodford, D.A.: Reply to Comments on " The Mechanism of Cavity Formation on High-Temperature Oxidation of Nickel", *Scripta Metallurgica*, **16** (1982) 761

587 Caplan, D.; Woodford, D.A.; Sproule, G.I.; Graham, M.J.: Effect of Carbon on Cavity Formation during High-Temperature Oxidation of Ni, *Oxidation of Metals*, **14** (1980) 279

588 Bakker, H.: A Curvature in the ln D versus $1/T$ Plot for Self-Diffusion in Nickel at Temperatures from 980 to 1400°C, *Physica Status Solidi*, **28** (1968) 569

589 Needleman, A.: A Continuum Model for Void Nucleation by Inclusion Debonding, *Journal of Applied Mechanics*, **54** (1987) 525

590 Vitek, V.: Yielding on Inclined Planes at the Tip of a Crack Loaded in Uniform Tension, *Journal of the Mechanics and Physics of Solids*, **24** (1976) 263

591 Muthiah, R.C.: Diffusion-Controlled Brittle Intergranular Fracture in Cu-Alloy and Polycrystals, Dissertation, University of Pennsylvania (1997)

Subject Index

a
ab initio methods 25
acoustic emission technique 51
adsorption-induced dislocation emission (AIDE) 189
alloy 718 182, 197, 246
alloy IN100 184
alternating current potential drop 50
aluminium alloy 127, 138, 141, 165, 167, 174, 212
aluminium-lithium alloy 225
aluminium-magnesium alloy 143
anisotropy 75
– anisotropy factor 79, 109, 113
– elastic 78, 107, 122
– strain 116
– stress 117 ff.
annihilation 105
– of dislocations 96
arcsin $\sqrt{P}$ method 15
ARMCO iron 150
aspect ratio a/c 173 ff.
atomic distance potential 76
atomic force microscopy (AFM) 72
austenitic steel 116, 119, 126 f., 143, 174, 235
automated EBSD 62 f.

b
backscattered electrons 58
Barkhausen emission 73
barrier strength 150. 213, 215
Basquin relationship 5, 15
Bauschinger effect 5
BCS crack 33, 214, 219
beach marks 57
bicrystals 64, 206 ff.
blunting 148
boundary conditions 228
boundary-element method (BEM) 28, 226
Bragg angle Θ 60 f.
Bragg condition 60
Brandon criterion 69
brittle–ductile transition temperature (BDTT) 57
Burgers vector 80, 226

c
carbon steel 127
carburization 123
cast iron 163
cell structures 97
cellular metals 129 ff.
– open-cell 45, 129
– closed-cell 45, 129
CMOD *see* crack mouth opening displacement
coherent 103
cohesive energy 245
cohesive strength 25
cohesive zone 188, 196
– dislocation density 248
cohesive-zone model 247
coincident site lattice (CSL) 68
compact-tension specimen 52
compatibility
– condition 28
– plastic 144
compliance tensor 108, 110 ff.
compression modulus 75
computer tomography 55, 73
conjugate slip system 92
convergent beam electron diffraction (CBED) 63
copper 116, 119, 141, 147, 248
corrosion 123
counting methods 17
crack-arrest lines 58

Fatigue Crack Propagation in Metals and Alloys: Microstructural Aspects and Modelling Concepts. Ulrich Krupp

ISBN: 978-3-527-31537-6

crack branching 175
crack closure 7, 17, 22, 138, 211, 228 f., 232
– development in the short-crack regime 164 ff.
– effects 51, 152 ff., 239
– fluid-induced 155
– geometry-induced 155, 233, 241
– measurements 47
– oxidation-induced 155, 162 ff.
– plasticity-induced 154, 156 ff., 241
– roughness-induced 155, 161 f., 165
– stress-induced 154
– steady state 165
– transformation-induced 155, 162 ff.
– transient regime 169
– two-parameter approach 160, 163
crack coalescence 173, 179 ff., 239
crack-course lines 58
crack deflection 143, 158, 165
crack initiation 99 ff., 117, 119 ff.
– elastic anisotropy 107 ff.
– elevated temperature 123 ff.
– environmental effects 123 ff.
– grain boundaries 107, 120
– inclusions 102 ff.
– intercrystalline 119, 123
– persistant slip bands 104 ff.
– pores 102 ff.
– transcrystalline 119
– transgranular 126 ff.
– twin boundaries 117
crack mouth opening displacement (CMOD) 166 f., 176
crack propagation 135 ff., 139 ff.
– brittle 186
– crystallographic 139, 143
– intercrystalline 181 ff., 186
– multiple-slip 169
– oscillating 141 ff., 212
– rate 138, 177, 182
– scatter 173
– single-slip 169
– statistics 217
– time-dependent 242
– transgranular 139 ff.
– transition, mode II to mode I 171 ff.
– velocity 194
crack-opening modes 27
crack-opening stress-intensity factor 22
crack-propagation
– measurements 48 ff.
– rate 224, 226
crack-resistance curve 51
crack-tip
– blunting 34
– opening displacement (CTOD) 24, 34, 60, 216
– plasticity 8
crack-tip-slide displacement (CTSD) 165, 216
creep 124, 194, 242, 247 f.
– corrosion 182
– pores 189
critical plane 18
critical shear stress 80, 231
cross slip 87
crystal coordinate system 65, 110, 116
crystal plasticity 114
crystallographic misorientation 218, 235
crystallographic orientation (distribution) 59 ff., 108, 165
– factor 218, 222
– relationship 65 ff.
CSL grain boundaries 120, 125, 197, 247
CSSC *see* cyclic stress–strain curve
CTOD *see* crack-tip opening displacement
CTSD *see* crack-tip-slide displacement
ΔCTSD 221, 223 f., 226, 228
cutting stress 89
cutting dislocations 83
cycle duration 182
cyclic bending 18
cyclic crack-tip opening 8
cyclic creep 12, 132
cyclic deformation 94 ff.
cyclic exponent of strain hardening 14
cyclic hardening 12, 42, 95
cyclic plastic zone 34
cyclic relaxation 12
cyclic softening 12, 42
cyclic strength coefficient 14
cyclic stress–strain behavior 11
cyclic stress–strain curve 13, 16, 41
cyclic torsion 18
cyclic yield strength 234

d

damage
– accumulation approach 1
– parameters 17
– sum 6
damage-tolerant approach 19 ff., 208
decohesion 244
deformation
– bands 131
– mechanisms maps 89

diffusion
– bonding 204
– differential equation 244
direct current potential-drop (DCPD) 49
dislocation
– climb 181
– dipole 83
– dynamics 218
– mathematical 226
– pairing 126, 140
– pileup 92
– source 93, 221, 231
– structures 94
– theory 6, 80 ff.
dislocation-density distribution 219 ff.
distributed dislocations 33
double slip 139
Dugdale approach 32 ff., 216
duplex steel 126, 139, 146 f., 150, 168, 174, 232 ff., 235, 237
dwell time 182, 192, 196
– cracking, modeling 242 ff.
dynamic embrittlement 181 ff., 186, 243 ff.
– alloy 718 192 ff.
– governing equation 245
– mechanism 187, 191, 243 ff.
– model 243 ff.
– process zone 247

e
EBSD *see* electron backscattered diffraction
eddy-current testing 49
edge dislocation 81
effective stress-intensity factor 155, 170
EGM model 105
elastic anisotropy 113, 122, 174
– analytical calculation 116 ff.
elastic constants 109 ff.
elastic deformation 76 ff.
elastic–plastic fracture mechanics (EPFM) 8, 24
electrochemical fatigue sensor (EFS) 73
electro-discharge machining 45
electrolytic thinning 70
electron backscattered diffraction (EBSD) 55 f., 58 ff., 110, 115, 226, 232
electron channeling contrast 58
electron diffraction 60
electron microscopy 6
electropolishing 44, 62
energy release rate 51
energy-dispersive X-ray spectroscopy (EDS) 55
environmental effects 242
epitactic NiO 184
equilibrium conditions 27, 118
equilibrium crack depth *a* 179
Euler angles 67
extrusions 105

f
fail-safe 19
fatigue-crack-propagation
– approach 207
– rate 184, 221
fatigue life 100
– assessment 207 ff.
– prediction 15, 24
fatigue limit 15, 127 ff., 137, 143, 210
fatigue notch factor 16
fatigue-loading spectra 17
ferritic steel 150, 180, 235
field emission gun 56
finite element method 28, 113 ff., 158
foam 129
focused ion beam milling (FIB) 48, 55, 70
foreign object damage (FOD) 101
forest dislocations 84, 141
four-point bending 192
fracture mechanics 25 ff.
fracture surfaces 56, 195
fracture toughness 51 ff.
Frank–Read source 86
friction stress 88 f., 215, 219, 225, 228, 235
frictional corrosion 162

g
γ' phase 182, 184
geometry functions *Y* 29
glancing incidence 61
grain boundary 150
– design 197
– energy 68
– engineering 197 ff.
– geometry 68
– plane 144
– sliding 124
– width 125
grain-boundary diffusion 125, 190, 199, 245
– coefficient 247
– stress-induced 187
– sulfur 190
grain-boundary facets 183
grain-boundary structure 244
grain size 165, 235
– effects 149 ff., 197

Green functions 228
Griffith approach 25 ff.

h
Hall–Petch relationship 93, 141, 231, 233 ff.
HCF (high-cycle fatigue) 12, 99, 102, 120, 135, 141, 161, 180, 208
heat affected zone (HAZ) 189
hexagons 217
high-temperature corrosion 124, 182
homogenization 129
Hooke's law 10, 78, 108
Hough transformation 63
Hull–Rimmer model 189
hydrogen embrittlement 188
hydrogen-enhanced localized plasticity (HELP) 188
hydrogen-induced failure 57

i
in situ observation 54 ff.
inclusions 127
incompatibility 120
– crack faces 158
– elastic 120
– plastic 120, 122, 126
incremental step test (IST) 41
incubation time 192
inhomogeneous materials 129 ff.
inspection intervals 21
intercrystalline corrosion 198
intercrystalline crack
– initiation 114
– oxygen-induced 243
interferometric strain/displacement gauge (ISDG) 47
internal oxidation 123 f.
interstitial dipoles 106
intrusions 105
ion-thinning 70
irreversibility 12
– limit 127
– plastic deformation 80, 148
– slip 145, 213
Irwin approach 31
ISDG 110, 166, 176
– technique 46 ff.

j
J integral 8, 34, 51, 53, 180, 207
– cyclic 24, 36
jog 83

k
K concept 22, 179
– of LEFM 27 ff.
ΔK concept 211
$\Delta K^*/K_{max}$ thresholds 163 f.
Kachanov-type damage functions 248
Kikuchi
– lines 61
– pattern 61
kink 83
Kitagawa–Takahashi diagram 209, 222
Kossel cones 60, 62
Kurdjumov–Sachs relationship 150

l
Laird model 148
laser-shock peening 101
lattice spacing 80
Laue camera 64
LCF (low-cycle fatigue) 11, 120, 172, 182, 192
linear-elastic fracture mechanics (LEFM) 7, 23, 179
liquid-metal embrittlement (LME) 188
load cell 40
load history effects 240
load relaxation 192, 201, 205
load sharing 151 ff.
load-shedding method 51
local-strain approach 16
local-stress approach 207
Lomer–Cottrell locks 86
long cracks 136
low-carbon steel 212
Lüders band formation 12

m
Manson–Coffin relationship 6, 15, 42
martensite formation 126, 163
Masing
– behavior 14
– model 98
mean strain 12
mean stress 12, 22
metallic bonding 76
microstructural barriers 137
microstructural fracture mechanics (MFM) 24
microstructurally short fatigue cracks 20, 99 ff., 135 ff.
Miner rule 6, 17
misorientation 65, 144
– angles 65
– axes 65

– crack factor 119
– matrix 65
mixed-mode loading 27
modeling, three-dimensional 239 ff.
multiple slip 146 f.
– plastic blunting 237

n
Nabarro–Herring creep 89
Nabarro–Peierls stress 80, 85 f., 88
Navarro–de los Rios model 141, 218 ff.
Neuber's rule 16
Neumann model 147
Newman–Raju model 175
nickel 119
nickel-base alloy IN718 150, 163
nickel-base alloy IN783 123, 162, 181 ff.
niobium carbides 184, 186
nitridation 123 f.
nominal-stress approach 15
noncoherent 103
Norton exponent *n* 37, 248
Norton's creep law 89
notch 11
– approach 207
– factors 6
– stress concept 16

o
orientation matrix 65
Orowan
– equation 84
– mechanism 152
oscillating crack-propagation velocity 190, 244
overloads 153 ff.
– compressive 159 f., 168
– tensile 160, 168, 171
oxygen effect 183 ff.
oxygen partial pressures 192
oxygen solubility 196

p
packing density 90
pairing of dislocations 126, 140
Palumbo–Aust criterion 206
parabolic oxidation 186
Paris law 7, 19, 21, 51, 136, 207, 211, 242
passing stress 88
Peach–Koehler equation 82
Peierls–Nabarro stress 141
persistent slip bands 7, 59, 96
phase boundaries 150
PID controlling 41
pileup 122, 126, 140, 213
plain carbon steel 168, 212
planar slip 140, 153
plane strain 34, 102, 127, 157, 174
plastic deformation 80 ff.
plastic-strain amplitude 12
plastic-strain control 41
plastic wake 154, 159, 161, 164
plastic zone 31
– correction 31
– size 34, 210, 216
plateau stress 96
Poisson ratio 28, 32, 75
pore formation 125
positron annihilation 73
potential-drop technique 49
power-law creep 37
precipitates 103, 122, 149 ff., 152
principal stresses 79
protrusions 96, 106

r
rainflow-counting approach 17
Ramberg–Osgood equation 14, 36
random-grain-boundary network 203
replica technique 49, 54
reproducibility 74 f.
residual stresses 100 ff.
resolved shear stress 91, 171
Reuss theory 111
ripples 148
roller hardening 101
rotating bending 5, 39
rotation matrix 79

s
Sachs factor 94, 222
safe-life design 19
saturation 13
scanning electron microscopy (SEM) 55
Schmid factor 91, 115 ff., 222
screw dislocation 81
segregation 188
selected area diffraction (SAD) 63
semi-coherent 103
sensor elements 231
SENB *see* single-edge notched bend
service-life prediction 9, 19
servohydraulic testing machines 40
shallow notch 43
shear modulus 28, 75, 80
Shockley partial dislocations 87

short cracks 8
– aspect ratio a/c 173 ff.
– chemically 136
– definition 136 ff.
– mechanically 136
– microstructurally 136 f., 149 ff.
– model, Navarro–de los Rios 217, 218 ff.
– models 211 ff.
– physically 136
– propagation, numerical modeling 226 ff.
– three-dimensional 173
shot peening 100
single-edge notched bend (SENB) 45, 49, 192
single-slip crack mechanism 145
skin effect 50
slip
– crystallographic 80
– irreversibility 102, 104
– planarity 122, 188
– planes 83
– steps 104, 122
– systems 90
– transmission 121, 126, 140, 149 ff.
slip-band crack 114, 142
– facets 143
– propagation mechanism 145
slip-blocking model 217
Smith–Watson–Topper parameter 17
Sneddon equations 28
solid-metal embrittlement (SME) 188
special grain boundaries 69, 197
specimen coordinate system 65, 110, 116
specimen geometries 43 ff.
SRR99 124
stacking fault energy (SFE) 88
stage Ib 139, 149
stage II 139, 149
staircase method 15
steady-state crack velocity 245
steel
– A508 189
– CrMoV 163
stiffness tensor 77, 108
strain
– control 41
– delocalization 106
– gauge 40, 46
– hardening 87
– tensor 77
strengthening mechanisms 89
stress
– control 41
– intensity factor 19, 29, 140, 192, 207
– raisers 11
– ratio 12, 77, 159, 162
stress-assisted grain boundary oxidation (SAGBO) 184, 243
stress-corrosion cracking 184
stress-intensity factor 7
– effective 22
– threshold 23
stress-relief cracking 189, 243
stress–strain
– analysis 47
– hysteresis 12 ff.
striations 7, 147, 183, 195
structure-stress approach 207
superalloys 182, 242
surface
– diffusion 249
– energy 26
– roughness 100, 161
synchrotron radiation 55, 68
synthetic microstructure 232 ff., 238

t
Taylor factor 94, 222
TEM *see* transmission electron microscopy
temperature dependence of plastic deformation 86
tensile test 10
thermomechanical fatigue (TMF) 54
thermomechanical processing (TMP) 199 ff.
threshold 156, 214
– stress-intensity factor 20
tilt misorientation angle Φ 144
titanium alloy
– LCB 44, 111, 120, 126, 142, 146, 150, 153, 167, 174 f., 177, 180, 224, 231
– Ti-2.5 Cu 161
– Ti-6A-2Sn-4Zr-6Mo 176
– Ti-6Al-V4 143, 159
total-life design 15 ff
transition
– long-crack propagation 210, 224
– mode II to mode I 236
– short-crack propagation 210, 224
– short to long crack behavior 172, 236
transmission electron microscopy (TEM) 55, 70 f., 141
– diffraction 63 f.
traveling microscope 54
Tresca stress 32
triple points 126
twin boundaries 116
twinning 90

twist misorientation 121
– angle ξ 144
two-parameter approach 164

u
ultrahigh-cycle fatigue (UHCF) 102, 127
ultrasonic detection 21
ultrasonic measurements 109
– Young's modulus 109
ultrasonic testing 49
unbounded solution 220

v
vacancy dipoles 106
Versailles accident 4
very high-cycle fatigue (VHCF) 102, 120, 127, 135, 208
vibro-polishing 62
Voigt theory 111
von Mises stress 32
Voronoi algorithm 218, 233

w
Waspalloy 61
water vapor 183 ff.
– effect 194
Wöhler diagram (S/N curve) 4, 15, 39, 128, 207
work hardening 148

x
X-ray diffraction 63 f., 72

y
yield strength 10
Young's modulus 10, 75 f., 107, 109

z
Z integral 36
zirconium alloy Zircaloy 179

Related Titles

K. U. Kainer (Ed.)

Metal Matrix Composites

Custom-made Materials for Automotive and Aerospace Engineering

2006

ISBN 978-3-527-31360-0

H. Baltes, O. Brand, G. K. Fedder, C. Hierold, J. G. Korvink, O. Tabata, D. Löhe, J. Haußelt (Eds.)

Microengineering of Metals and Ceramics

Part II: Special Replication Techniques, Automation, and Properties

2005

ISBN 978-3-527-31493-5

H. Baltes, O. Brand, G. K. Fedder, C. Hierold, J. G. Korvink, O. Tabata, D. Löhe, J. Haußelt (Eds.)

Microengineering of Metals and Ceramics

Part I: Design, Tooling, and Injection Molding

2005

ISBN 978-3-527-31208-5

A. Hazotte (Ed.)

Solid State Transformation and Heat Treatment

2005

ISBN 978-3-527-31007-4

D. Raabe, F. Roters, F. Barlat, L.-Q. Chen (Eds.)

Continuum Scale Simulation of Engineering Materials

Fundamentals – Microstructures – Process Applications

2004

ISBN 978-3-527-30760-9

M. J. Zehetbauer, R. Z. Valiev (Eds.)

Nanomaterials by Severe Plastic Deformation

2004

ISBN 978-3-527-30659-6

K. U. Kainer (Ed.)

Magnesium

Proceedings of the 6th International Conference Magnesium Alloys and Their Applications

2004

ISBN 978-3-527-30975-7

C. Leyens, M. Peters (Eds.)

Titanium and Titanium Alloys

Fundamentals and Applications

2003

ISBN 978-3-527-30534-6